D1758747

Books are to be returned on or before
the last date below.

Diffraction, Fourier Optics and Imaging

THE WILEY BICENTENNIAL—KNOWLEDGE FOR GENERATIONS

$\mathcal{E}$ach generation has its unique needs and aspirations. When Charles Wiley first opened his small printing shop in lower Manhattan in 1807, it was a generation of boundless potential searching for an identity. And we were there, helping to define a new American literary tradition. Over half a century later, in the midst of the Second Industrial Revolution, it was a generation focused on building the future. Once again, we were there, supplying the critical scientific, technical, and engineering knowledge that helped frame the world. Throughout the 20th Century, and into the new millennium, nations began to reach out beyond their own borders and a new international community was born. Wiley was there, expanding its operations around the world to enable a global exchange of ideas, opinions, and know-how.

For 200 years, Wiley has been an integral part of each generation's journey, enabling the flow of information and understanding necessary to meet their needs and fulfill their aspirations. Today, bold new technologies are changing the way we live and learn. Wiley will be there, providing you the must-have knowledge you need to imagine new worlds, new possibilities, and new opportunities.

Generations come and go, but you can always count on Wiley to provide you the knowledge you need, when and where you need it!

WILLIAM J. PESCE
PRESIDENT AND CHIEF EXECUTIVE OFFICER

PETER BOOTH WILEY
CHAIRMAN OF THE BOARD

Diffraction, Fourier Optics and Imaging

OKAN K. ERSOY

WILEY-INTERSCIENCE

A JOHN WILEY & SONS, INC., PUBLICATION

Published by John Wiley & Sons, Inc., Hoboken, New Jersey
Published simultaneously in Canada

For general information on our other products and services or for technical support, please contact our
Customer Care Department within the United States at (800) 762-2974, outside the United States at (317)
572-3993 or fax (317) 572-4002.

Wiley also publishes its books in a variety of electronic formats. Some content that appears in
print may not be available in electronic formats. For more information about Wiley products, visit
our web site at www.wiley.com.

Library of Congress Cataloging-in-Publication Data

Ersoy, Okan K.
  Diffraction, fourier optics, and imaging / by Okan K. Ersoy.
    p. cm.
  Includes bibliographical references and index.
  ISBN-13: 978-0-471-23816-4
  ISBN-10: 0-471-23816-3
1. Diffraction. 2. Fourier transform optics. 3. Imaging systems. I. Title.

QC415.E77 2007
535′.42--dc22   2006048263

Printed in the United States of America

10 9 8 7 6 5 4 3 2 1

Contents

Preface

Diffraction and imaging are central topics in many modern and scientific fields. Fourier analysis and sythesis techniques are a unifying theme throughout this subject matter. For example, many modern imaging techniques have evolved through research and development with the Fourier methods.

This textbook has its origins in courses, research, and development projects spanning a period of more than 30 years. It was a pleasant experience to observe over the years how the topics relevant to this book evolved and became more significant as the technology progressed. The topics involved are many and an highly multidisciplinary.

Even though Fourier theory is central to understanding, it needs to be supplemented with many other topics such as linear system theory, optimization, numerical methods, imaging theory, and signal and image processing. The implementation issues and materials of fabrication also need to be coupled with the theory. Consequently, it is difficult to characterize this field in simple terms. Increasingly, progress in technology makes it of central significance, resulting in a need to introduce courses, which cover the major topics together of both science and technology. There is also a need to help students understand the significance of such courses to prepare for modern technology.

This book can be used as a textbook in courses emphasizing a number of the topics involved at both senior and graduate levels. There is room for designing several one-quarter or one-semester courses based on the topics covered.

The book consists of 20 chapters and three appendices. The first three chapters can be considered introductory discussions of the fundamentals. Chapter 1 gives a brief introduction to the topics of diffraction, Fourier optics and imaging, with examples on the emerging techniques in modern technology.

Chapter 2 is a summary of the theory of linear systems and transforms needed in the rest of the book. The continous-space Fourier transform, the real Fourier transform and their properties are described, including a number of examples. Other topics involved are covered in the appendices: the impulse function in Appendix A, linear vector spaces in Appendix B, the discrete-time Fourier transform, the discrete Fourier transform, and the fast Fourier transform (FFT) in Appendix C.

Chapter 3 is on fundamentals of wave propagation. Initially waves are described generally, covering all types of waves. Then, the chapter specializes into electromagnetic waves and their properties, with special emphasis on plane waves.

The next four chapters are fundamental to scalar diffraction theory. Chapter 4 introduces the Helmholtz equation, the angular spectrum of plane waves, the Fresnel-Kirchoff and Rayleigh-Sommerfeld theories of diffraction. They represent wave propagation as a linear integral transformation closely related to the Fourier transform.

Chapter 5 discusses the Fresnel and Fraunhofer approximations that allow diffraction to be expressed in terms of the Fourier transform. As a special application area for these approximations, diffraction gratings with many uses are described.

Diffraction is usually discussed in terms of forward wave propagation. Inverse diffraction covered in Chapter 6 is the opposite, involving inverse wave propagation. It is important in certain types of imaging as well as in iterative methods of optimization used in the design of optical elements. In this chapter, the emphasis is on the inversion of the Fresnel, Fraunhofer, and angular spectrum representations.

The methods discussed so far are typically valid for wave propagation near the z-axis, the direction of propagation. In other words, they are accurate for wave propagation directions at small angles with the z-axis. The Fresnel and Fraunhofer approximations are also not valid at very close distances to the diffraction plane. These problems are reduced to a large extent with a new method discussed in Chapter 7. It is called the near and far field approximation (NFFA) method. It involves two major topics: the first one is the inclusion of terms higher than second order in the Taylor series expansion; the second one is the derivation of equations to determine the semi-irregular sampling point positions at the output plane so that the FFT can still be used for the computation of wave propagation. Thus, the NFFA method is fast and valid for wide-angle, near and far field wave propagation applications.

When the diffracting apertures are much larger than the wavelength, geometrical optics discussed in Chapter 8 can be used. Lens design is often done by using geometric optics. In this chapter, the rays and how they propagate are described with equations for both thin and thick lenses. The relationship to waves is also addressed.

Imaging with lenses is the most classical type of imaging. Chapters 9 and 10 are reserved to this topic in homogeneous media, characterizing such imaging as a linear system. Chapter 9 discusses imaging with coherent light in terms of the 2-D Fourier transform. Two important applications, phase contrast microscopy and scanning confocal microscopy, are described to illustrate how the theory is used in practice.

Chapter 10 is the continuation of Chapter 9 to the case of quasimonochromatic waves. Coherent imaging and incoherent imaging are explained. The theoretical basis involving the Hilbert transform and the analytic signal is covered in detail. Optical aberrations and their evaluation with Zernike polynomials are also described.

The emphasis to this point is on the theory. The implementation issues are introduced in Chapter 11. There are many methods of implementation. Two major ones are illustrated in this chapter, namely, photographic films and plates and electron-beam lithography for diffractive optics.

In Chapters 9 and 10, the medium of propagation is assumed to be homogeneous (constant index of refraction). Chapter 12 discusses wave propagation in inhomogeneous media. Then, wave propagation becomes more difficult to compute numerically. The Helmholtz equation and the paraxial wave equation are generalized to inhomogeneous media. The beam propagation method (BPM) is introduced as a powerful numerical method for computing wave propagation in inhomogenous media. The theory is illustrated with the application of a directional coupler that allows light energy to be transferred from one waveguide to another.

Holography as the most significant 3-D imaging technique is the topic of Chapter 13. The most basic types of holographic methods including analysis of holographic imaging, magnification, and aberrations are described in this chapter.

In succeeding chapters, diffractive optical elements (DOEs), new modes of imaging, and diffraction in the subwavelength scale are considered, with extensive emphasis on numerical methods of computation. These topics are also related to signal/image processing and iterative optimization techniques discussed in Chapter 14. These techniques are also significant for the topics of previous chapters, especially when optical images are further processed digitally.

The next two chapters are devoted to diffractive optics, which is creation of holograms, more commonly called DOEs, in a digital computer, followed by a recording system to create the DOE physically. Generation of a DOE under implementational constraints involves coding of amplitude and phase of an incoming wave, a topic borrowed from communication engineering. There are many such methods. Chapter 15 starts with Lohmann's method, which is the first such method historically. This is followed by two methods, which are useful in a variety of waves such as 3-D image generation, and a method called one-image-only holography, which is capable of generating only the desired image while suppressing the harmonic images due to sampling and nonlinear coding of amplitude and phase. The final section of the chapter is on the binary Fresnel zone plate, which is a DOE acting as a flat lens.

Chapter 16 is a continuation of Chapter 15, and covers new methods of coding DOEs and their further refinements. The method of projections onto convex sets (POCS) discussed in Chapter 14 is used in several ways for this purpose. The methods discussed are virtual holography, which makes implementation easier, iterative interlacing technique (IIT), which makes use of POCS for optimizing a number of subholograms, the ODIFIIT, which is a further refinement of IIT by making use of the decimation-in-frequency property of the FFT, and the hybrid Lohmann–ODIFIIT method, resulting in considerably higher accuracy.

Chapters 17 and 18 are on computerized imaging techniques. The first such technique is synthetic aperture radar (SAR) covered in Chapter 17. In a number of ways, a raw SAR image is similar to the image of a DOE. Only further processing, perhaps more appropriately called decoding, results in a reconstructed image of a terrain of the earth. The images generated are very useful in remote sensing of the earth. The principles involved are optical and diffractive, such as the use of the Fresnel approximation.

In the second part of computerized imaging, computed tomography (CT) is covered in Chapter 18. The theoretical basis for CT is the Radon transform, a cousin of the Fourier transform. The projection slice theorem shows how the 1-D Fourier transforms of projections are used to generate slices of the image spectrum in the 2-D Fourier transform plane. CT is highly numerical, as evidenced by a number of algorithms for image reconstruction in the rest of the chapter.

Optical Fourier techniques have become very important in optical communications and networking. One such area covered in Chapter 19 is arrayed waveguide gratings (AWGs) used in dense wavelength division multiplexing (DWDM). AWG is also called phased array (PHASAR). It is an imaging device in which an array of waveguides are used. The waveguides are different in length by an integer m times the central wavelength so that a large phase difference is achieved from one waveguide to the next. The integer m is quite large, such as 30, and is responsible for the large resolution capability of the phasar device, meaning that the small changes in wavelength can be resolved in the output plane. This is the reason why waveguides are used rather than free space. However, it is diffraction that is used past the waveguides to generate images of points at different wavelengths at the output plane. This is similar to a DOE, which is a sampled device. Hence, the images repeat at certain intervals. This limits the number of wavelengths that can be imaged without interference from other wavelengths. A method called irregularly sampled zero-crossings (MISZCs) is discussed to avoid this problem. The MISZC has its origin in one-image-only holography discussed in Chapter 15.

Scalar diffraction theory becomes less accurate when the sizes of the diffracting apertures are smaller than the wavelength of the incident wave. Then, the Maxwell equations need to be solved by numerical methods. Some emerging approaches for this purpose are based on the method of finite differences, the Fourier modal analysis, and the method of finite elements. The first two approaches are discussed in Chapter 20. First, the paraxial BPM method discussed in Section 12.4 is reformulated in terms of finite differences using the Crank-Nicholson method. Next, the wide-angle BPM using the Pade approximation is discussed. The final sections highlight the finite difference time domain and the Fourier modal method.

Many colleagues, secretaries, friends, and students across the globe have been helpful toward the preparation of this manuscript. I am especially grateful to them for keeping me motivated under all circumstances for a lifetime. I am also very fortunate to have worked with John Wiley & Sons, on this project. They have been amazingly patient with me. Without such patience, I would not have been able to finish the project. Special thanks to George Telecki, the editor, for his patience and support throughout the project.

1

Diffraction, Fourier Optics and Imaging

1.1 INTRODUCTION

When wave fields pass through "obstacles," their behavior cannot be simply described in terms of rays. For example, when a plane wave passes through an aperture, some of the wave deviates from its original direction of propagation, and the resulting wave field is different from the wave field passing initially through the aperture, both in size and shape [Sommerfeld, 2006]. This type of phenomenon is called *diffraction*.

Wave propagation involves diffraction. Diffraction occurs with all types of waves, such as electromagnetic waves, acoustic waves, radio waves, ultrasonic waves, acoustical waves, ocean swells, and so on. Our main concern will be electromagnetic (EM) waves, even though the results are directly applicable to other types of waves as well.

In the past, diffraction was considered a nuisance in conventional optical design. This is because the resolution of an optical imaging system is determined by diffraction. The developments of analog holography (demonstrated in the 1940s and made practical in the 1960s), synthetic aperture radar (1960s), and computer-generated holograms and kinoforms, more generally known as diffractive optical elements (DOE's) (late 1960s) marked the beginning of the development of optical elements based on diffraction. More recently, combination of diffractive and refractive optical elements, such as a refractive lens corrected by diffractive optics, showed how to achieve new design strategies.

Fourier optics involves those topics and applications of optics that involve continuous-space as well as discrete-space Fourier transforms. As such, scalar diffraction theory is a part of Fourier optics. Among other significant topics of Fourier optics, we can cite Fourier transforming and imaging properties of lenses, frequency analysis of optical imaging systems, spatial filtering and optical information processing, analog and computer-generated holography, design and analysis of DOE's, and novel imaging techniques.

The modern theories of diffraction, imaging, and other related topics especially based on Fourier analysis and synthesis techniques have become essential for

Diffraction, Fourier Optics and Imaging, by Okan K. Ersoy
Copyright © 2007 John Wiley & Sons, Inc.

1

understanding, analyzing, and synthesizing modern imaging, optical communications and networking, and micro/nanotechnology devices and systems. Some typical applications include tomography, magnetic resonance imaging, synthetic aperture radar (SAR), interferometric SAR, confocal microscopy, devices used in optical communications and networking such as directional couplers in fiber and integrated optics, analysis of very short optical pulses, computer-generated holograms, analog holograms, diffractive optical elements, gratings, zone plates, optical and microwave phased arrays, and wireless systems using EM waves.

Micro/nanotechnology is poised to develop in many directions and to result in novel products. In this endeavor, diffraction is a major area of increasing significance. All wave phenomena are governed by diffraction when the wavelength(s) of interest is/are of the order of or smaller than the dimensions of diffracting sources. It is clear that technology will aim at smaller and smaller devices/systems. As their complexity increases, a major approach, for example, for testing and analyzing them would be achieved with the help of diffraction.

In advanced computer technology, it will be very difficult to continue system designs with the conventional approach of a clock and synchronous communications using electronic pathways at extremely high speeds. The necessity will increase for using optical interconnects at increasing complexity. This is already happening in the computer industry today with very complex chips. It appears that the day is coming when only optical technologies will be able to keep up with the demands of more and more complex microprocessors. This is simply because photons do not suffer from the limitations of the copper wire.

Some microelectronic laboratory equipment such as the scanning electron microscope and reactive ion etching equipment will be more and more crucial for micro/nanodevice manufacturing and testing. An interesting application is to test devices obtained with such technologies by diffractive methods. Together with other equipment for optical and digital information processing, it is possible to digitize such images obtained from a camera and further process them by image processing. This allows processing of images due to diffraction from a variety of very complex micro- and nanosystems. In turn, the same technologies are the most competitive in terms of implementing optical devices based on diffraction.

In conclusion, Fourier and related transform techniques and diffraction have found significant applications in diverse areas of science and technology, especially related to imaging, communications, and networking. Linear system theory also plays a central role because the systems involved can often be modeled as linear, governed by convolution, which can be analyzed by the Fourier transform.

1.2   EXAMPLES OF EMERGING APPLICATIONS WITH GROWING SIGNIFICANCE

There are numerous applications with increasing significance as the technology matures while dimensions shrink. Below some specific examples that have recently emerged are discussed in more detail.

1.2.1 Dense Wavelength Division Multiplexing/Demultiplexing (DWDM)

Modern techniques for multispectral communications, networking, and computing have been increasingly optical. Topics such as *dense wavelength division multiplexing/demultiplexing* (DWDM) are becoming more significant in the upcoming progress for communications and networking, and as the demand for more and more number of channels (wavelengths) is increasing.

DWDM provides a new direction for solving capacity and flexibility problems in communications and networking. It offers a very large transmission capacity and new novel network architectures. Major components in DWDM systems are the wavelength multiplexers and demultiplexers. Commercially available optical components are based on fiber-optic or microoptic techniques. Research on integrated-optic (de)multiplexers has increasingly been focused on grating-based and phased-array (PHASAR)-based devices (also called arrayed waveguide gratings). Both are imaging devices, that is, they image the field of an input waveguide onto an array of output waveguides in a dispersive way. In grating-based devices, a vertically etched reflection grating provides the focusing and dispersive properties required for demultiplexing. In phased-array-based devices, these properties are provided by an array of waveguides, the length of which has been chosen such as to obtain the required imaging and dispersive properties. As phased-array-based devices are realized in conventional waveguide technology and do not require the vertical etching step needed in grating-based devices, they appear to be more robust and fabrication tolerant. Such devices are based on diffraction to a large degree.

1.2.2 Optical and Microwave DWDM Systems

Technological interest in optical and wireless microwave phased-array dense wavelength division multiplexing systems is also fast increasing. Microwave array antennas providing multiplexing/demultiplexing are becoming popular for wireless communications. For this purpose, use of optical components leads to the advantages of extreme wide bandwidth, miniaturization in size and weight, and immunity from electromagnetic interference and crosstalk.

1.2.3 Diffractive and Subwavelength Optical Elements

Conventional optical devices such as lenses, mirrors, and prisms are based on refraction or reflection. By contrast, *diffractive optical elements* (DOE's), for example, in the form of a phase relief are based on diffraction. They are becoming increasingly important in various applications.

A major factor in the practical implementation of diffractive elements at optical wavelengths was the revolution undergoing in electronic integrated circuits technology in the 1970s. Progress in optical and electron-beam lithography allowed complex patterns to be generated into resist with high precision. Phase control through surface relief with fine-line features and sharp sidewalls was made possible by dry etching techniques. Similar progress in diamond turning machines and laser writers also provided new ways of fabricating diffractive optical elements with high precision.

More recently, the commercial introduction of wafer-based nanofabrication techniques makes it possible to create a new class of optical components called *subwavelength optical elements* (SOEs). With physical structures far smaller than the wavelength of light, the physics of the interaction of these fine-scale surface structures with light yields new arrangements of optical-processing functions. These arrangements have greater density, more robust performance, and greater levels of integration when compared with many existing technologies and could fundamentally alter approaches to optical system design.

1.2.4 Nanodiffractive Devices and Rigorous Diffraction Theory

The physics of such devices depends on rigorous application of the boundary conditions of Maxwell's equations to describe the interaction of light with the structures. For example, at the wavelengths of light used in telecommunications – 980 through 1800 nm – the structures required to achieve those effects have some dimensions on the order of tens to a few hundred nanometers. At the lower end of the scale, single-electron or quantum effects may also be observed. In many applications, subwavelength structures act as a nanoscale diffraction grating whose interaction with incident light can be modeled by rigorous application of diffraction-grating theory and the above-mentioned boundary conditions of Maxwell's equations.

Although these optical effects have been researched in the recent past, cost-effective manufacturing of the optical elements has not been available. Building subwavelength grating structures in a research environment has generally required high-energy techniques, such as electron-beam (E-beam) lithography. E-beam machines are currently capable of generating a beam with a spot size around 5–10 nm. Therefore, they are capable of exposing patterns with lines of widths less than 0.1 µm or 100 nm.

The emergence of all-optical systems is being enabled in large part by new technologies that free systems from the data rate, bandwidth, latencies, signal loss, cost, and protocol dependencies inherent in optical systems with electrical conversion. In addition, the newer technologies, for example, microelectromechanical systems (MEMS)-based micromirrors allow external control of optical switching outside of the optical path. Hence, the electronics and optical parameters can be adjusted independently for optimal overall results. Studies of diffraction with such devices will also be crucial to develop new systems and technologies.

1.2.5 Modern Imaging Techniques

If the source in an imaging system has a property called spatial coherence, the source wavefield is called coherent and can be described as a spatial distribution of complex-valued field amplitude. For example, holography is usually a coherent imaging technique. When the source does not have spatial coherence, it is called incoherent and can be described as a spatial distribution of real-valued intensity. Laser and microwave sources usually represent sources for coherent imaging. Then, the Fourier transform and diffraction are central for the understanding of imaging. Sunlight represents an incoherent source. Incoherent imaging can also be analyzed by Fourier techniques.

A number of computerized modern imaging techniques rely heavily on the Fourier transform and related computer algorithms for image reconstruction. For example, synthetic aperture radar, image reconstruction from projections including computerized tomography, magnetic resonance imaging, confocal microscopy, and confocal scanning microscopy are among such techniques.

2

Linear Systems and Transforms

2.1 INTRODUCTION

Diffraction as well as imaging can often be modeled as *linear systems*. First of all, a system is an input–output mapping. Thus, given an input, the system generates an output. For example, in a diffraction or imaging problem, the input and output are typically a wave at an input plane and the corresponding diffracted wave at a distance from the input plane.

Optical systems are quite analogous to communication systems. Both types of systems have a primary purpose of collecting and processing information. Speech signals processed by communication systems are 1-D whereas images are 2-D. One-dimensional signals are typically temporal whereas 2-D signals are typically spatial. For example, an optical system utilizing a laser beam has spatial coherence. Then, the signals can be characterized as 2-D or 3-D complex-valued *field amplitudes*. Spatial coherence is necessary in order to observe diffraction. Illumination such as ordinary daylight does not have spatial coherence. Then, the signals can be characterized as 2-D spatial, real-valued intensities.

Linear time-invariant and space-invariant communication and optical systems are usually analyzed by *frequency analysis* using the Fourier transform. Nonlinear optical elements such as the photographic film and nonlinear electronic components such as diodes have similar input–output characteristics.

In both types of systems, Fourier techniques can be used for *system synthesis* as well. An example is two-dimensional filtering. Theoretically optical matched filters, optical image processing techniques are analogous to matched filters and image processing techniques used in communications and signal processing.

In this chapter, linear system theory and Fourier transform theory as related especially to diffraction, optical imaging, and related areas are discussed. The chapter consists of eight sections. The properties of linear systems with emphasis on convolution and shift invariance are highlighted in Section 2.2. The 1-D Fourier transform and the continuous-space Fourier transform (simply called the Fourier transform (FT) in the rest of the book) are introduced in Section 2.3. The conditions

Diffraction, Fourier Optics and Imaging, by Okan K. Ersoy
Copyright © 2007 John Wiley & Sons, Inc.

for the existence of the Fourier transform are given in Section 2.4. The properties of the Fourier transform are summarized in Section 2.5.

The Fourier transform discussed so far has a complex exponential kernel. It is actually possible to define the Fourier transform as a real transform with cosine and sine kernel functions. The resulting real Fourier transform is sometimes more useful. The 1-D real Fourier transform is discussed in Section 2.6. Amplitude and phase spectra of the 1-D Fourier transform are defined in Section 2.7.

Especially in optics and wave propagation applications, the 2-D signals sometimes have circular symmetry. In that case, the Fourier transform becomes the Hankel transform in cylindrical coordinates. The Hankel transform is discussed in Section 2.8.

2.2 LINEAR SYSTEMS AND SHIFT INVARIANCE

Linearity allows the decomposition of a complex signal into elementary signals often called basis signals. In Fourier analysis, basis signals or functions are sinusoids.

In a linear system, a given input maps into a unique output. However, more than one input may map into the same output. Thus, the mapping may be one-to-one, or many-to-one.

A 2-D system is shown in Figure 2.1, where $u(x, y)$ is the input signal, and $g(x, y)$ is the output signal. Mathematically, the system can be written as

$$g(x, y) = O[u(x, y)] \qquad (2.2\text{-}1)$$

in the continuous-space case. $O[\bullet]$ is an operator, mapping the input to the output. In the discrete-space case, the point (x, y) is sampled as $[\Delta x \bullet m, \ \Delta y \bullet n]$, where Δx and Δy are the sampling intervals along the two directions. $[\Delta x \bullet m, \ \Delta y \bullet n]$ can be simply represented as $[m, n]$, and the system can be written as

$$g[m, n] = O[u[m, n]] \qquad (2.2\text{-}2)$$

Below the continuous-space case is considered. The system is called linear if any *linear* combination of two inputs $u_1(x, y)$, and $u_2(x, y)$ generates the same combination of their respective outputs $g_1(x, y)$ and $g_2(x, y)$. This is called *superposition principle* and written as

$$O[a_1 u_1(t_1, t_2) + a_2 u_2(x, y)] = a_1 \, O[u_1(x, y)] + a_2 \, O[u_2(x, y)] \qquad (2.2\text{-}3)$$

Figure 2.1. A system diagram.

where a_1 and a_2 are scalars. Above (x, y) is replaced by $[m, n]$ in the case of a linear discrete-space system.

Suppose that the input at (x_1, y_1) is the delta function $\delta(x_1, y_1)$ (see Appendix A for a discussion of the delta function). The output at location (x, y) is defined as

$$h(x, y; x_1, y_1) = O[\delta(x - x_1, y - y_1)] \tag{2.2-4}$$

$h(x_2, y_2; x_1, y_1)$ is called the *impulse response* (*point-spread function*) of the system.

The sifting property of the delta function allows an arbitrary input $u(x, y)$ to be expressed as

$$u(x, y) = \int\limits_{-\infty}^{\infty} \int u(x_1, y_1)\delta(x - x_1, y - y_1) dx_1 dy_1 \tag{2.2-5}$$

Now the output can be written as

$$\begin{aligned} g(x, y) &= O[u(x, y)] \\ &= \int\limits_{-\infty}^{\infty} \int u(x_1, y_1) O[\delta(x - x_1, y - y_1)] dx_1 dy_1 \\ &= \int\limits_{-\infty}^{\infty} u(x_1, y_1) h(x, y; x_1, y_1) dx_1 dy_1 \end{aligned} \tag{2.2-6}$$

This result is known as the *superposition integral*. Physically, the delta function corresponds to a point source. The superposition integral implies that all we need to know is the response of the system to point sources throughout the field of interest in order to characterize the system.

A linear imaging system is called *space invariant* or *shift invariant* if a translation of the input causes the same translation of the output. For a point source at the origin, the output of a shift-invariant system can be written as

$$h(x, y; 0, 0) = O[\delta(x, y)] \tag{2.2-7}$$

If the input is shifted as $\delta(-x_1, -y_1)$, the output of the shift-invariant system must be $h(x - x_1, y - y_1; 0, 0)$. This is usually written simply as

$$h(x, y; x_1, y_1) = h(x - x_1, y - y_1) \tag{2.2-8}$$

Then, the superposition integral becomes

$$g(x, y) = \int\limits_{-\infty}^{\infty} \int u(x_1, y_1) h(x - x_1, y - y_1) dx_1 dy_1 \tag{2.2-9}$$

By a change of variables, this can also be written as

$$g(x, y) = \int\limits_{-\infty}^{\infty} \int h(x_1, y_1) f(x - x_1, y - y_1) dx_1 dy_1 \qquad (2.2\text{-}10)$$

This is the same as the *2-D convolution* of $h(x, y)$ with $u(x, y)$ to yield $g(x, y)$. It is often written symbolically as

$$g(x, y) = h(x, y) * u(x, y) \qquad (2.2\text{-}11)$$

The significance of this result is that a linear shift-invariant (LSI) system is governed by convolution. Hence, the convolution theorem can be used to express the input–output relationship as

$$G(f_x, f_y) = H(f_x, f_y) U(f_x, f_y) \qquad (2.2\text{-}12)$$

where $G(f_x, f_y)$, $H(f_x, f_y)$, and $U(f_x, f_y)$ are the Fourier transforms of $g(x, y)$, $h(x, y)$ and $u(x, y)$, respectively. The Fourier transform is discussed in the next section. $H(f_x, f_y)$ given by

$$H(f_x, f_y) = \int\limits_{-\infty}^{\infty} \int h(x, y) e^{-j 2\pi (f_x x + f_y y)} dx dy, \qquad (2.2\text{-}13)$$

is called the *transfer function* of the system.

In the case of a discrete-space system, the superposition integral becomes the superposition sum, given by

$$g[m, n] = \sum_{m_1=-\infty}^{\infty} \sum_{n_1=-\infty}^{\infty} u[m_1, n_1] h[m, n; m_1, n_1] \qquad (2.2\text{-}14)$$

In the case of a discrete-space LSI system, the convolution integral becomes the convolution sum, given by

$$g[m, n] = \sum_{m_1=-\infty}^{\infty} \sum_{n_1=-\infty}^{\infty} u[m_1, n_1] h[m - m_1, n - n_1], \qquad (2.2\text{-}15)$$

which can also be written as

$$g[m, n] = \sum_{m_1=-\infty}^{\infty} \sum_{m_2=-\infty}^{\infty} h[m_1, n_1] u[m - m_1, n - n_1] \qquad (2.2\text{-}16)$$

The transfer function of a discrete-space LSI system is the discrete-space Fourier transform of the impulse response, given by

$$H(f_x, f_y) = \sum_{m_1=-\infty}^{\infty} \sum_{n_1=-\infty}^{\infty} h[m_1, n_1] e^{-j 2\pi (f_x m_1 \Delta_x + f_2 n_1 \Delta_y)} \tag{2.2-17}$$

The convolution theorem is stated by Eq. (2.2-12) in this case as well.

2.3   CONTINUOUS-SPACE FOURIER TRANSFORM

The property of linearity allows the decomposition of a complex signal into elementary signals often called *basis signals*. In Fourier analysis, basis signals or functions are sinusoids.

The 1-D Fourier transform of a signal $u(t)$, $-\infty \leq t \leq \infty$ is defined as

$$U_c(f) = \int_{-\infty}^{\infty} u(t) e^{-j 2\pi ft} df \tag{2.3-1}$$

The inverse Fourier transform is the representation of $u(t)$ in terms of the basis functions $e^{j 2\pi ft}$ and is given by

$$u(t) = \int_{-\infty}^{\infty} U_c(f) e^{j 2\pi ft} df \tag{2.3-2}$$

Equation (2.3-1)is also referred to as the *analysis equation*. Equation (2.3-2) is the corresponding *synthesis equation*.

The multidimensional (MD) Fourier transform belongs to the set of separable unitary transforms as the transformation kernel is separable along each direction. For example, the 2-D transform kernel $b(x, y, f_x, f_y)$ can be written as

$$b(x, y, f_x, f_y) = b_1(x, f_x) b_2(y, f_y) \tag{2.3-3}$$

$b_i(\bullet, \bullet)$ for i equal to 1 or 2 is the 1-D transform kernel.

The 2-D Fourier transform of a signal $u(x, y)$, $-\infty < x, y < \infty$ is defined as

$$U(f_x, f_y) = \int_{-\infty}^{\infty} \int u(x, y) e^{-j 2\pi (x f_x + y f_y)} dx dy, \tag{2.3-4}$$

where f_x and f_y are the spatial frequencies corresponding to the x- and y-directions, respectively.

The inverse transform is given by

$$u(x,y) = \int\limits_{-\infty}^{\infty} \int U(f_x,f_y)e^{j2\pi(xf_x+yf_y)}\,df_x df_y \tag{2.3-5}$$

2.4   EXISTENCE OF FOURIER TRANSFORM

Sufficient conditions for the existence of the Fourier transform are summarized below:

A. The signal must be absolutely integrable over the infinite space.
B. The signal must have only a finite number of discontinuities and a finite number of maxima and minima in any finite subspace.
C. The signal must have no infinite discontinuities.

Any of these conditions can be weakened if necessary. For example, a 2-D strong, narrow pulse is often represented by a 2-D impulse (Dirac delta) function, defined by

$$\delta(x,y) = \lim_{N\to\infty} N^2 e^{-N^2\pi(x^2+y^2)} \tag{2.4-1}$$

This function fails to satisfy condition C. Two other functions that fail to satisfy condition A are

$$\begin{aligned} u(x,y) &= 1 \\ u(x,y) &= \cos 2\pi f_x x \end{aligned} \tag{2.4-2}$$

With such functions, the Fourier transform is still defined by incorporating generalized functions such as the delta function above and defining the Fourier transform in the limit. The resulting transform is often called the *generalized Fourier transform*.

EXAMPLE 2.1 Find the FT of the 2-D delta function, using its definition.
Solution: Let

$$u(x,y) = N^2 e^{-N^2\pi(x^2+y^2)}$$

Then,

$$\delta(x,y) = \lim_{N\to\infty} f(x,y)$$

Now, we can write

$$U(f_x, f_y) = e^{-\pi(f_x^2 + f_y^2)/N^2}$$

Denoting $\Delta(f_x, f_y)$ as the Fourier transform of $\delta(x, y)$ in the limit as $N \to \infty$, we find

$$\Delta(f_x, f_y) = \lim_{N \to \infty} U(f_x, f_y) = 1$$

2.5   PROPERTIES OF THE FOURIER TRANSFORM

The properties of the 2-D Fourier transform are generalizations of the 1-D Fourier transform. We will need the definitions of even and odd signals in 2-D. A signal $u(x, y)$ is *even* (*symmetric*) if

$$u(x, y) = u(-x, -y) \tag{2.5-1}$$

$u(x, y)$ is *odd* (*antisymmetric*) if

$$u(x, y) = -u(-x, -y) \tag{2.5-2}$$

These definitions really indicate two-fold symmetry. It is possible to extend them to four-fold symmetry in 2-D.

Below we list the properties of the Fourier transform.

Property 1: Linearity
  If $g(x, y) = au_1(x, y) + bu_2(x, y)$, then

$$G(f_x, f_y) = aU_1(f_x, f_y) + bU_2(f_x, f_y) \tag{2.5-3}$$

Property 2: Convolution
  If $g(x, y) = u_1(x, y) * u_2(x, y)$, then

$$G(f_x, f_y) = U_1(f_x, f_y)U_2(f_x, f_y) \tag{2.5-4}$$

Property 3: Correlation is similar to convolution, and the correlation between $u_1(x, y)$ and $u_2(x, y)$ is given by

$$g(x, y) = \iint\limits_{-\infty}^{\infty} u_1(x_1, y_1)u_2(x + x_1, y + y_1)dx_1 dy_1 \tag{2.5-5}$$

If $g(x, y) = u_1(x, y) \circ u_2(x, y)$, where $\circ$ denotes 2-D correlation, then

$$U(f_x, f_y) = U_1(f_x, f_y)U_2^*(f_x, f_y) \tag{2.5-6}$$

Property 4: Modulation
  If $g(x, y) = u_1(x, y)u_2(x, y)$, then

$$G(f_x, f_y) = F_1(f_x, f_y) * F_2(f_x, f_y) \tag{2.5-7}$$

Property 5: Separable function
  If $g(x, y) = u_1(x)u_2(y)$, then

$$G(f_x, f_y) = U_1(f_x)U_2(f_y) \tag{2.5-8}$$

Property 6: Space shift
  If $g(x, y) = u(x - x_0, y - y_0)$, then

$$G(f_x, f_y) = e^{-j2\pi(f_x x_0 + f_y y_0)} \bullet U(f_x, f_y) \tag{2.5-9}$$

Property 7: Frequency shift
  If $g(x, y) = e^{j2\pi(f_{x0}x + f_{y0}y)} \bullet u(x, y)$, then

$$G(f_x, f_y) = U(f_x - f_{x0}, f_2 - f_{y0}) \tag{2.5-10}$$

Property 8: Differentiation in space domain
  If $g(x, y) = \left(\partial^k / \partial x^k\right)\left(\partial^\ell / \partial y^\ell\right)u(x, y)$, then

$$G(f_x, f_y) = (2\pi j f_x)^k (2\pi j f_y)^\ell U(f_x, f_y) \tag{2.5-11}$$

Property 9: Differentiation in frequency domain
  If $g(x, y) = (-j2\pi x)^k (-j2\pi y)^\ell u(x, y)$, then

$$G(f_x, f_y) = \frac{\partial^k}{\partial f_x^k} \frac{\partial^\ell}{\partial f_y^\ell} U(f_x, f_y) \tag{2.5-12}$$

Property 10: Parseval's theorem

$$\int\limits_{-\infty}^{\infty} \int u(x, y)g^*(x, y)\mathrm{d}x\mathrm{d}y = \int\limits_{-\infty}^{\infty} \int U(f_x, f_y)G^*(f_x, f_y)\mathrm{d}f_x\mathrm{d}f_y \tag{2.5-13}$$

Property 11: Real $u(x, y)$

$$U(f_x, f_y) = U^*(-f_x, -f_y) \tag{2.5-14}$$

Property 12: Real and even $u(x, y)$

$$U(f_x, f_y) \text{ is real and even}$$

Property 13: Real and odd $u(x, y)$

$$U(f_x, f_y) \text{ is imaginary and odd}$$

Property 14: Laplacian in the space domain

If $g(x, y) = \left(\dfrac{\partial^2}{\partial x^2} + \dfrac{\partial^2}{\partial y^2} \right) u(x, y)$, then

$$G(f_x, f_y) = -4\pi^2 (f_x^2 + f_y^2) U(f_x, f_y) \qquad (2.5\text{-}15)$$

Property 15: Laplacian in the frequency domain
If $g(x, y) = -4\pi^2 (x^2 + y^2) u(x, y)$, then

$$G(f_x, f_y) = \left(\dfrac{\partial^2}{\partial f_x^2} + \dfrac{\partial^2}{\partial f_y^2} \right) U(f_x, f_y) \qquad (2.5\text{-}16)$$

Property 16: Square of signal
If $g(x, y) = |u(x, y)|^2$, then

$$G(f_x, f_y) = U(f_x, f_y) * U^*(f_x, f_y) \qquad (2.5\text{-}17)$$

Property 17: Square of spectrum
If $g(x, y) = u(x, y) * u^*(x, y)$, then

$$G(f_x, f_y) = |U(f_x, f_y)|^2 \qquad (2.5\text{-}18)$$

Property 18: Rotation of axes
If $g(x, y) = u(\pm x, \pm y)$, then

$$G(f_x, f_y) = U(\pm f_x, \pm f_y) \qquad (2.5\text{-}19)$$

The important properties of the FT are summarized in Table 2-1.

EXAMPLE 2.2 Find the 1-D FT of

$$g(x) = \int\limits_{-\infty}^{\infty} u(x, y) \, dy$$

as a function of the FT of $u(x, y)$.

Table 2.1. **Properties of the Fourier transform (a, b, f_{x0} and f_{y0} are real nonzero constants; k and l are nonnegative integers).**

Property	$g(x,y)$	$G(f_x,f_y)$
Linearity	$au_1(x,y) + bu_2(x,y)$	$aU_1(f_x,f_y) + bU_2(f_x,f_y)$
Convolution	$u_1(x,y) * u_2(x,y)$	$U_1(f_x,f_y)U_2(f_x,f_y)$
Correlation	$u_1(x,y) \circ u_2(x,y)$	$U_1(f_x,f_y)U_2^*(f_x,f_y)$
Modulation	$u_1(x,y)u_2(x,y)$	$U_1(f_x,f_y) * U_2(f_x,f_y)$
Separable function	$u_1(x)u_2(y)$	$U_1(f_x)U_2(f_y)$
Space shift	$u(x - x_0, y - y_0)$	$e^{-j2\pi(f_x x_0 + f_y y_0)} \cdot U(f_x,f_y)$
Frequency shift	$g(x,y) = e^{j2\pi(f_{x0}x + f_{y0}y)} \cdot u(x,y)$	$G(f_x,f_y) = U(f_x - f_{x0}, f_2 - f_{y0})$
Differentiation in space domain	$\dfrac{\partial^k}{\partial x^k}\dfrac{\partial^l}{\partial y^l} u(x,y)$	$(2\pi j f_x)^k (2\pi j f_y)^l U(f_x,f_y)$
Differentiation in frequency domain	$(-j2\pi x)^k (-j2\pi y)^l u(x,y)$	$\dfrac{\partial^k}{\partial f_x^k}\dfrac{\partial^l}{\partial f_y^l} U(f_x,f_y)$
Laplacian in the space domain	$\left(\dfrac{\partial^2}{\partial x^2} + \dfrac{\partial^2}{\partial y^2}\right) u(x,y)$	$-4\pi^2(f_x^2 + f_y^2)U(f_x,f_y)$
Laplacian in the frequency domain	$-4\pi^2(x^2 + y^2)u(x,y)$	$\left(\dfrac{\partial^2}{\partial f_x^2} + \dfrac{\partial^2}{\partial f_y^2}\right) U(f_x,f_y)$
Square of signal	$|u(x,y)|^2$	$U(f_x,f_y) * U^*(f_x,f_y)$
Square of spectrum	$u(x,y) * u^*(x,y)$	$|U(f_x,f_y)|^2$
Rotation of axes	$u(\pm x, \pm y)$	$U(\pm f_x, \pm f_y)$
Parseval's theorem	$\displaystyle\int_{-\infty}^{\infty}\int u(x,y)g^*(x,y)\mathrm{d}x\mathrm{d}y = $	$\displaystyle\int_{-\infty}^{\infty}\int U(f_x,f_y)G^*(f_x,f_y)\mathrm{d}f_x\mathrm{d}f_y$
Real $u(x,y)$		$U(f_x,f_y) = U^*(-f_x, -f_y)$
Real and even $u(x,y)$		$U(f_x,f_y)$ is real and even
Real and odd $u(x,y)$		$U(f_x,f_y)$ is imaginary and odd

Solution: $g(x)$ and $\delta(x)$ are considered as 2-D functions $g(x) \bullet 1$ and $\delta(x) \bullet 1$, respectively. Then, $g(x)$ can be written as

$$g(x) \bullet 1 = u(x,y) * [\delta(x) \bullet 1]$$

$$= \iint_{-\infty}^{\infty} u(x_1, y_1)\delta(x_1 - x)\mathrm{d}x_1\mathrm{d}y_1$$

$$= \int_{-\infty}^{\infty} u(x,y)\mathrm{d}y \tag{2.5-20}$$

Computing the FT of both sides of Eq. (2.5-20) and using the convolution theorem and property 5 of separable functions gives

$$G(f_x)\delta(f_y) = U(f_x,f_y)\delta(f_y)$$

or

$$G(f_x) = U(f_x,0)$$

EXAMPLE 2.3 Find the FT of

$$g(x,y) = u(a_1x + b_1y + c_1, a_2x + b_2y + c_2)$$

as a function of the Fourier transform of $u(x,y)$.

Solution:

$$G(f_x,f_y) = \int\!\!\!\int_{-\infty}^{\infty} u(a_1x + b_1y + c_1, a_2x + b_2y + c_2)e^{-j2\pi(f_1x+f_2y)}dxdy$$

Letting $x_1 = a_1x + b_1y + c_1$ gives

$$G(f_x,f_y) = \int\!\!\!\int_{-\infty}^{\infty} u\left(x_1, a_2\left(\frac{x_1 - b_1y - c_1}{a_1}\right) + b_2y + c_2\right)e^{-j2\pi\left[f_x\left(\frac{x_1-a_1y-c_1}{a_1}\right)+f_yy\right]}\frac{dx_1dy}{|a_1|}$$

Also letting $y_1 = a_2\left(\frac{x_1-b_1y-c_1}{a_1}\right) + b_2y + c_2$ gives

$$G(f_x,f_y) = \frac{1}{|D|}\int\!\!\!\int_{-\infty}^{\infty} u(x_1,y_1)e^{-j2\pi[x_1f_x'+y_1f_y'+T_1f_x+T_2f_y]}dx_1dy_1 \qquad (2.5\text{-}21)$$

where

$$D = a_2b_1 - a_1b_2$$

$$f_x' = \frac{1}{D}(-a_1f_x + a_2f_y)$$

$$f_y' = \frac{1}{D}(b_1f_x - a_1f_y)$$

$$T_1 = \frac{b_1c_2 - b_2c_1}{D}$$

$$T_2 = \frac{a_2c_1 - a_1c_2}{D}$$

Equation (2.5-21) is the same as

$$G(f_x, f_y) = \frac{1}{|D|} e^{j2\pi(T_1 f_x + T_2 f_y)} U(f_x', f_y')$$

In other words, when the input signal is scaled, shifted, and skewed, its transform is also scaled, skewed, and linearly phase-shifted, but not shifted in position.

EXAMPLE 2.4 Find the FT of the "one-zero" function defined by

$$u(x, y) = \begin{cases} 1 & h_1(x) < y < h_2(x) \\ 0 & \text{otherwise} \end{cases},$$

where $h_1(x)$ and $h_2(x)$ are given single-valued functions of x. Determine $U(f_x, f_y)$ and $U(f_x, 0)$.

Solution: $u(x, y)$ is as shown in Figure 2.2. Its FT is given by

$$U(f_x, f_y) = \iint\limits_{-\infty}^{\infty} u(x, y) e^{-j2\pi(f_x x + f_y y)} \, dx dy$$

$$= \int\limits_{-\infty}^{\infty} e^{-j2\pi f_x x} dx \int\limits_{h_1(x)}^{h_2(x)} e^{-j2\pi f_y y} dy$$

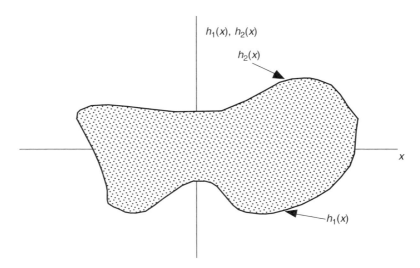

Figure 2.2.   A typical zero-one function.

When $f_y = 0$,

$$U(f_x, 0) = \int_{-\infty}^{\infty} [h_2(x) - h_1(x)]e^{-j2\pi f_x x}dx$$

$$= H_2(f_x) - H_1(f_x)$$

Thus, $U(f_x, 0)$ is the difference of the 1-D FT of the functions $h_2(x)$ and $h_1(x)$.

2.6   REAL FOURIER TRANSFORM

Sometimes it is more convenient to represent the Fourier transform with real sine and cosine basis functions. Then, it is referred to as the real Fourier transform (RFT). What was discussed as the Fourier transform before would then be the complex Fourier transform (CFT) [Ersoy, 1994]. For example, the analytic signal representation of nonmonochromatic wave fields can be more effectively derived using the RFT, as discussed in Section 9.4. In this and next sections, we will discuss the 1-D transforms only.

The RFT of a signal $u(x)$ can be defined as

$$U(f) = 2w(f) \int_{-\infty}^{\infty} u(t) \cos(2\pi ft + \theta(f))dx \qquad (2.6-1)$$

where

$$w(f) = \begin{cases} 1 & f \neq 0 \\ 1/2 & f = 0 \end{cases} \qquad (2.6-2)$$

$$\theta(f) = \begin{cases} 0 & f \geq 0 \\ \pi/2 & f < 0 \end{cases} \qquad (2.6-3)$$

The *inverse RFT* is given by

$$u(t) = \int_{-\infty}^{\infty} U(f) \cos(2\pi ft + \theta(f))df \qquad (2.6-4)$$

It is observed that $\cos(2\pi ft + \theta(f))$ equals $\cos(2\pi ft)$ for $f = 0$ and $\sin(2\pi|f|t)$ for $f < 0$. This is a "trick" used to cover both the cosine and sine basis functions in a single integral. Negative frequencies are used for this purpose. It is interesting

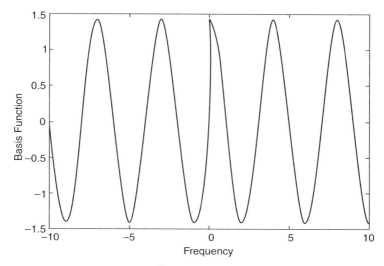

Figure 2.3. The basis function $\sqrt{2}\cos(2\pi ft + \theta(f))$ for $t = 0.25$ and $-10 < f < 10$.

to observe that $\sqrt{2}\cos(2\pi ft + \theta(f))$ is orthonormal with respect to f because Eq. (2.6-4) is true. Thus,

$$2\int_{0}^{\infty}\cos(2\pi ft + \theta(f))\cos(2\pi f\tau + \theta(f))df = \delta(t - \tau) \qquad (2.6\text{-}5)$$

where $\delta(\bullet)$ is the Dirac-delta function.

The basis function $\sqrt{2}\cos(2\pi ft + \theta(f))$ is shown in Figure 2.3 for $t = 0.25$ and $-10 < f < 10$.

Equations (2.6-1) and (2.6-4) can also be written for $f = 0$ as

$$U_1(f) = 2w(f)\int_{-\infty}^{\infty} u(t)\cos(2\pi ft)dt \qquad (2.6\text{-}6)$$

$$U_0(f) = 2\int_{-\infty}^{\infty} u(t)\sin(2\pi ft)dt \qquad (2.6\text{-}7)$$

and

$$u(t) = \int_{0}^{\infty}[U_1(f)\cos(2\pi ft) + U_0(f)\sin(2\pi ft)]df \qquad (2.6\text{-}8)$$

Thus, $U(f)$ equals $U_1(f)$ for $f = 0$ and $X_0(|f|)$ for $f < 0$. $X_1(f)$ and $X_0(f)$ will be referred to as *the cosine* and *sine parts*, respectively. Equations (2.6-1), (2.6-6), and (2.6-7) are also referred to as the *analysis equations*. Equations (2.6-4) and (2.6-8) are the corresponding *synthesis equations*.

The relationship between the CFT and the RFT can be expressed for $f \geq 0$ as

$$U_c(0) = U(0)$$

$$\begin{bmatrix} U_c(f) \\ U_c(-f) \end{bmatrix} = \frac{1}{2} \begin{bmatrix} 1 & -j \\ 1 & j \end{bmatrix} \begin{bmatrix} U_1(f) \\ U_0(f) \end{bmatrix} f > 0 \qquad (2.6\text{-}9)$$

Equation (2.6-9) reflects the fact that $U_1(f)$ and $U_0(f)$ are even and odd functions, respectively.

The inverse of Eq. (2.6-9) for $f \geq 0$ is given by

$$U_1(0) = U_c(0)$$

$$\begin{bmatrix} U_1(f) \\ U_0(f) \end{bmatrix} = \begin{bmatrix} 1 & 1 \\ j & -j \end{bmatrix} \begin{bmatrix} U_c(f) \\ U_c(-f) \end{bmatrix} f > 0 \qquad (2.6\text{-}10)$$

Equations (2.6-9) and (2.6-10) are useful to convert from one representation to the other. When $x(t)$ is real, $U_1(f)$ and $U_0(f)$ are also real. Then, Eq. (2.6-9) shows that $U_c(f)$ and $U_c(-f)$ are complex conjugates of each other.

2.7 AMPLITUDE AND PHASE SPECTRA

$U_c(f)$ can be written as

$$U_c(f) = U_a(f)\, e^{j\phi(f)}, \qquad (2.7\text{-}1)$$

where the *amplitude (magnitude) spectrum* $U_a(f)$ and the *phase* spectrum $\phi(f)$ of the signal $x(t)$ are defined as

$$U_a(f) = |U_c(f)| = \frac{1}{2} \left[|U_1(f)|^2 + |U_0(f)|^2 \right]^{1/2} \qquad (2.7\text{-}2)$$

$$\phi(f) = \tan^{-1} \left[\frac{\text{Imaginary }[U_c(f)]}{\text{Real }[U_c(f)]} \right] \qquad (2.7\text{-}3)$$

$U_a(f)$ is an even function. With real signals, $\phi(f)$ is an odd function and can be written as

$$\phi(f) = \tan^{-1}[-U_0(f)/U_1(f)] \qquad (2.7\text{-}4)$$

Equation (1.2.2) for the inverse CFT can be written in terms of the amplitude and phase spectra as

$$u(t) = \int_{-\infty}^{\infty} U_a(f) e^{j[2\pi ft + \phi(f)]} \mathrm{d}f \tag{2.7-5}$$

When $u(t)$ is real, this equation reduces to

$$u(t) = 2 \int_{0}^{\infty} U_a(f) \cos[2\pi ft + \phi(f)] \mathrm{d}f \tag{2.7-6}$$

because $U_c(f) = U_c^*(-f)$, and the integrand in

$$j \int_{-\infty}^{\infty} U_a(f) \sin[2\pi ft + \phi(f)] \mathrm{d}f$$

is an odd function and integrates to zero.

Equation (2.7-6) will be used in representing nonmonochromatic wave fields in Section 9.4.

2.8   HANKEL TRANSFORMS

Functions having radial symmetry are easier to handle in polar coordinates. This is often the case, for example, in optics where lenses, aperture stops, and so on are often circular in shape.

Let us first consider the Fourier transform in polar coordinates. The rectangular and polar coordinates are shown in Figure 2.4. The transformation to polar

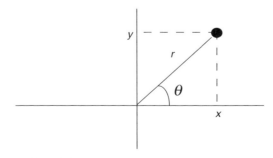

Figure 2.4.   The rectangular and polar coordinates.

coordinates is given by

$$r = [x^2 + y^2]^{1/2}$$

$$\theta = \tan^{-1}\left(\frac{y}{x}\right)$$

$$\rho = [f_x^2 + f_y^2]^{1/2} \tag{2.8-1}$$

$$\phi = \tan^{-1}\left(\frac{f_y}{f_x}\right)$$

The FT of $u(x, y)$ is given by

$$U(f_x, f_y) = \iint\limits_{-\infty}^{\infty} u(x, y)e^{-j2\pi\,(f_x x + f_y y)}\,dxdy \tag{2.8-2}$$

$f(x, y)$ in polar coordinates is $f(r, \theta)$. $F(f_x, f_y)$ in polar coordinates is $F(\rho, \phi)$, given by

$$U(\rho, \phi) = \int\limits_{0}^{2\pi} d\theta \int\limits_{0}^{\infty} u(r, \theta)e^{-j2\pi\,r\rho(\cos\theta\cos\phi + \sin\theta\sin\phi)}\,rdr \tag{2.8-3}$$

When $u(x, y)$ is circularly symmetric, it can be written as

$$u(r, \theta) = u(r) \tag{2.8-4}$$

Then, Eq. (2.6-3) becomes

$$U(\rho, \phi) = \int\limits_{0}^{\infty} u(r)rdr \int\limits_{0}^{2\pi} e^{-j2\pi r\rho\cos(\theta - \phi)}\,d\theta \tag{2.8-5}$$

The Bessel function of the first kind of zero order is given by

$$J_0(t) = \frac{1}{2\pi}\int\limits_{0}^{2\pi} e^{-jt\cos(\theta - \phi)}d\theta \tag{2.8-6}$$

It is observed that $J_0(t)$ is the same for all values of ϕ. Substituting this identity in Eq. (2.8-5) and incorporating an extra factor of 2π gives

$$U(\rho, \phi) = U(\rho) = 2\pi\int\limits_{0}^{\infty} u(r)J_0(2\pi r\rho)rdr \tag{2.8-7}$$

Thus, the 2-D FT of a circularly symmetric function is itself circularly symmetric and is given by Eq. (2.8-7). This relation is called the *Hankel transform of zero order* or the *Fourier-Bessel transform*.

It can be easily shown that the inverse Hankel transform of zero order is given by the same type of integral as

$$u(r) = 2\pi \int_0^\infty U(\rho)J_0(2\pi r\rho)\rho \mathrm{d}\rho \tag{2.8-8}$$

EXAMPLE 2.5 Derive Eq. (2.8-5).

Solution: x, y and f_x, f_y are given by

$$x = r\cos\theta \qquad y = r\sin\theta$$
$$f_x = \rho\cos\phi \qquad f_y = \rho\sin\phi$$

Equation (2.8-2) becomes

$$U(\rho, \phi) = \int_0^{2\pi} \mathrm{d}\theta \int_0^\infty u(r, \theta)e^{-j2\pi r\rho(\cos\theta\cos\phi + \sin\theta\sin\phi)}J\mathrm{d}r \tag{2.8-9}$$

where J is the *Jacobian* given by

$$J = \begin{vmatrix} \dfrac{\partial x}{\partial r} & \dfrac{\partial x}{\partial \theta} \\ \dfrac{\partial y}{\partial r} & \dfrac{\partial y}{\partial \theta} \end{vmatrix} = \begin{vmatrix} \cos\theta & -r\sin\theta \\ \sin\theta & r\cos\theta \end{vmatrix} = r \tag{2.8-10}$$

where $|\bullet|$ indicates determinant. Substituting $J = r$ in Eq. (2.8-9) gives the desired result.

EXAMPLE 2.6 (a) Find the Hankel transform of the cylinder function $\mathrm{cyl}(r)$ defined by

$$\mathrm{cyl}(r) = \begin{cases} 1 & 0 \le r < \dfrac{1}{2} \\ \dfrac{1}{2} & r = \dfrac{1}{2} \\ 0 & r > \dfrac{1}{2} \end{cases} \tag{2.8-11}$$

(b) Find the Hankel transform of $\mathrm{cyl}(r/D)$.

Solution: (a) Let

$$u(r) = \text{cyl}(r)$$

Then

$$U(\rho) = 2\pi \int_0^\infty u(r)J_0(2\pi r\rho)r\,\mathrm{d}r$$

$$= 2\pi \int_0^{1/2} J_0(2\pi r\rho)r\,\mathrm{d}r$$

The Bessel function of the first kind of first order is given by [Erdelyi]

$$J_1(\pi\rho) = 4\pi\rho \int_0^{1/2} J_0(2\pi r\rho)r\,\mathrm{d}r \qquad (2.8\text{-}12)$$

Hence,

$$U(\rho) = \frac{J_1(\pi\rho)}{2\rho}$$

The sombrero function $\text{somb}(r)$ is defined by

$$\text{somb}(r) = \frac{2J_1(\pi r)}{\pi r} \qquad (2.8\text{-}13)$$

$U(\rho)$ is related to $\text{somb}(\rho)$ by

$$U(\rho) = \frac{\pi}{4}\text{somb}(\rho)$$

(b) We have

$$U'(\rho) = 2\pi \int_0^{D/2} J_0(2\pi r\rho)r\,\mathrm{d}r$$

Let $r' = r/D$. Then, $\mathrm{d}r = D\mathrm{d}r'$, and

$$U'(\rho) = 2\pi D^2 \int_0^{1/2} r'J_0(2\pi rD\rho)\mathrm{d}r'$$

$$= D^2 U(\rho) = \frac{D^2\pi}{4}\text{somb}(D\rho)$$

3

Fundamentals of Wave Propagation

3.1 INTRODUCTION

In this chapter and Chapter 4, waves are considered in 3-D, in general. However, in some applications such as in integrated optics in which propagation of waves on a surface is often considered, 2-D waves are of interest. For example, see Chapter 19 on dense wavelength division multiplexing. Two-dimensional equations are simpler because one of the space variables, say, y is omitted from the equations. Hence, the results discussed in 3-D in what follows can be easily reduced to the 2-D counterparts.

Electromagnetic (EM) waves will be of main concern. They are generated when a time-varying electric field $\mathbf{E}(\mathbf{r}, t)$ produces a time-varying field $\mathbf{H}(\mathbf{r}, t)$. EM waves propagate through *unguided media* such as free space or air and in *guided media* such as an optical fiber or the medium between the earth's surface and the ionosphere. In this chapter, we will be mainly concerned with unbounded media.

Spherical waves result when a source such as an antenna emits EM energy as shown in Figure 3.1(a). At a far away distance from the source, the spherical wave appears like a plane wave with uniform properties at all points of the wavefront, as seen in Figure 3.1(b). Another example would be an electric dipole directed along the z-axis, located at the origin, and oscillating with the circular frequency w. It generates electric and magnetic fields with a complicated expression, but far from the origin where the fields look like plane waves. A perfect plane wave does not exist physically, but it is a component that is very useful in modeling all kinds of waves.

Waves propagate in a medium. In the case of optical waves, the optical medium is characterized by a quantity n called the *refractive index*. It is the ratio of the speed of light in free space to that of the speed of light in the medium. The medium is *homogeneous* if n is constant, otherwise, it is *inhomogeneous*. In this chapter, we will assume that the medium is homogeneous.

The chapter consists of seven sections. How waves come about and some of their fundamental properties are discussed in Section 3.2. The fundamental properties of EM waves and the Kirchoff equations that characterize them are discussed in Section 3.3. The phasor representation is reviewed in Section 3.4. Wave equations,

Diffraction, Fourier Optics and Imaging, by Okan K. Ersoy
Copyright © 2007 John Wiley & Sons, Inc.

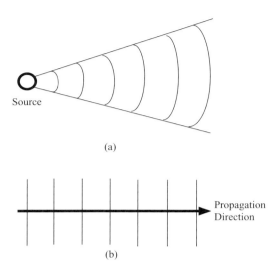

(a)

Propagation
Direction

(b)

Figure 3.1. (a) Spherical wave generated by a source; (b) plane wave with uniform properties along the direction of propagation.

the wave equation in a source free medium as well as the plane wave solution with wave number and direction cosines, are described in Section 3.5. Wave equations in phasor representation in a charge-free medium are discussed in Section 3.6. Plane waves are fundamental components of EM waves. They are described in more detail in Section 3.7, including their polarization properties.

3.2   WAVES

Nature is rich in a large variety of waves, such as electromagnetic, acoustical, water, and brain waves. A wave can be considered as a disturbance of some kind that can travel with a fixed velocity and is unchanged in form from point to point.

Let $u(x, t)$ denote a 1-D wave in the x-direction in a homogeneous medium. If v is its velocity, $u(x, t)$ satisfies

$$u(x, t) = u(x - vt, 0) \tag{3.2-1}$$

if it is traveling to the right and

$$u(x, t) = u(x + vt, 0) \tag{3.2-2}$$

if it is traveling to the left.

Assuming the wave is traveling to the left, let s be given by

$$s = x + vt \tag{3.2-3}$$

Then, the following can be computed:

$$\frac{\partial u}{\partial x} = \frac{\partial u}{\partial s}$$

$$\frac{\partial^2 u}{\partial x^2} = \frac{\partial^2 u}{\partial s^2}$$

$$\frac{\partial u}{\partial t} = \frac{\partial s}{\partial t}\frac{\partial u}{\partial s} = v\frac{\partial u}{\partial s} \qquad (3.2\text{-}4)$$

$$\frac{\partial^2 u}{\partial t^2} = v^2\frac{\partial^2 u}{\partial s^2}$$

Hence,

$$\frac{\partial^2 u(x,t)}{\partial x^2} = \frac{1}{v^2}\frac{\partial^2 u(x,t)}{\partial t^2} \qquad (3.2\text{-}5)$$

This equation is known as the *nondispersive wave equation*. A particular solution, which is also a solution for all other wave equations, is the simple harmonic solution given by

$$\Psi(x,t) = A\cos(k(x+vt)) = A\cos(kx+wt) = A\cos\left(2\pi\left(\frac{x}{\lambda}+ft\right)\right), \qquad (3.2\text{-}6)$$

where $\omega = 2\pi f$ and $k = 2\pi/\lambda$. λ is the *wavelength*, f is the *time frequency*, $1/\lambda$ is the *spatial frequency*, and ω and k are the corresponding *angular frequencies*. k is also known as the *wave number*. $\Psi(x,t)$ given by Eq. (3.2-6), is also known as the *1-D plane wave*.

Because of mathematical ease, Eq. (3.2-6) is often written as

$$u(x,t) = \text{Re}[Ae^{j(kx+\omega t)}] \qquad (3.2\text{-}7)$$

or simply as

$$u(x,t) = Ae^{j(kx+\omega t)} \qquad (3.2\text{-}8)$$

"real part" being understood from the content. We reassert that the wave number k is the *radian spatial frequency* along the x-direction. It can also be written as

$$k = 2\pi f_x \qquad (3.2\text{-}9)$$

where f_x is the *spatial frequency* in the x-direction in cycles/unit length.

Substituting Eq. (3.2-8) into Eq. (3.2-5) gives

$$v = \frac{\omega}{k} \qquad (3.2\text{-}10)$$

v is also known as the *phase velocity* and can be larger than the vacuum speed of light in the case of light waves.

More generally, a 1-D wave given by Eq. (3.2-8) may satisfy another partial differential equation with respect to x and t, in addition to Eq. (3.2-5). A general wave equation that has a simple harmonic solution can be written as

$$f\left(\frac{\partial}{\partial x}, \frac{\partial}{\partial t}\right) u(x, t) = 0, \qquad (3.2\text{-}11)$$

where f is a polynomial function of $\frac{\partial}{\partial x}$ and $\frac{\partial}{\partial t}$. For example,

$$\frac{\partial^3 u}{\partial x^3} + \alpha \frac{\partial^3 u}{\partial x^2 \partial t} - \mathcal{B}\frac{\partial u}{\partial t} = 0 \qquad (3.2\text{-}12)$$

Using Eq. (3.2-8) for the simple harmonic solution allows the substitution of $\frac{\partial}{\partial t} \to j\omega$, and Eq. (3.2-12) becomes

$$k^3 - \alpha k^2 \omega - \mathcal{B}\omega = 0 \qquad (3.2\text{-}13)$$

Such an equation relating ω and k is known as *the dispersion equation*. The phase velocity v in this case is given by

$$v = \frac{\omega}{k} = \frac{k^2}{\alpha k^2 + \mathcal{B}} \qquad (3.2\text{-}14)$$

A wave with a phase velocity, which is a function of wave number or wavelength other than Eq. (3.2-10), is known as a *dispersive wave*. Hence, Eq. (3.2-14) is a *dispersive wave equation*.

Some examples of waves are the following:

A. Mechanical waves, such as longitudinal sound waves in a compressible fluid, are governed by the wave equation

$$\frac{\partial^2 u(x, t)}{\partial x^2} = \frac{\rho}{K_c} \frac{\partial^2 u(x, t)}{\partial t^2}, \qquad (3.2\text{-}15)$$

where ρ is the fluid density, and K_c is the compressibility. Hence, the phase velocity is given by

$$v = \sqrt{\frac{K_c}{\rho}} \qquad (3.2\text{-}16)$$

B. Heat diffuses under steady-state conditions according to the wave equation

$$\frac{K_h}{s}\frac{\partial^2 u(x,t)}{\partial x^2} = \frac{\partial u(x,t)}{\partial t},\qquad(3.2\text{-}17)$$

where K_h is the thermal conductivity, s is the specific heat per unit volume, and $u(x,t)$ is the local temperature.

C. The Schrodinger's wave equation in quantum mechanics is given by

$$-\frac{h^2}{4\pi^2 m}\frac{\partial^2 u(x,t)}{\partial x^2} + V(x)u(x,t) = -j\frac{h}{2\pi}\frac{\partial u(x,t)}{\partial t},\qquad(3.2\text{-}18)$$

where m is mass, $V(x)$ is potential energy of the particle, h is Planck's constant, and $u(x,t)$ is the state of the particle. The dispersion relation can be written as

$$\hbar\omega = \frac{\hbar^2 k^2}{2m} + V(x),\qquad(3.2\text{-}19)$$

where $\hbar$ equals $h/2\pi$.

D. Time-varying imagery can be considered to be a 3-D function $g(x,y,t)$. Its 3-D Fourier representation can be written as

$$g(x,y,t) = \int\!\!\!\int\!\!\!\int_{-\infty}^{\infty} G(f_x,f_y,f)e^{j2\pi(f_x x + f_y y + ft)}\,df_x df_y df,\qquad(3.2\text{-}20)$$

where

$$\begin{aligned} G(f_x,f_y,f) &= \int\!\!\!\int\!\!\!\int_{-\infty}^{\infty} g(x,y,t)e^{-j2\pi(f_x x + f_y y + ft)}\,dxdydt\\ &= A(f_x,f_y,f)e^{j\Theta(f_x,f_y,f)} \end{aligned}\qquad(3.2\text{-}21)$$

For real $g(x,y,t)$, $G(f_x,f_y,f) = G^*(-f_x,-f_y,-f)$ so that Eq. (3.2-21) can be written as

$$g(x,y,t) = 2\int_{0}^{\infty}\left[\int_{-\infty}^{\infty}\int_{-\infty}^{\infty} A(f_x,f_y,f)\cos(2\pi(f_x x + f_y y + ft) + \Theta(f_x,f_y,f))df_x df_y df\right]$$

$$(3.2\text{-}22)$$

$A(f_x,f_y,f)\cos(2\pi(f_x x + f_y y + ft) + \Theta(f_x,f_y,f))$ is a 2-D plane wave with amplitude $A(f_x,f_y,f)$, phase $\Theta(f_x,f_y,f)$, spatial frequencies f_x and f_y, and time frequency f. In this case, the wave vector $\mathbf{k}$ is given by

$$\mathbf{k} = 2\pi(f_x \mathbf{e_x} + f_y \mathbf{e_y}) = k_x \mathbf{e_x} + k_y \mathbf{e_y}, \qquad (3.2\text{-}23)$$

where $\mathbf{e_x}$ and $\mathbf{e_y}$ are the unit vectors along the x- and y-directions, respectively. The components k_x and k_y are the radian spatial frequencies along the x-direction and y-direction, respectively. f_x and f_y are the corresponding spatial frequencies.

The 2-D plane wave can be written as

$$u(\mathbf{r},t) = A(\mathbf{k},\omega)\cos(\mathbf{k} \cdot \mathbf{r} + \omega t + \Theta(\mathbf{k},\omega)), \qquad (3.2\text{-}24)$$

where

$$\mathbf{r} = x\mathbf{e_x} + y\mathbf{e_y} \qquad (3.2\text{-}25)$$

If we assume $u(\mathbf{r},t)$ moves in the direction $\mathbf{k}$ with velocity v, $u(\mathbf{r},t)$ can be shown to be a solution to the 2-D nondispersive wave equation given by

$$\nabla^2 u(\mathbf{r},t) = \frac{1}{v^2}\frac{\partial^2 u(\mathbf{r},t)}{\partial t^2}, \qquad (3.2\text{-}26)$$

where

$$\nabla^2 = \frac{\partial^2}{\partial x^2} + \frac{\partial^2}{\partial y^2} \qquad (3.2\text{-}27)$$

In conclusion, time-varying imagery can be represented as the sum of a number of 2-D plane waves. This conclusion can be extended to any multidimensional signal in which one of the dimensions is treated as time.

Equation (3.2-26) is the generalization of Eq. (3.2-5) to two spatial coordinates x and y. If three spatial coordinates x, y, and z are considered with the same wave property, Eqs. (3.2-23) and (3.2-25) remain the same, with the following definitions of $\mathbf{r}$ and ∇^2:

$$\mathbf{r} = x\mathbf{e_x} + y\mathbf{e_y} + z\mathbf{e_z}, \qquad (3.2\text{-}28)$$

where $\mathbf{e_z}$ is the additional unit vector along the z-direction, and

$$\nabla^2 = \frac{\partial^2}{\partial x^2} + \frac{\partial^2}{\partial y^2} + \frac{\partial^2}{\partial z^2} \qquad (3.2\text{-}29)$$

The ∇^2 operator is known as the *Laplacian*. The 3-D wave vector used in Eq. (3.2-24) is given by

$$\mathbf{k} = 2\pi(f_x\mathbf{e_x} + f_y\mathbf{e_y} + f_z\mathbf{e_z}) = k_x\mathbf{e_x} + k_y\mathbf{e_y} + k_z\mathbf{e_z} \qquad (3.2\text{-}30)$$

Thus, k_z is the additional radian frequency along the z-direction, with f_z being the corresponding spatial frequency. The plane for which $u(\mathbf{r}, t)$ is constant is called the *wavefront*. It moves in the direction of $\mathbf{k}$ with velocity v. This is discussed further in Section 3.5.

In what follows, we will assume linear, homogeneous, and isotropic media unless otherwise specified. We will also assume that metric units are used.

EXAMPLE 3.1 Consider $u(x_1, t_1) = u(x_2, t_2)$, $x_2 x_1$, $t_2 t_1$ for a right-traveling wave. Show how x_1 and x_2 are related.
Solution: $u(x_1, t_1)$ can be considered to be moving to the right with velocity v. Letting $\Delta t = t_2 - t_1$, $u(x_1, t_1)$ and $u(x_2, t_2)$ will be the same when

$$x_2 - x_1 = v\Delta t$$

3.3 ELECTROMAGNETIC WAVES

Electromagnetic waves are 3-D waves, with three space dimensions and one time dimension. There are four quantities $\mathbf{D}$, $\mathbf{B}$, $\mathbf{H}$, and $\mathbf{E}$, which are vectors in space coordinates and whose components are functions of x, y, z, and t. In other words, $\mathbf{D}$, $\mathbf{B}$, $\mathbf{H}$, and $\mathbf{E}$ are directional in space. $\mathbf{D}$ is the *electric displacement (flux) (vector) field* (C/m^2), $\mathbf{E}$ is the *electric (vector) field* (V/m), $\mathbf{B}$ is the *magnetic density (vector) field* (Wb/m^2), $\mathbf{H}$ is the *magnetic (vector) field* (A/m), ρ is the *charge density* (C/m^3), and $\mathbf{J}$ is the *current density (vector) field* (A/m^2).

Electromagnetic waves are governed by four *Maxwell's equations*. In metric units, they are given by

$$\nabla \cdot \mathbf{D} = \rho \qquad (3.3\text{-}1)$$

$$\nabla \cdot \mathbf{B} = 0 \qquad (3.3\text{-}2)$$

$$\nabla \times \mathbf{H} = \frac{\partial \mathbf{D}}{\partial t} + \mathbf{J} \qquad (3.3\text{-}3)$$

$$\nabla \times \mathbf{E} = -\frac{\partial \mathbf{B}}{\partial t}, \qquad (3.3\text{-}4)$$

where the divergence and curl of a vector $\mathbf{A}$ with components A_x, A_y, and A_z are given by

$$\nabla \cdot \mathbf{A} = \frac{\partial A_x}{\partial x} + \frac{\partial A_y}{\partial y} + \frac{\partial A_z}{\partial z} \qquad (3.3\text{-}5)$$

$$\nabla \times \mathbf{A} = \begin{vmatrix} \mathbf{e_x} & \mathbf{e_y} & \mathbf{e_z} \\ \dfrac{\partial}{\partial x} & \dfrac{\partial}{\partial y} & \dfrac{\partial}{\partial z} \\ A_x & A_y & A_z \end{vmatrix}$$

$$= \left(\frac{\partial A_z}{\partial y} - \frac{\partial A_y}{\partial z} \right) \mathbf{e_x} + \left(\frac{\partial A_x}{\partial z} - \frac{\partial A_z}{\partial x} \right) \mathbf{e_y} + \left(\frac{\partial A_y}{\partial x} - \frac{\partial A_x}{\partial y} \right) \mathbf{e_z} \qquad (3.3\text{-}6)$$

Two more relations are

$$\mathbf{D} = \varepsilon \mathbf{E} \qquad (3.3\text{-}7)$$

$$\mathbf{B} = \mu \mathbf{H}, \qquad (3.3\text{-}8)$$

where ε is the *permittivity*, and μ is the *permeability*. Their values in free space or vacuum are given by

$$\varepsilon_0 = \frac{10^{-9}}{36\pi} \ \text{F/m}$$

$$\mu_0 = 4\pi \times 10^{-7} \ \text{H/m}$$

In dielectrics, ε and μ are greater than ε_0 and μ_0, respectively. In such media, we also consider *dipole moment density* $\mathbf{P}$ (C/m^2), which is related to the electric field $\mathbf{E}$ by

$$\mathbf{P} = \chi \varepsilon_0 \mathbf{E}, \qquad (3.3\text{-}9)$$

where χ is called the *electric susceptibility*. χ can be considered as a measure for electric dipoles in the medium to align themselves with the electric field.

We also have

$$\mathbf{D} = \varepsilon_0 \mathbf{E} + \mathbf{P} = \varepsilon_0 (1 + \chi) \mathbf{E} = \varepsilon \mathbf{E}, \qquad (3.3\text{-}10)$$

where

$$\varepsilon = \varepsilon_0 (1 + \chi) \qquad (3.3\text{-}11)$$

In a uniform isotropic dielectric medium in which space charge ρ and current density $\mathbf{J}$ are zero, Maxwell's equations become

$$\nabla \cdot \mathbf{D} = \varepsilon \nabla \cdot \mathbf{E} = 0 \qquad (3.3\text{-}12)$$

$$\nabla \cdot \mathbf{B} = \mu \nabla \cdot \mathbf{H} = 0 \qquad (3.3\text{-}13)$$

$$\nabla \times \mathbf{H} = \frac{\partial \mathbf{D}}{\partial t} = \varepsilon \frac{\partial \mathbf{E}}{\partial t} \qquad (3.3\text{-}14)$$

$$\nabla \times \mathbf{E} = -\frac{\partial \mathbf{B}}{\partial t} = -\mu \frac{\partial \mathbf{H}}{\partial t} \qquad (3.3\text{-}15)$$

3.4 PHASOR REPRESENTATION

The electric and magnetic fields we consider are usually sinusoidal with a time-varying dependence in the form

$$u(\mathbf{r}, t) = A(\mathbf{r}) \cos(\mathbf{k} \bullet \mathbf{r} \pm wt) \qquad (3.4\text{-}1)$$

We can express $u(\mathbf{r}, t)$ as

$$u(\mathbf{r}, t) = \text{Real}[A'(\mathbf{r})e^{jwt}], \qquad (3.4\text{-}2)$$

where

$$A'(\mathbf{r}) = A(\mathbf{r})e^{j\mathbf{k}\bullet\mathbf{r}} \qquad (3.4\text{-}3)$$

$A'(\mathbf{r})$ is called the phasor (representation) of $u(\mathbf{r}, t)$. It is time independent. We note the following:

$$\frac{\mathrm{d}}{\mathrm{d}t}A'(\mathbf{r})e^{jwt} = jwA'(\mathbf{r})e^{jwt} \qquad (3.4\text{-}4)$$

$$\int A'(\mathbf{r})e^{jwt}\mathrm{d}t = \frac{1}{jw}A'(\mathbf{r})e^{jwt} \qquad (3.4\text{-}5)$$

Hence, differentiation and integration are equivalent to multiplying $A'(\mathbf{r})$ by jw and $1/jw$, respectively.

The electric and magnetic fields can be written in the phasor representation as

$$\mathbf{E}(\mathbf{r}, t) = \text{Real}[\tilde{\mathbf{E}}(\mathbf{r})e^{jwt}] \qquad (3.4\text{-}6)$$

$$\mathbf{H}(\mathbf{r}, t) = \text{Real}[\tilde{\mathbf{H}}(\mathbf{r})e^{jwt}], \qquad (3.4\text{-}7)$$

where $\tilde{\mathbf{E}}(\mathbf{r})$ and $\tilde{\mathbf{H}}(\mathbf{r})$ are the phasors. The corresponding phasors for $\mathbf{D}$, $\mathbf{B}$, and $\mathbf{J}$ are defined similarly.

Maxwell's equations in terms of $\tilde{\mathbf{E}}(\mathbf{r})$ and $\tilde{\mathbf{H}}(\mathbf{r})$ can be written as

$$\nabla \bullet \tilde{\mathbf{D}} = \rho \qquad (3.4\text{-}8)$$

$$\nabla \bullet \tilde{\mathbf{B}} = 0 \qquad (3.4\text{-}9)$$

$$\nabla \times \tilde{\mathbf{H}} = jw\varepsilon\tilde{\mathbf{E}} + \tilde{\mathbf{J}} \qquad (3.4\text{-}10)$$

$$\nabla \times \tilde{\mathbf{E}} = -jw\tilde{\mathbf{B}} \qquad (3.4\text{-}11)$$

3.5 WAVE EQUATIONS IN A CHARGE-FREE MEDIUM

Taking the curl of both sides of Eq. (3.3-3) gives

$$\nabla \times \nabla \times \mathbf{H} = \varepsilon \frac{\partial}{\partial t} \nabla \times \mathbf{E} = -\varepsilon\mu \frac{\partial^2 \mathbf{H}}{\partial t^2} \qquad (3.5\text{-}1)$$

$\nabla \times \nabla \times \mathbf{H}$ can be expanded as

$$\nabla \times \nabla \times \mathbf{H} = \nabla(\nabla \bullet \mathbf{H}) - \nabla \bullet \nabla\mathbf{H}, \qquad (3.5\text{-}2)$$

where $\nabla\mathbf{H}$ is the gradient of $\mathbf{H}$. This means $\nabla H_i, i = x, y, z$ is given by

$$\nabla H_i = \frac{\partial H_i}{\partial x}\mathbf{e_x} + \frac{\partial H_i}{\partial y}\mathbf{e_y} + \frac{\partial H_i}{\partial z}\mathbf{e_z} \qquad (3.5\text{-}3)$$

$\nabla \cdot \nabla H_i = \nabla^2 H_i$ is given by

$$\nabla^2 H_i = \frac{\partial^2 H_i}{\partial x^2}\mathbf{e_x} + \frac{\partial^2 H_i}{\partial y^2}\mathbf{e_y} + \frac{\partial^2 H_i}{\partial z^2}\mathbf{e_z} \qquad (3.5\text{-}4)$$

Thus, $\nabla \cdot \nabla\mathbf{H}$ is a vector whose components along the three directions are $\nabla^2 H_i, i = x, y, z$, respectively.

It can be shown that $\nabla(\nabla \cdot \mathbf{H})$, which is the gradient vector of $\nabla \cdot \mathbf{H}$, equals zero. Hence,

$$\nabla \times \nabla \times \mathbf{H} = -\nabla \cdot \nabla\mathbf{H} = -\nabla^2\mathbf{H} \qquad (3.5\text{-}5)$$

Substituting this result in Eq. (3.5-1) gives

$$\nabla^2\mathbf{H} = \varepsilon\mu \frac{\partial^2 \mathbf{H}}{\partial t^2} \qquad (3.5\text{-}6)$$

Similarly, it can be shown that

$$\nabla^2\mathbf{E} = \varepsilon\mu \frac{\partial^2 \mathbf{E}}{\partial t^2} \qquad (3.5\text{-}7)$$

Equations (3.5-6) and (3.5-7) are called the *homogeneous wave equations* for $\mathbf{E}$ and $\mathbf{H}$, respectively.

In conclusion, each component of the electric and magnetic field vectors satisfies the nondispersive wave equation with phase velocity v equal to $1/\sqrt{\varepsilon\mu}$. In free space, we get

$$c = v = \frac{1}{\sqrt{\varepsilon_0\mu_0}} \cong 3 \times 10^8 \text{ m}/\text{sec} \qquad (3.5\text{-}8)$$

Let us consider one such field component as $u(\mathbf{r}, t)$. It satisfies

$$\nabla^2 u(\mathbf{r}, t) = \left(\frac{\partial^2}{\partial x^2} + \frac{\partial^2}{\partial y^2} + \frac{\partial^2}{\partial z^2}\right) u(\mathbf{r}, t) = \frac{1}{c^2}\frac{\partial^2 u(\mathbf{r}, t)}{\partial t^2} \qquad (3.5\text{-}9)$$

A 3-D plane wave solution of this equation is given by

$$u(\mathbf{r}, t) = A(\mathbf{k}, \omega)\cos(\mathbf{k}\bullet\mathbf{r} - \omega t), \qquad (3.5\text{-}10)$$

where the wave vector $\mathbf{k}$ is given by

$$\mathbf{k} = k_x + k_y\mathbf{e_y} + k_z\mathbf{e_z} \qquad (3.5\text{-}11)$$

and the phase velocity v is related to $\mathbf{k}$ and ω by

$$v = \frac{\omega}{|\mathbf{k}|} \qquad (3.5\text{-}12)$$

A *phase front* is defined by

$$\mathbf{k}\cdot\mathbf{r} - \omega t = \text{constant} \qquad (3.5\text{-}13)$$

This is a plane whose normal is in the direction of $\mathbf{k}$. When ωt changes, the plane changes, and the wave propagates in the direction of $\mathbf{k}$ with velocity c. We reassert that $\cos(\mathbf{k}\bullet\mathbf{r} - \omega t)$ can also be chosen as $\cos(\mathbf{k}\bullet\mathbf{r} + \omega t)$. Then, the wave travels in the direction of $-\mathbf{k}$.

The components of the wave vector $\mathbf{k}$ can be written as

$$k_i = \frac{2\pi}{\lambda}\alpha_i = k\alpha_i \quad i = x, y, z, \qquad (3.5\text{-}14)$$

where $k = |\mathbf{k}|$, and α_i is the *direction cosine* in the ith direction. If the spatial frequency along the ith direction is denoted by f_i equal to $k_i/2\pi$, then the direction cosines can be written as

$$\alpha_x = \lambda f_x \qquad (3.5\text{-}15)$$
$$\alpha_y = \lambda f_y \qquad (3.5\text{-}16)$$
$$\alpha_z = \sqrt{1 - \lambda^2 f_x^2 - \lambda^2 f_y^2} \qquad (3.5\text{-}17)$$

as

$$\alpha_x^2 + \alpha_y^2 + \alpha_z^2 = 1 \qquad (3.5\text{-}18)$$

3.6   WAVE EQUATIONS IN PHASOR REPRESENTATION
IN A CHARGE-FREE MEDIUM

This time let us start with the curl of Eq. (3.4-11):

$$\nabla \times \nabla \times \tilde{\mathbf{E}} = -jw\mu\tilde{\mathbf{H}}, \tag{3.6-1}$$

where $\tilde{\mathbf{B}} = \mu\tilde{\mathbf{H}}$ is used. Utilizing Eq. (3.4-10) with $\tilde{\mathbf{J}} = 0$ in Eq. (3.6-1) yields

$$\nabla \times \nabla \times \tilde{\mathbf{E}} = w^2\mu\varepsilon\tilde{\mathbf{E}} \tag{3.6-2}$$

As in Section 3.4, we have

$$\nabla \times \nabla \times \tilde{\mathbf{E}} = \nabla(\nabla \bullet \tilde{\mathbf{E}}) - \nabla \bullet \nabla E \tag{3.6-3}$$

As $\nabla \bullet \tilde{\mathbf{E}} = 0$ by Eq. (3.4-8), we have

$$\nabla^2\tilde{\mathbf{E}} + w^2\mu\varepsilon\tilde{\mathbf{E}} = 0 \tag{3.6-4}$$

We define the wave number k by

$$k = w\sqrt{\mu\varepsilon} \tag{3.6-5}$$

Then, Eq. (3.6-4) becomes

$$\nabla^2\tilde{\mathbf{E}} + k^2\tilde{\mathbf{E}} = 0 \tag{3.6-6}$$

This is called the *homogeneous wave equation* for $\tilde{\mathbf{E}}$. The homogeneous wave
equation for $\tilde{\mathbf{H}}$ can be similarly derived as

$$\nabla^2\tilde{\mathbf{H}} + k^2\tilde{\mathbf{H}} = 0 \tag{3.6-7}$$

Let $\tilde{\mathbf{E}}$ be written as

$$\tilde{\mathbf{E}} = \tilde{E}_x\mathbf{e_x} + \tilde{E}_y\mathbf{e_y} + \tilde{E}_z\mathbf{e_z} \tag{3.6-8}$$

Equation (3.6-6) can now be written for each component $\tilde{E}_i$ of $\tilde{\mathbf{E}}$ as

$$\left[\frac{\partial^2}{\partial x^2} + \frac{\partial^2}{\partial y^2} + \frac{\partial^2}{\partial z^2} + k^2\right]\tilde{E}_i = 0 \tag{3.6-9}$$

EXAMPLE 3.2 (a) Simplify Eq. (3.6-9) for a uniform plane wave moving in the
z-direction, (b) show that the z-component of $\tilde{\mathbf{E}}$ and $\tilde{\mathbf{H}}$ of a uniform plane wave equals
zero, using the phasor representation, (c) repeat part (b) using Maxwell's equations.

Solution: (a) A uniform plane wave is characterized by

$$\frac{\delta \tilde{E}_i}{\delta x} = \frac{\delta \tilde{E}_i}{\delta y} = \frac{\delta \tilde{H}_i}{\delta x} = \frac{\delta \tilde{H}_i}{\delta y} = 0$$

Hence, Eq. (3.6-9) simplifies to

$$\left[\frac{\partial^2}{\partial z^2} + k^2 \right] \tilde{E}_i = 0$$

(b) Consider the z-component of Eq. (3.4-10):

$$\frac{\delta \tilde{H}_y}{\delta x} - \frac{\delta \tilde{H}_x}{\delta y} = jw\varepsilon \tilde{E}_z$$

As $\frac{\delta \tilde{H}_y}{\delta x} = \frac{\delta \tilde{H}_x}{\delta y} = 0$, $\tilde{E}_z = 0$. We can similarly show that $\tilde{H}_z = 0$ by using Eq. (3.4-11).
(c) We write

$$\mathbf{E} = E_x e^{j(kz+wt)} \mathbf{e_x} + E_y e^{j(kz+wt)} \mathbf{e_y} + E_z e^{j(kz+wt)} \mathbf{e_z}$$

Substituting $\mathbf{E}$ in $\nabla \bullet \mathbf{E} = 0$, we get

$$\frac{\partial}{\partial z} E_z e^{j(kz+wt)} = 0$$

implying $E_z = 0$. We can similarly consider the magnetic field $\mathbf{H}$ as

$$\mathbf{H} = H_x e^{j(kz+wt)} \mathbf{e_x} + H_y e^{j(kz+wt)} \mathbf{e_y} + H_z e^{j(kz+wt)} \mathbf{e_z}$$

Substituting $\mathbf{H}$ in $\nabla \bullet \mathbf{H} = 0$, we get

$$\frac{\partial}{\partial z} H_z e^{j(kz+wt)} = 0$$

implying $H_z = 0$.

3.7   PLANE EM WAVES

Consider a plane wave propagating along the z-direction. The electric and magnetic fields can be written as

$$\mathbf{E} = E_x e^{j(kz-wt)} \hat{e}_x + E_y e^{j(kz-wt)} \hat{e}_y$$
$$\mathbf{H} = H_x e^{j(kz-wt)} \hat{e}_x + H_y e^{j(kz-wt)} \hat{e}_y,$$

$$(3.7\text{-}1)$$

where the real parts are actually the physical solutions. They can be more generally written as

$$\mathbf{E}(\mathbf{r}, t) = \mathbf{E}_0 \cos(\mathbf{k} \bullet \mathbf{r} - wt)$$
$$\mathbf{H}(\mathbf{r}, t) = \mathbf{H}_0 \cos(\mathbf{k} \bullet \mathbf{r} - wt), \tag{3.7-2}$$

where $\mathbf{E}_0$ and $\mathbf{H}_0$ have components (E_x, E_y), (H_x, H_y), and $\mathbf{k}, \mathbf{r}$ in this case are simply $k = 2\pi/\lambda$ and z, respectively.

Substituting Eqs. (3.7-1) in to $\nabla \times \mathbf{E} = -\mu \frac{\partial \mathbf{H}}{\partial t}$ gives the following:

$$kE_y \hat{e}_\mathbf{x} - kE_x \hat{e}_y = -\mu w[H_x \hat{e}_\mathbf{x} + H_y \hat{e}_y] \tag{3.7-3}$$

Hence,

$$H_x = -\frac{1}{\eta} E_y$$
$$H_y = \frac{1}{\eta} E_x, \tag{3.7-4}$$

where η is called the *characteristic impedance* of the medium. It is given by

$$\eta = \frac{w}{k} \mu = v\mu = \sqrt{\mu/\varepsilon} \tag{3.7-5}$$

Note that

$$\mathbf{E} \bullet \mathbf{H} = (E_x H_x + E_y H_y) e^{j(kz+wt)} = 0 \tag{3.7-6}$$

Thus, the electric and magnetic fields are orthogonal to each other.

The *Poynting vector* $\mathbf{S}$ is defined by

$$\mathbf{S} = \mathbf{E} \times \mathbf{H}, \tag{3.7-7}$$

which has units of W/m^2, indicating power flow per unit area in the direction of propagation.

Polarization indicates how the electric field vector varies with time. Again assuming the direction of propagation to be z, the electric field vector including time dependence can be written as

$$\mathbf{E} = \text{Real}[(E_x \mathbf{e}_x + E_y \mathbf{e}_y) e^{j(kz+wt)}] \tag{3.7-8}$$

E_x and E_y can be chosen relative to each other as

$$E_x = E_{x_0} \tag{3.7-9}$$
$$E_y = E_{y_0} e^{j\phi}, \tag{3.7-10}$$

where E_{x_0} and E_{y_0} are positive scalars, and ϕ is the relative phase, which decides the direction of the electric field.

Linear polarization is obtained when $\phi = 0 \ or \ \pi$. Then, we get

$$\mathbf{E} = (E_{x_0}\mathbf{e_x} \pm E_{y_0}\mathbf{e_y})\cos(kz + wt) \tag{3.7-11}$$

In this case, $(E_{x_0}\mathbf{e_x} \pm E_{y_0}\mathbf{e_y})$ can be considered as a vector that does not change in direction with time or propagation distance.

Circular polarization is obtained when $\phi = \pm\pi/2, \ and \ E_0 = E_{x_0} = E_{y_0}$. Then, we get

$$\mathbf{E} = E_0\cos(kz + wt)\mathbf{e_x} \pm E_0\sin(kz + wt)\mathbf{e_y} \tag{3.7-12}$$

We note that $|\mathbf{E}| = E_0$. For $\phi = -\pi/2$, $\mathbf{E}$ describes a circle rotating clockwise during propagation. For $\phi = +\pi/2$, $\mathbf{E}$ describes a circle rotating counterclockwise during propagation.

Elliptic polarization corresponds to an arbitrary ϕ. Equation (3.7-1) for the electric field can be written as

$$\mathbf{E} = E_x\mathbf{e_x} + E_y\mathbf{e_y} \tag{3.7-13}$$

where

$$\begin{aligned} E_x &= E_{x_0}\cos(kz + wt) \\ E_y &= E_{y_0}\cos(kz + wt - \phi) \end{aligned} \tag{3.7-14}$$

Equation (3.7-14) can be written as

$$\begin{aligned} \frac{E_y}{E_{y_0}} &= \cos(kz + wt)\cos\phi + \sin(kz + wt)\sin\phi \\ &= \frac{E_x}{E_{x_0}}\cos\phi + \left[1 - \left(\frac{E_x}{E_{x_0}}\right)^2\right]\sin\phi \end{aligned} \tag{3.7-15}$$

This equation can be further written as

$$E_x'^2 - 2E_x'E_y'\cos\phi + E_y'^2 = \sin^2\phi, \tag{3.7-16}$$

where

$$E_x' = \frac{E_x}{E_{x_0}}, \quad E_y' = \frac{E_y}{E_{y_0}} \tag{3.7-17}$$

Equation (3.7-17) is the equation of an ellipse.

Suppose the field is linearly polarized and $\mathbf{E}$ is along the x-direction. Then, we can write

$$\mathbf{E} = E_0 e^{j(kz+wt)} \mathbf{e}_x$$
$$\mathbf{H} = -\frac{E_0}{\eta} e^{j(kz+wt)} \mathbf{e}_y,$$

(3.7-18)

where Real [●] is assumed from the context. *Intensity or irradiance I* is defined as the time-averaged power given by

$$I = \frac{w}{2\pi} \int\limits_0^{2\pi/w} |\mathbf{S}| dt = \varepsilon\mu \frac{E_0^2}{2},$$

(3.7-19)

where $\mathbf{S}$ is the Poynting vector. It is observed that the intensity is proportional to the square of the field magnitude. This will be assumed to be true in general unless otherwise specified.

4

Scalar Diffraction Theory

4.1 INTRODUCTION

When the wavelength of a wave field is larger than the "aperture" sizes of the diffraction device used to control the wave, the *scalar diffraction theory* can be used. Even when this is not true, scalar diffraction theory has been found to be quite accurate [Mellin and Nordin, 2001]. Scalar diffraction theory involves the conversion of the *wave equation*, which is a partial differential equation, into an integral equation. It can be used to analyze most types of diffraction phenomena and imaging systems within its realm of validity. For example, Figure 4.1 shows the diffraction pattern from a double slit illuminated with a monochromatic plane wave. The resulting wave propagation can be quite accurately described with scalar diffraction theory.

In this chapter, scalar diffraction theory will be first derived for monochromatic waves with a single wavelength. Then, the results will be generalized to nonmonochromatic waves by using Fourier analysis and synthesis in the time direction.

This chapter consists of eight sections. In Section 4.2, the Helmholtz equation is derived. It characterizes the spatial variation of the wave field, by characterizing the time variation as a complex exponential factor. In Section 4.3, the solution of the Helmholtz equation in homogeneous media is obtained in terms of the angular spectrum of plane waves. This formulation also characterizes wave propagation in a homogeneous medium as a linear system. The FFT implementation of the angular spectrum of plane waves is discussed in Section 4.4.

Diffraction can also be treated by starting with the Helmholtz equation and converting it to an integral equation using *Green's theorem*. The remaining sections cover this topic. In Section 4.5, the Kirchoff theory of diffraction results in one formulation of this approach. The Rayleigh–Sommerfeld theory of diffraction covered in Sections 4.6 and 4.7 is another formulation of the same approach. The Rayleigh–Sommerfeld theory of diffraction for nonmonochromatic waves is treated in Section 4.8.

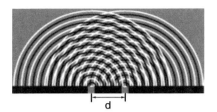

Figure 4.1. Diffraction from a double slit [Wikipedia].

4.2 HELMHOLTZ EQUATION

Monochromatic waves have a single time frequency f. As

$$k = \frac{w}{c} = \frac{2\pi f}{c} \tag{4.2-1}$$

the wavelength λ is also fixed. For such waves, the plane wave solution discussed in Section 3.7 can be written as

$$u(\mathbf{r}, t) = A(\mathbf{r}) \cos(\omega t + \Theta(\mathbf{r})) \tag{4.2-2}$$

where $A(\mathbf{r})$ is the amplitude, and $\Theta(\mathbf{r})$ is the phase at $\mathbf{r}$. In phasor representation, this becomes

$$u(\mathbf{r}, t) = \mathrm{Re}[U(\mathbf{r})e^{j\omega t}] \tag{4.2-3}$$

where the phasor $U(\mathbf{r})$ also called the *complex amplitude* equals $A(\mathbf{r})e^{j\Theta(r)}$. Equation (4.2-3) is often written without explicitly writing "the real part" for the sake of simplicity as well as simplicity of computation as in the next equation.

Substituting $u(\mathbf{r},t)$ into the wave equation (3.5-7) yields

$$(\nabla^2 + k^2)U(\mathbf{r}) = 0 \tag{4.2-4}$$

where $k = \omega/c$. This is called the *Helmholtz equation*. It is same as Eq. (3.6-9). Thus, in the case of EM plane waves, $U(\mathbf{r})$ represents a component of the electric field or magnetic field phasor, as discussed in Chapter 3. The Helmholtz equation is valid for all waves satisfying the nondispersive wave equation. For example, with acoustical and ultrasonic waves, $U(\mathbf{r})$ is the pressure or velocity potential.

If $U(\mathbf{r}, t)$ is not monochromatic, it can be represented in terms of its time Fourier transform as

$$U(\mathbf{r}, t) = \int_{-\infty}^{\infty} U_f(\mathbf{r}, f)e^{j2\pi ft}\mathrm{d}f \tag{4.2-5}$$

where

$$U_f(\mathbf{r},f) = \int_{-\infty}^{\infty} U(\mathbf{r},t)e^{-j2\pi ft} dt \qquad (4.2\text{-}6)$$

Substituting Eq. (4.2-6) into the wave equation (3.5-7) again results in the Helmholtz equation for $U_f(\mathbf{r},f)$.

EXAMPLE 4.1 Show that a spherical wave, defined by $e^{\pm jkr}/r$ where $r = \sqrt{x^2 + y^2 + z^2}$, is a solution of the Helmholtz equation.
Solution: ∇^2 in spherical coordinates is given by

$$\nabla^2 = \frac{\partial^2}{\partial r^2} + \frac{1}{r}\frac{\partial}{\partial r} + \frac{1}{r^2}\frac{\partial^2}{\partial \phi^2} + \frac{\partial^2}{\partial \theta^2}$$

$e^{\pm jkr}/r$ is spherically symmetric. In this case, ∇^2 in polar coordinates is simplified to

$$\nabla^2 = \frac{1}{r^2}\frac{\partial}{\partial r}\left(r^2 \frac{\partial}{\partial r}\right)$$

The Helmholtz equation with a spherically symmetric wave function $U(r)$ becomes

$$\frac{1}{r}\frac{\partial}{\partial r}\left(r^2 \frac{\partial}{\partial r}\right)U(r) + k^2 U(r) = 0$$

Let $U(r)$ be of the form $f(r)/r$. It is not difficult to show by partial differentiation that

$$\frac{1}{r^2}\frac{\partial}{\partial r}\left(r^2 \frac{\partial}{\partial r}\right)\frac{f(r)}{r} = \frac{1}{r^2}\frac{\partial^2 f(r)}{\partial r^2}$$

The Helmholtz equation becomes

$$\frac{1}{r^2}\frac{\partial^2 f(r)}{\partial r^2} + K^2 \frac{f(r)}{r} = 0$$

The solution of this equation is given by

$$f(r) = e^{\pm jkr}$$

Thus, $U(r) = e^{\pm jkr}/r$.

EXAMPLE 4.2 Find a similar solution in cylindrical coordinates.
Solution: In cylindrical coordinates, ∇^2 is given by

$$\nabla^2 = \frac{\partial^2}{\partial r^2} + \frac{1}{r}\frac{\partial}{\partial r} + \frac{1}{r^2}\frac{\partial^2}{\partial \phi^2} + \frac{\partial^2}{\partial z^2}$$

Assuming a solution with cylindrical symmetry, $U(r, \phi, z)$ becomes $U(r, z)$. Cylindrical symmetry means the wavefronts are circles for constant z. In this case, there are no simple solutions. The exact solution has a Bessel-function type of dependence on r. It can also be shown that

$$U(r, z) \cong \frac{C}{\sqrt{r}} e^{\pm jkr}$$

approximately satisfies the wave equation. Such a wave is called a *cylindrical wave*.

4.3   ANGULAR SPECTRUM OF PLANE WAVES

We will consider the propagation of the wave field $U(x, y, z)$ in the z-direction. The wave field is assumed to be at a wavelength λ such that $k = 2\pi/\lambda$. Let $z = 0$ initially. The 2-D Fourier representation of $U(x,y,0)$ is given in terms of its Fourier transform $A(f_x, f_y, 0)$ by

$$U(x, y, 0) = \iint\limits_{-\infty}^{\infty} A(f_x, f_y, 0)e^{j2\pi(f_x x + f_y y)}df_x df_y \qquad (4.3\text{-}1)$$

where

$$A(f_x, f_y, 0) = \iint\limits_{-\infty}^{\infty} U(x, y, 0)e^{-j2\pi(f_x x + f_y y)}dx dy \qquad (4.3\text{-}2)$$

Including time variation, $A(f_x, f_y, 0)e^{j2\pi(f_x x + f_y y + ft)}$ is a plane wave at $z = 0$, propagating with direction cosines α_i given by Eqs. (3.5-15)–(3.5-17). $A(f_x, f_y, 0)$ is called the *angular spectrum* of $U(x, y, 0)$.

Consider next the wave field $U(x, y, z)$. Its angular spectrum $A(f_x, f_y, z)$ is given by

$$A(f_x, f_y, z) = \iint\limits_{-\infty}^{\infty} U(x, y, z)e^{-j2\pi(f_x x + f_y y)}dx dy \qquad (4.3\text{-}3)$$

and its Fourier representation in terms of its angular spectrum is given by

$$U(x, y, z) = \iint\limits_{-\infty}^{\infty} A(f_x, f_y, z)e^{j2\pi(f_x x + f_y y)}df_x df_y \qquad (4.3\text{-}4)$$

$U(x, y, z)$ satisfies the Helmholtz equation at all points without sources, namely,

$$\nabla^2 U(x, y, z) + k^2 U(x, y, z) = 0 \tag{4.3-5}$$

Substitution of $U(x, y, z)$ from Eq. (4.3-4) into Eq. (4.3-5) yields

$$\int\!\!\!\int_{-\infty}^{\infty} \left[\frac{d^2}{dz^2} A(f_x, f_y, z) + (k^2 - 4\pi^2(f_x^2 + f_y^2)) A(f_x, f_y, z) \right] e^{j2\pi(f_x x + f_y y)} df_x df_y = 0$$

$$\tag{4.3-6}$$

This is true for all waves only if the integrand is zero:

$$\frac{d^2}{dz^2} A(f_x, f_y, z) + (k^2 - 4\pi^2(f_x^2 + f_y^2)) A(f_x, f_y, z) = 0 \tag{4.3-7}$$

This differential equation has the solution

$$A(f_x, f_y, z) = A(f_x, f_y, 0) e^{j\mu z} \tag{4.3-8}$$

where

$$\mu = \sqrt{k^2 - 4\pi^2(f_x^2 + f_y^2)} = k_z \tag{4.3-9}$$

If $4\pi^2(f_x^2 + f_y^2) k^2$, μ is real, and each angular spectrum component is just modified by a phase factor $e^{j\mu z}$. Plane wave components satisfying this condition are known as *homogeneous waves*.

If $4\pi^2(f_x^2 + f_y^2) k^2$, then μ can be written as

$$\mu = j\sqrt{4\pi^2(f_x^2 + f_y^2) - k^2} \tag{4.3-10}$$

and Eq. (4.3-8) becomes

$$A(f_x, f_y, z) = A(f_x, f_y, 0) e^{-\mu z} \tag{4.3-11}$$

This result indicates that the amplitudes of such plane wave components are strongly attenuated by propagation in the z-direction. They are called *evanescent waves*.

If $4\pi^2(f_x^2 + f_y^2) = k^2$, $A(f_x, f_y, z)$ is the same as $A(f_x, f_y, 0)$. Such components correspond to plane waves traveling perpendicular to the z-axis.

Knowing $A(f_x, f_y, z)$ in terms of $A(f_x, f_y, 0)$ allows us to find the wave field at (x, y, z) by using Eq. (4.3-11) in Eq. (4.3-4):

$$U(x, y, z) = \int\!\!\!\int_{-\infty}^{\infty} A(f_x, f_y, 0) e^{jz\sqrt{k^2 - 4\pi^2(f_x^2 + f_y^2)}} e^{j2\pi(f_x x + f_y y)} df_x df_y \tag{4.3-12}$$

Thus, if $U(x,y,0)$ is known, $A(f_x,f_y,0)$ can be computed, followed by the computation of $U(x,y,z)$. The limits of integration in Eq. (4.3-12) can be limited to a circular region given by

$$4\pi^2(f_x^2 + f_y^2) \leq k^2 \tag{4.3-13}$$

provided that the distance z is at least several wavelengths long so that the evanescent waves may be neglected.

Under these conditions, Eq. (4.3-8) shows that wave propagation in a homogeneous medium is equivalent to a linear 2-D spatial filter with the transfer function given by

$$H(f_x,f_y) = \begin{cases} e^{jz\sqrt{k^2-4\pi^2(f_x^2+f_y^2)}} & 4\pi^2(f_x^2 + f_y^2)k^2 \\ 0 & \text{otherwise} \end{cases} \tag{4.3-14}$$

This is schematically shown in Figure 4.1. The impulse response, which is the inverse Fourier transform of this transfer function, is discussed in Example 4.6.

Propagation of a wavefield in the z-direction in a source-free space is correctly described by the propagation of the angular spectrum in the near field as well as the far field. Two other ways to characterize such propagation are in terms of the Fresnel and Fraunhofer approximations, discussed in Chapter 5. However, they are valid only under certain constraints.

Let $F[\bullet]$ and $F^{-1}[\bullet]$ denote the forward and inverse Fourier transform operators, respectively. In terms of these operators, Eq. (4.3-12) can be written as

$$U(x,y,z) = F^{-1}\left[F[U(x,y,0)]e^{jkz\sqrt{1-\alpha_x^2-\alpha_y^2}}\right] \tag{4.3-15}$$

$A(f_x,f_y,0) = F[U(x,y,0)]$ for particular values of f_x and f_y is the complex amplitude of a plane wave traveling in the direction specified by the direction cosines $\alpha_x = 2\pi f_x$, $\alpha_y = 2\pi f_y$, and $\alpha_z = 2\pi f_z$, where $f_z = \frac{1}{2\pi}[1 - \alpha_x^2 - \alpha_y^2]^{1/2}$. The effect of propagation is to modify the relative phases of the various plane waves by $e^{jkz\sqrt{1-\alpha_x^2-\alpha_y^2}}$ without changing their amplitudes.

EXAMPLE 4.3 Find the angular spectrum of a wave field at plane $z = a$ (constant) if the wave field results from a plane wave $Be^{jk_{x_0}x}e^{jk_{z_0}z}$ passing through a circular aperture of diameter d at $z = 0$.

Solution: The wave at $z = 0$ can be written as

$$U(x,y,0) = Be^{jk_{x_0}x}\text{cyl}\left(\frac{r}{d}\right)$$

$$= Be^{j2\pi f_{x_0}x}\text{cyl}\left(\frac{\sqrt{x^2+y^2}}{d}\right)$$

The FTs of the two factors above are

$$Be^{j2\pi f_{x_0}x} \leftrightarrow B\delta(f_x - f_{x_0}, f_y)$$

$$\mathrm{cyl}\left(\frac{r}{d}\right) \leftrightarrow \frac{d^2\pi}{4}\,\mathrm{somb}(d\rho)$$

By convolution theorem, the angular spectrum of $U(x, y, 0)$ is given by

$$A(f_x, f_y, 0) = B\delta(f_x - f_{x_0}, f_y) * \frac{d^2\pi}{4}\,\mathrm{somb}(d\rho)$$

$$= B\delta(f_x - f_{x_0}, f_y) * \frac{d^2\pi}{4}\,\mathrm{somb}\left(d\sqrt{f_x^2 + f_y^2}\right)$$

$$= \frac{d^2\pi B}{4}\,\mathrm{somb}\left(d\sqrt{(f_x - f_{x_0})^2 + f_y^2}\right)$$

It is observed that the single plane wave component before the circular aperture is changed to a spectrum with an infinite number of such components after the aperture. They are mostly propagating in the direction of the incident wave field, with a spread given by the sombrero function. This spread is inversely proportional to d, the diameter of the circular aperture.

At a distance z, the angular spectrum becomes

$$A(f_x, f_y, z) = \frac{d^2\pi B}{4}\,\mathrm{somb}\left(d\sqrt{(f_x - f_{x_0})^2 + f_y^2}\right)e^{jz\sqrt{k^2 - 4\pi^2(f_x^2 + f_y^2)}}$$

4.4   FAST FOURIER TRANSFORM (FFT) IMPLEMENTATION OF THE ANGULAR SPECTRUM OF PLANE WAVES

The angular spectrum of plane waves relating $U(x, y, z)$ to $U(x, y, 0)$ can be implemented by the fast Fourier transform (FFT) algorithm [Brigham, 1974] after discretizing and truncating the space variables and the spacial frequency variables. For a discussion of the discrete Fourier transform and the FFT, see Appendix B. The discretized and truncated variables in the space domain and the spatial frequency domain are given by

$$x = \Delta x \cdot n_1$$
$$y = \Delta y \cdot n_2$$
$$f_x = \Delta f_x \cdot m_1 \quad\quad\quad (4.4\text{-}1)$$
$$f_y = \Delta f_y \cdot m_2$$

where Δx, Δy, Δf_x, Δf_y are the sampling intervals, and n_1, m_1, as well as n_2, m_2 are integers satisfying

$$
\begin{aligned}
-M_1 n_1, m_1 &\leq M_1 \\
-M_2 n_2, m_2 &\leq M_2
\end{aligned}
\tag{4.4-2}
$$

We will write $U(\Delta x n_1, \Delta y n_2, z)$ and $A(\Delta f_x m_1, \Delta f_y m_2, z)$ as $U[n_1, n_2, z]$ and $A[m_1, m_2, z]$, respectively. In terms of discretized space and frequency variables, Eqs. (4.3-2) and (4.3-8) are approximated by

$$
A[m_1, m_2, 0] = \Delta x \Delta y \sum_{n_1} \sum_{n_2} U[n_1, n_2, 0] e^{-j2\pi(\Delta f_x \Delta x n_1 m_1 + \Delta f_y \Delta y n_2 m_2)}
\tag{4.4-3}
$$

$$
U[n_1, n_2, z] = \Delta f_x \Delta f_y \sum_{m_1} \sum_{m_2} A[m_1, m_2, 0] e^{jz\sqrt{k^2 - 4\pi^2(\Delta f_x^2 m_1^2 + \Delta f_y^2 m_2^2)}}
$$
$$
\times e^{j2\pi(\Delta f_x \Delta x n_1 m_1 + \Delta f_y \Delta y n_2 m_2)}
\tag{4.4-4}
$$

M_1 and M_2 should be chosen such that the inequality (4.3-13) is satisfied if evanescent waves are to be neglected. An approximation is to choose a rectangular region in the Fourier domain such that

$$
|k_x| = 2\pi |f_x| \leq k
\tag{4.4-5}
$$

$$
|k_y| = 2\pi |f_y| \leq k
\tag{4.4-6}
$$

Then, the following must be satisfied:

$$
|f_{x\,max}| = |f_{y\,max}| = \frac{1}{\lambda}
\tag{4.4-7}
$$

Choosing $M = M_1 = M_2$, $\Delta f = \Delta f_x = \Delta f_y$ gives

$$
f_{max} = \Delta f M = \frac{1}{\lambda}
\tag{4.4-8}
$$

Hence,

$$
M = \frac{1}{\Delta f \lambda}
\tag{4.4-9}
$$

In practice, it is often true that $f_{x\,max}$ and $f_{y\,max}$ are less than $1/\lambda$. If they are known, Eq. (4.4-8) becomes

$$
M_1 = \frac{f_{x\,max}}{\Delta f_x}
\tag{4.4-10}
$$

in the x-direction, and

$$M_2 = \frac{f_{y\,\max}}{\Delta f_y} \qquad (4.4\text{-}11)$$

in the y-direction. Assuming $f_{x\,\max} = f_{y\,\max}$, $\Delta f_x = \Delta f_y = \Delta f$ gives $M = M_x = M_y$.

FFTs of length N will be assumed to be used along the two directions so that $\Delta = \Delta x = \Delta y$, and $\Delta f = \Delta f_x = \Delta f_y$. How N is related to M is discussed below. In order to be able to use the FFT, the following must be valid:

$$\Delta\Delta f = \frac{1}{N} \qquad (4.4\text{-}12)$$

$U[n_1, n_2, z]$ and $A[m_1, m_2, z]$ for any z are also assumed to be periodic with period N. Hence, for m_1, m_2, n_1, $n_2 0$, they satisfy

$$
\begin{aligned}
U[-n_1, n_2, z] &= U[N - n_1, n_2, z] \\
U[n_1, -n_2, z] &= U[n_1, N - n_2, z] \\
U[-n_1, -n_2, z] &= U[N - n_1, N - n_2, z] \\
A[-m_1, m_2, z] &= A[N - m_1, m_2, z] \\
A[m_1, -m_2, z] &= A[m_1, N - m_2, z] \\
A[-m_1, -m_2, z] &= A[N - m_1, N - m_2, z]
\end{aligned}
\qquad (4.4\text{-}13)
$$

Mapping of negative indices to positive indices is shown in Figure 4.2. Using Eq. (4.4-13), Eqs. (4.4-3) and (4.4-4) can be written as follows:

$$A[m_1, m_2, 0] = \Delta^2 \sum_{n_1=0}^{N-1} \sum_{n_2=0}^{N-1} U[n_1, n_2, 0] e^{-j\frac{2\pi}{N}(n_1 m_1 + n_2 m_2)} \qquad (4.4\text{-}14)$$

$$U[n_1, n_2, z] = (\Delta f')^2 \sum_{m_1=0}^{N-1} \sum_{m_2=0}^{N-1} A[m_1, m_2, z] e^{j\frac{2\pi}{N}(m_1 n_1 + m_2 n_2)} \qquad (4.4\text{-}15)$$

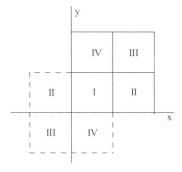

Figure 4.2. Mapping of negative index regions to positive index regions to use the FFT.

where

$$A[m_1, m_2, z] = A[m_1, m_2, 0] e^{jz \sqrt{k^2 - 4\pi^2 (\Delta f_x)^2 (m_1^2 + m_2^2)}} \qquad (4.4\text{-}16)$$

where m_1 and m_2 satisfy condition (4.4-2). $A[m_1, m_2, z]$ is also assumed to be periodic with period N:

$$A[-m_1, m_2, z] = A[N - m_1, m_2, z]$$
$$A[m_1, -m_2, z] = A[m_1, N - m_2, z] \qquad (4.4\text{-}17)$$
$$A[-m_1, -m_2, z] = A[N - m_1, N - m_2, z]$$

Note that the periodicity condition due to the use of the FFT causes aliasing because circular convolution rather than linear convolution is computed. In order to reduce aliasing, the input and output apertures can be zero padded to a size, say, $M' = 2M$. Then, N can be chosen as follows:

$$N = \begin{cases} 2M' & N \text{ even} \\ 2M' + 1 & N \text{ odd} \end{cases} \qquad (4.4\text{-}18)$$

In addition, FFT algorithms are usually more efficient when N is a power of 2. Hence, N may actually be chosen larger than the value computed above to make it a power of 2.

Equations (4.4-15) and (4.4-16) can now be written as

$$U[n_1, n_2, z] = (\Delta f)^2 \sum_{m_1=0}^{N-1} \sum_{m_2=0}^{N-1} \left[A[m_1, m_2, 0] e^{jz \sqrt{k^2 - 4\pi^2 (\Delta f_x)^2 (m_1^2 + m_2^2)}} \right] e^{j\frac{2\pi}{N}(n_1 m_1 + n_2 m_2)}$$

$$(4.4\text{-}19)$$

Equation (4.4-19) is in the form of an inverse 2-D DFT except for a normalization factor.

As $(\Delta \Delta f)^2 = \frac{1}{N^2}$, Δ^2 and $(\Delta f)^2$ in Eqs. (4.4-14) and (4.4-15) can be omitted, and only one of these equations, say, Eq. (4.4-14) is multiplied by $\frac{1}{N^2}$.

In summary, the procedure to obtain $U(x, y, z)$ from $U(x, y, 0)$ with the angular spectrum method and the FFT is as follows:

1. Generate $U[n_1, n_2, 0]$ as discussed above.
2. Compute $A[m_1, m_2, 0]$ by FFT according to Eq. (4.4-14).
3. Compute $A[m_1, m_2, z]$ according to Eq. (4.4-16).
4. Compute $U[n_1, n_2, z]$ by FFT according to Eq. (4.4-19).
5. Arrange $U[n_1, n_2, z]$ according to Eq. (4.4-13) so that negative space coordinates are regenerated.

The results discussed above can be easily generalized to different number of data points along the x- and y-directions.

EXAMPLE 4.4 Suppose that the following definitions are made:

fftshift: operations defined by Eqs. (4.4.13) and (4.417)

fft2, ifft2: routines for 2-D FFT and inverse FFT, respectively

H: transfer function given by Eq. (4.3.14)

u, U: input and output fields, respectively

Using these, an ASM program can be written as

$u1 = fftshift\ ((fft2(fftshift(u))))$

$u2 = H \bullet u1$

$U = fftshift(ifft2(fftshift(u2)))$

Show that this program can be reduced to

$u1 = fftshift((fft2(u)))$

$u2 = H \bullet u1$

$U = ifft2(fftshift(u2))$

Solution: The difference between the two programs is reduction of *fftshift* in the first and third steps of the program. This is possible because of the following property of the DFT shown in 1-D:

If $u'[n] = u\left[n - \frac{N}{2}\right]$, then $U'[k] = U[k]e^{-j\pi k}$. When the *fftshift* in the first step of the program is skipped, the end result is the phase shift $e^{-j\pi k}$. When the *fftshift* in the third step is skipped, the same phase shift causes the output U to be correctly generated.

EXAMPLE 4.5 Suppose the parameters are given as follows:

$$\text{Input size: } 32 \times 32 \qquad N = 512$$
$$\lambda = 0.0005 \qquad z = 1000$$

The units above can be in mm. A Gaussian input field is given by

$$U(x, y, 0) = \exp[-0.01\pi(x^2 + y^2)]$$

The input field intensity is shown in Figure 4.3. The corresponding output field intensity computed with the ASM is shown in Figure 4.4. If z is changed to 100, the corresponding output field is as shown in Figure 4.5. Thus, the ASM can be used for any distance z. As expected, if the input field is Gaussian, the output field also remains Gaussian.

EXAMPLE 4.6 Determine the impulse response function corresponding to the transfer function given by Eq. (4.3.14).

Solution: The impulse response is the inverse Fourier transform of the transfer function:

$$h(x, y) = \int_{-\infty}^{\infty} \int_{-\infty}^{\infty} e^{jkz}[1 - \lambda^2 f_x^2 - \lambda^2 f_y^2]^{1/2} e^{j2\pi(f_x x + f_y y)} df_x df_y \qquad (4.4\text{-}20)$$

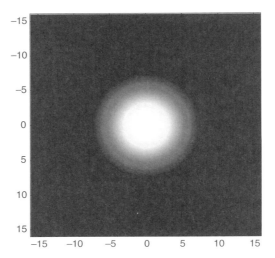

Figure 4.3. The input Gaussian field intensity.

where $\alpha_x^2 + \alpha_y^2 = \lambda^2(f_x^2 + f_y^2)$ is used. We note that $H(f_x, f_y)$ has cylindrical symmetry. For this reason, it is better to use cylindrical coordinates by letting

$$x = r\cos\theta, \; y = r\sin\theta, f_x = \rho\cos\phi, \quad \text{and} \quad f_y = \rho\sin\phi$$

In cylindrical coordinates, Eq. (4.4-20) becomes [Stark, 1982]:

$$h(r) = \frac{1}{2\pi} \int_0^\infty e^{-z(t^2 - k^2)^{1/2}} J_0(2\pi r\rho)\rho\,\mathrm{d}\rho$$

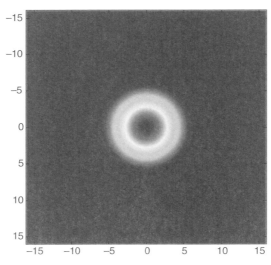

Figure 4.4. The output field intensity when $z = 1000$.

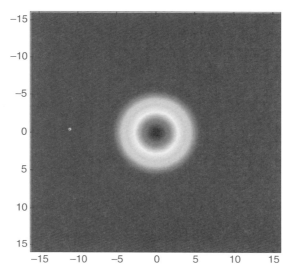

Figure 4.5. The output field intensity when $z = 100$.

where $J_0(\bullet)$ is the Bessel function of the first kind of zeroth order. The integral above can be further evaluated as

$$h(r) = \frac{e^{jk[z^2 + r^2]^{1/2}}}{j\lambda[z^2 + r^2]^{1/2}} \frac{z}{[z^2 + r^2]^{1/2}} \left[1 + \frac{1}{jk[z^2 + r^2]^{1/2}} \right] \tag{4.4-21}$$

EXAMPLE 4.7 Determine the wave field due to a point source modeled by $U(x, y, 0) = \delta(x - 3)\delta(y - 4)$.

Solution: The output wave field is given by

$$U(x, y, z) = U(x, y, 0) * h(x, y, z)$$

$$\simeq \delta(x - 3)\delta(y - 4) * \frac{1}{j\lambda z} e^{jk[z^2 + x^2 + y^2]^{1/2}}$$

$$\simeq \frac{1}{j\lambda z} e^{jk[z^2 + (x-3)^2 + (y-4)^2]^{1/2}}$$

4.5 THE KIRCHOFF THEORY OF DIFFRACTION

The propagation of the angular spectrum of plane waves as discussed in Section 4.5 does characterize diffraction. However, diffraction can also be treated by starting

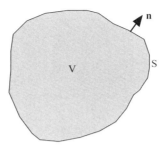

Figure 4.6.   The volume and its surface used in Green's theorem.

with the Helmholtz equation and converting it to an integral equation using *Green's theorem*.

Green's theorem involves two complex-valued functions $U(\mathbf{r})$ and $G(\mathbf{r})$ (*Green's function*). We let S be a closed surface surrounding a volume V, as shown in Figure 4.6.

If U and G, their first and second partial derivatives respectively, are single-valued and continuous, without any singular points within and on S, Green's theorem states that

$$\iiint_V (G\nabla^2 U - U\nabla^2 G)\,dv = \iint_S \left(G\frac{\partial U}{\partial n} - U\frac{\partial G}{\partial n} \right) ds \qquad (4.5\text{-}1)$$

where $\partial/\partial n$ indicates a partial derivative in the outward normal direction at each point of S. In our case, U corresponds to the wave field.

Consider the propagation of an arbitrary wave field incident on a screen at an initial plane at $z = 0$ to the observation plane at some z. The geometry to be used is shown in Figure 4.7. In this figure, P_0 is a point in the observation plane, and P_1, which can be an arbitrary point in space, is shown as a point at the initial plane. The closed surface is the sum of the surfaces S_0 and S_1. There are two choices of the Green's function that lead us to a useful integral representation of diffraction. They are discussed below.

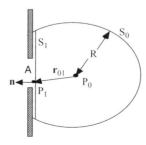

Figure 4.7.   The geometry used in Kirchoff formulation of diffraction.

4.5.1   Kirchoff Theory of Diffraction

The Green function chosen by Kirchhoff is a spherical wave given by

$$G(\mathbf{r}) = \frac{e^{jkr_{01}}}{r_{01}} \tag{4.5-2}$$

where $\mathbf{r}$ is the position vector pointing from P_0 to P_1, and r_{01} is the corresponding distance, given by

$$r_{01} = [(x_0 - x)^2 + (y_0 - y)^2 + z^2]^{1/2} \tag{4.5-3}$$

$U(\mathbf{r})$ satisfies the Helmholtz equation. As $G(\mathbf{r})$ is an expanding spherical wave, it also satisfies the Helmholtz equation:

$$\nabla^2 G(\mathbf{r}) + k^2 G(\mathbf{r}) = 0 \tag{4.5-4}$$

The left hand side of Eq. (4.5-1) can now be written as

$$\iiint_V [G\nabla^2 U - U\nabla^2 G]\mathrm{d}V = \iiint_V k^2 [UG - UG]\mathrm{d}v = 0$$

Hence, Eq. (4.5-1) becomes

$$\iint_S \left[G\frac{\partial U}{\partial n} - U\frac{\partial G}{\partial n} \right] \mathrm{d}s = 0 \tag{4.5-5}$$

On the surface S_0 of Figure 4.7, we have

$$G(\mathbf{r}) = \frac{e^{jkR}}{R}$$

$$\frac{\partial G(\mathbf{r})}{\partial n} = \left(jk - \frac{1}{R} \right) \frac{e^{jkR}}{R} \left(\frac{\partial U}{\partial n} - jkU \right)$$

The part of the last integral in Eq. (4.5-1) over S_0 becomes

$$I_{S_2} = \int_{S_0} G(\mathbf{r}) \left(\frac{\partial U}{\partial n} - jkU \right) \mathrm{d}s$$

$$= \int_{\Omega} G(\mathbf{r}) \left(\frac{\partial U}{\partial n} - jkU \right) R^2 \mathrm{d}w$$

$$= \int_{\Omega} e^{jkR} \left(R\left(\frac{\partial U}{\partial n} - jkU \right) \right) \mathrm{d}w$$

where Ω is the solid angle subtended by S_0 at P_0. The last integral goes to zero as $R \to \infty$ if

$$\lim_{R \to \infty} R\left(\frac{\partial U}{\partial n} - jkU\right) = 0 \qquad (4.5\text{-}6)$$

This condition is known as *the Sommerfeld radiation condition* and leads us to results in agreement with experiments.

What remains is the integral over S_1, which is an infinite opaque plane except for the open aperture to be denoted by A. Two commonly used approximations to the boundary conditions for diffraction by apertures in plane screens are the *Debye approximation* and the *Kirchhoff approximation* [Goodman, 2004]. In the Debye approximation, the angular spectrum of the incident field has an abrupt cutoff such that only those plane waves traveling in certain directions passed the aperture are maintained. For example, the direction of travel from the aperture may be toward a focal point. Then, the field in the focal region is a superposition of plane waves whose propagation directions are inside the geometrical cone whose apex is the focal point and whose base is the aperture.

The Kirchhoff approximation is more commonly used. If an aperture A is on plane $z = 0$, as shown in Figure 4.7, the Kirchhoff approximation on the plane $z = 0^+$ is given by the following equations:

$$U(x, y, 0^+) = \begin{cases} U(x, y, 0) & \text{inside } A \\ 0 & \text{outside } A \end{cases} \qquad (4.5\text{-}7)$$

$$\frac{\partial}{\partial z} U(x, y, z)\bigg|_{z=0^+} = \begin{cases} \dfrac{\partial}{\partial z} U(x, y, z)\bigg|_{z=0}. & \text{inside } A \\ 0 & \text{outside } A \end{cases} \qquad (4.5\text{-}8)$$

It is observed that $U(x, y, z)$ and its partial derivative in the z-direction are discontinuous outside the aperture and continuous inside the aperture at $z = 0$ according to the Kirchhoff approximation. These conditions are also called Kirchoff boundary conditions. They lead us to the following solution for $U(P_0)$:

$$U(P_0) = \frac{1}{4\pi} \iint_A \left[G\frac{\partial U}{\partial n} - U\frac{\partial G}{\partial n} \right] ds \qquad (4.5\text{-}9)$$

4.5.2 Fresnel–Kirchoff Diffraction Formula

Assuming r_{01} is many optical wavelengths, the following approximation can be made:

$$\frac{\partial G(P_1)}{\partial n} = \frac{\partial}{\partial n}\left[\frac{e^{jkr_{01}}}{r_{01}}\right] = \cos(\theta)\left[jk - \frac{1}{r_{01}}\right]G(P_1) \cong jk\cos(\theta)G(P_1) \qquad (4.5\text{-}10)$$

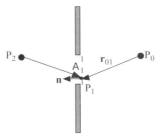

Figure 4.8.   Spherical wave illumination of a plane aperture.

Hence, Eq. (4.5-9) becomes

$$U(P_0) = \frac{1}{4\pi} \iint\limits_{A} \frac{e^{jkr_{01}}}{r_{01}} \left[\frac{\partial U}{\partial n} - jk\cos(\theta)U \right] ds$$

Let us apply this equation to the case of $U(P_1)$ being a spherical wave originating at a point P_2 as shown in Figure 4.8. Denoting the distance between P_1 and P_2 as r_{21}, and the angle between $\mathbf{n}$ and $\mathbf{r_{21}}$ by θ_2, we have

$$U(P_1) = G(r_{21}) \frac{e^{jkr_{21}}}{r_{21}}$$

$$\frac{\partial U(P_1)}{\partial n} \cong jk\cos(\theta_2)G(r_{21})$$

Hence, we get

$$U(P_0) = \frac{1}{j\lambda} \iint\limits_{A} \frac{e^{jk(r_{21}+r_{01})}}{r_{21}r_{01}} \left[\frac{\cos(\theta) - \cos(\theta_2)}{2} \right] ds \qquad (4.5\text{-}11)$$

This result is known as the *Fresnel–Kirchhoff diffraction formula*, valid for diffraction of a spherical wave by a plane aperture.

4.6   THE RAYLEIGH–SOMMERFELD THEORY OF DIFFRACTION

The use of the Green's function given by Eq. (4.5-2) together with a number of simplifying assumptions leads us to the *Fresnel–Kirchhoff diffraction formula* as discussed above. However, there are certain inconsistencies in this formulation. These inconsistencies were later removed by Sommerfeld by using the following

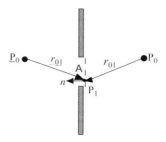

Figure 4.9. Rayleigh–Sommerfeld modeling of diffraction by a plane aperture.

Green's function:

$$G_2(\mathbf{r}) = \frac{e^{jkr_{01}}}{r_{01}} - \frac{e^{jk\underline{r_{01}}}}{\underline{r_{01}}} \tag{4.6-1}$$

where r_{01} is the distance from $\underline{P_0}$ to P_1, $\underline{P_0}$ being the mirror image of P_0 with respect to the initial plane. This is shown in Figure 4.9.

The use of the second Green's function leads us to the *first Rayleigh–Sommerfeld diffraction formula* given by

$$U(x_0, y_0, z) = \frac{1}{j\lambda} \iint\limits_{-\infty}^{\infty} U(x, y, 0) \frac{z}{r_{01}} \frac{e^{jkr_{01}}}{r_{01}} \, dx dy \tag{4.6-2}$$

Another derivation of this equation by using the convolution theorem is given in the next section.

Equation (4.6-2) shows that $U(x_0, y_0, z)$ may be interpreted as a linear superposition of diverging spherical waves, each of which emanates from a point $(x, y, 0)$ and is weighted by $\frac{1}{j\lambda} \frac{z}{r_{01}} U(x, y, 0)$. This mathematical statement is also known as the *Huygens–Fresnel principle*.

It is observed that Eq. (4.6-2) is a 2-D linear convolution with respect to x, y:

$$U(x_0, y_0, z) = U(x, y, 0) * h(x, y, z) \tag{4.6-3}$$

where the impulse response $h(x, y, z)$ is given by

$$h(x, y, z) = \frac{1}{j\lambda z} \frac{e^{jkz\left[1 + \frac{x^2 + y^2}{z^2}\right]^{1/2}}}{\left[1 + \frac{x^2 + y^2}{z^2}\right]} \tag{4.6-4}$$

The transfer function $H(f_1, f_2, z)$, the 2-D Fourier transform of $h(x, y, z)$, is given by

$$H(f_x, f_y, z) = e^{jz[k^2 - 4\pi^2(f_x^2 + f_y^2)]^{1/2}} \tag{4.6-5}$$

This is the same as the transfer function for the propagation of angular spectrum of plane waves discussed in Section 4.5. Thus, there is complete equivalence between the Rayleigh–Sommerfeld theory of diffraction and the method of the angular spectrum of plane waves.

4.6.1   The Kirchhoff Approximation

Incorporating the Kirchhoff approximation into the first Rayleigh–Sommerfeld integral yields

$$U(x, y, z) = \frac{1}{j\lambda} \iint\limits_{A} U(x, y, 0) \frac{z}{r_{01}} \frac{e^{jkr_{01}}}{r_{01}} \mathrm{d}x\mathrm{d}y \qquad (4.6\text{-}6)$$

z/r_{01} is sometimes written as $\cos(n, r_{01})$, indicating the cosine of the angle between the z-axis and r_{01} shown in Figure 4.9.

4.6.2   The Second Rayleigh–Sommerfeld Diffraction Formula

Another valid Green's function that can be used instead of $G_2(\mathbf{r})$ is given by

$$G_3(\mathbf{r}) = \frac{e^{jkr_{01}}}{r_{01}} + \frac{e^{jkr'_{01}}}{r'_{01}} \qquad (4.6\text{-}7)$$

$$U_2(x, y, z) = \frac{1}{2\pi} \iint\limits_{A} \frac{\delta U(x, y, 0)}{\delta n} \frac{e^{jkr_{01}}}{r_{01}} \mathrm{d}x\mathrm{d}y \qquad (4.6\text{-}8)$$

where $\frac{\delta U(x,y,0)}{\delta n}$ is the partial derivative of $U(x, y, 0)$ in the normal direction (z-direction).

It can be shown that the Kirchoff solution discussed in Section 4.5 is the arithmetic average of the first and second Rayleigh–Sommerfeld solutions [Goodman, 2004].

4.7   ANOTHER DERIVATION OF THE FIRST RAYLEIGH–SOMMERFELD DIFFRACTION INTEGRAL

We assume that the field $U(\mathbf{r})$ is due to sources in the half space $z0$ and that it is known in the plane $z = 0$. We want to determine $U(\mathbf{r})$ for $z0$. $U(x, y, 0)$ in terms of its angular spectrum is given by

$$U(x, y, 0) = \iint\limits_{-\infty}^{\infty} A(f_x, f_y, 0) e^{j2\pi(f_x x + f_y y)} \mathrm{d}f_x \mathrm{d}f_y \qquad (4.7\text{-}1)$$

In terms of the convolution theory, this equation can be interpreted as finding the output of a LSI system whose transfer function is unity. The impulse response of the system is given by

$$h(x, y, 0) = \iint\limits_{-\infty}^{\infty} e^{j2\pi(f_x x + f_y y)} df_x df_y \qquad (4.7\text{-}2)$$

What if $z \neq 0$? $h(x, y, z)$ consistent with Eq. (4.7-1) and other derivations of diffraction integrals as in Section 4.6 is given by

$$h(x, y, z) = \frac{1}{4\pi^2} \iint\limits_{-\infty}^{\infty} e^{j\mathbf{k}\cdot\mathbf{r}} dk_x dk_y \qquad (4.7\text{-}3)$$

where $\mathbf{k}$ and $\mathbf{r}$ are given by Eqs. (3.2-28) and (3.2-30), respectively. Equation (4.7-3) can be written as

$$h(x, y, z) = -\frac{1}{2\pi} \frac{\partial}{\partial z} \left[\frac{j}{2\pi} \iint\limits_{-\infty}^{\infty} \frac{e^{j\mathbf{k}\cdot\mathbf{r}}}{k_z} dk_x dk_y \right] \qquad (4.7\text{-}4)$$

The quantity inside the brackets can be shown to be the plane wave expansion of a spherical wave [Weyl]:

$$\frac{e^{jkr}}{r} = \frac{j}{2\pi} \iint\limits_{-\infty}^{\infty} \frac{e^{j\mathbf{k}\cdot\mathbf{r}}}{k_z} dk_x dk_y \qquad (4.7\text{-}5)$$

Substituting this result in Eq. (4.7-4) yields

$$h(x, y, z) = \frac{1}{2\pi} \frac{\partial}{\partial z} \left[\frac{e^{jkr}}{r} \right] \qquad (4.7\text{-}6)$$

The above equation is the same as

$$h(x, y, z) = \frac{1}{j\lambda} \frac{z}{r} \frac{e^{jkr}}{r} \qquad (4.7\text{-}7)$$

The diffracted field $U(x, y, z)$ is interpreted as the convolution of $U(x, y, 0)$ with $h(x, y, z)$. Thus,

$$U(x, y, z) = \frac{1}{j\lambda} \iint\limits_{-\infty}^{\infty} U(x, y, 0) \frac{z}{r_{01}} \frac{e^{jkr_{01}}}{r_{01}} dx dy \qquad (4.7\text{-}8)$$

Using Eq. (4.8-6), this result is sometimes written as

$$U(x, y, z) = \frac{1}{2\pi} \int\limits_{-\infty}^{\infty} \int\limits_{-\infty}^{\infty} U(x, y, 0) \frac{\delta}{\delta z} \frac{e^{jkr_{01}}}{r_{01}} dx dy \qquad (4.7\text{-}9)$$

4.8 THE RAYLEIGH–SOMMERFELD DIFFRACTION INTEGRAL FOR NONMONOCHROMATIC WAVES

In the case of nonmonochromatic waves, $U(\mathbf{r}, t)$ is represented in terms of its Fourier transform $U_f(\mathbf{r}, f) = U_f(x, y, z, f)$ as in Eq. (4.2-5). By substituting $f' = -f$, this equation can also be written as

$$U(x, y, z, t) = \int_{-\infty}^{\infty} U_f(x, y, z, -f') e^{-j2\pi f' t} df' \qquad (4.8\text{-}1)$$

$U_f(x, y, z, -f')$ satisfies the Rayleigh–Sommerfeld integral at an aperture:

$$U_f(x, y, z, -f') = \frac{1}{j\lambda} \iint_A U_f(x, y, 0, -f') \frac{z}{r_{01}} \frac{e^{jkr_{01}}}{r_{01}} dx dy \qquad (4.8\text{-}2)$$

Substituting this result in Eq. (4.8-1) yields

$$U(x, y, z, t) = \int_{-\infty}^{\infty} \left[\frac{1}{j\lambda} \iint_A U_f(x, y, 0, -f') \frac{z}{r_{01}} \frac{e^{jkr_{01}}}{r_{01}} dx dy \right] e^{-j2\pi f' t} df' \qquad (4.8\text{-}3)$$

Note that

$$\lambda f = c \qquad (4.8\text{-}4)$$

where c is the phase velocity of the wave. Using this relation and exchanging orders of integration in Eq. (4.9-3) results in

$$U(x, y, z, t) = \iint_A \frac{z}{2\pi c r_{01}^2} \left[\int_{-\infty}^{\infty} -j2\pi f' U_f(x, y, 0, -f') e^{-j2\pi f' (t - \frac{r_{01}}{c})} df' \right] dx dy \quad (4.8\text{-}5)$$

As

$$\frac{d}{dt} U(x, y, 0, t) = \frac{d}{dt} \int_{-\infty}^{\infty} U_f(x, y, 0, -f') e^{-j2\pi f' t} df'$$

$$= \int_{-\infty}^{\infty} -j2\pi f' U_f(x, y, 0, -f') e^{-j2\pi f' t} df'$$

Eq. (4.8-5) can also be written as

$$U(x, y, z, t) = \frac{z}{2\pi c} \iint\limits_{A} \frac{1}{r_{01}^2} \frac{d}{dt} \left[U\left(x, y, 0, t - \frac{r_{01}}{c}\right) \right] dx dy \qquad (4.8\text{-}6)$$

Thus, the wave at (x, y, z), $z0$, is related to the time derivative of the wave at $(x, y, 0)$ with a time delay of r_{01}/c, which is the time for the wave to propagate from P_1 to P_0.

5

Fresnel and Fraunhofer Approximations

5.1 INTRODUCTION

Fresnel and Fraunhofer approximations of the scalar diffraction integral allow simpler Fourier integral computations to be used for wave propagation. They also allow different input and output plane window sizes. However, they are valid only in certain regions, not very close to the input aperture plane. The valid regions for the Rayleigh–Sommerfeld integral, Fresnel, and Fraunhofer approximations are shown in Figure 5.1.

The Rayleigh–Sommerfeld region is observed to be the entire half-space to the right of the input diffraction plane. The Fresnel and Fraunhofer regions are parts of the Rayleigh–Sommerfeld region. Approximate bounds indicating where they start will be derived below as Eqs. (5.25) and (5.27) for the Fresnel region and Eq. (5.41) for the Fraunhofer region. However, these bounds need to be interpreted with care. See Chapter 7 for further explanation.

The term *far field* usually refers to the Fraunhofer region. The term *near field* can be considered to be the region between the input diffraction plane and the Fraunhofer region.

This chapter consists of seven sections. Section 5.2 describes wave propagation in the Fresnel region. The FFT implementation of wave propagation in the Fresnel region is covered in Section 5.3. The fact that the Fresnel approximation is actually the solution of the paraxial wave equation is shown in Section 5.4. Wave propagation in the Fraunhofer region is discussed in Section 5.5.

Diffraction gratings are periodic optical devices that have many significant uses in applications. They also provide excellent examples of how to analyze waves emanating from such devices in the Fresnel and Fraunhofer regions. Section 5.6 discusses the fundamentals of diffraction gratings. Sections 5.7, 5.8, and 5.9 highlight Fraunhofer diffraction with a sinusoidal amplitude grating, Fresnel diffraction with a sinusoidal amplitude grating, and Fraunhofer diffraction with a sinusoidal phase grating, respectively.

Diffraction, Fourier Optics and Imaging, by Okan K. Ersoy
Copyright © 2007 John Wiley & Sons, Inc.

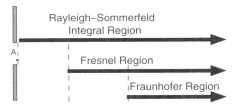

Figure 5.1.   The three diffraction regions.

5.2   DIFFRACTION IN THE FRESNEL REGION

Let the input wave field be restricted to a radial extent L_1:

$$U(x, y, 0) = 0 \quad \text{if} \quad \sqrt{x^2 + y^2} > L_1 \tag{5.2-1}$$

Similarly, the radial extent of the observed wave field $U(x,y,z)$ at the output plane is confined to a region L_2 so that

$$U(x_0, y_0, z_0) = 0 \quad \text{if} \quad \sqrt{x_0^2 + y_0^2} > L_2 \tag{5.2-2}$$

r_{01} in Eq. (4.6-3) is given by

$$r_{01} = z\left[1 + \frac{(x_0 - x)^2 + (y_0 - y)^2}{z^2}\right]^{1/2} \tag{5.2-3}$$

where $(x,y,0)$ are the coordinates of a point on the input plane and (x_0, y_0, z) are the coordinates of a point on the observation plane. With the two restrictions discussed above, the following is true:

$$[(x_0 - x)^2 + (y_0 - y)^2]_{\text{max}} \le (L_1 + L_2)^2 \tag{5.2-4}$$

The upper limit will be used below. If

$$|z| \gg L_1 + L_2, \tag{5.2-5}$$

the term $(i/j\lambda)(z/r_{01}^2)$ in Eq. (4.7-6) can be approximated by $1/j\lambda z$. However, more care is required with the phase. kr_{10} can be expanded in a binomial series as

$$kr_{10} = kz + \frac{k}{2z}[(x_0 - x)^2 + (y_0 - y)^2] - \frac{k}{8z^3}[(x_0 - x)^2 + (y_0 - y)^2]^2 + \dots$$

$$\tag{5.2-6}$$

The third term has maximum absolute value equal to $k(L_1 + L_2)^4/(8|z|^3)$. This will be much less than 1 radian if

$$|z|^3 \gg \frac{k(L_1 + L_2)^2}{8} \qquad (5.2\text{-}7)$$

With this constraint on $|z|$, the phase can be approximated as

$$kr_{10} \cong kz + \frac{k}{2z}[(x_0 - x)^2 + (y_0 - y)^2] \qquad (5.2\text{-}8)$$

The approximations made above are known as *Fresnel approximations*. The region decided by Eqs. (5.2-5) and (5.2-7) is known as the *Fresnel region*. In this region, Eq. (4.7-6) becomes

$$U(x_0, y_0, z) = \frac{e^{jkz}}{j\lambda z} \iint\limits_{-\infty}^{\infty} U(x, y, 0) e^{j\frac{\pi}{\lambda z}[(x_0 - x)^2 + (y_0 - y)^2]} \, dxdy \qquad (5.2\text{-}9)$$

this is a 2-D convolution with respect to x and y and can be written as

$$U(x_0, y_0, z) = U(x, y, 0) * h(x, y, z) \qquad (5.2\text{-}10)$$

where the impulse response is given by

$$h(x, y, z) = \frac{e^{jkz}}{j\lambda z} e^{j\frac{\pi}{\lambda z}(x^2 + y^2)} \qquad (5.2\text{-}11)$$

The corresponding transfer function is given by

$$H(f_x, f_y, z) = e^{jkz} e^{-j\pi\lambda z(f_x^2 + f_y^2)} \qquad (5.2\text{-}12)$$

The quadratic terms in Eq. (5.2-9) can be expanded such that Eq. (5.2-9) becomes

$$U(x_0, y_0, z) = \frac{e^{jkz}}{j\lambda z} e^{j\frac{k}{2z}(x_0^2 + y_0^2)} \iint\limits_{-\infty}^{\infty} U'(x, y, 0) e^{-j\frac{2\pi}{\lambda z}(x_0 x + y_0 y)} \, dxdy \qquad (5.2\text{-}13)$$

where

$$U'(x, y, 0) = U(x, y, 0) e^{j\frac{k}{2z}(x^2 + y^2)} \qquad (5.2\text{-}14)$$

Aside from multiplicative amplitude and phase factors, $U(x_0, y_0, z)$ is observed to be the 2-D Fourier transform of $U'(x, y, 0)$ at spatial frequencies $f_x = x_0/\lambda z$ and $f_y = y_0/\lambda z$.

EXAMPLE 5.1 Determine the impulse response function for Fresnel diffraction by starting with Eq. (4.4-21) of the ASM method.
Solution: When $z \gg r$ in Eq. (4.4-21), the second and third terms on the right hand side can be approximated by 1. Also expanding $[z^2 + r^2]^{1/2}$ in a Taylor's series with two terms kept, Eq. (4.4-21) becomes

$$h(x, y, z) = \frac{e^{jkz}}{j\lambda z} e^{j\frac{\pi}{\lambda z}(x^2 + y^2)}$$

which is the same as Eq. (5.2-11).

EXAMPLE 5.2 Determine the wave field in the Fresnel region due to a point source modeled by $U(x, y, 0) = \delta(x - 3)\delta(y - 4)$.
Solution: The output wave field is given by

$$U(x, y, z) = U(x, y, 0) * h(x, y, z)$$

$$\simeq \delta(x - 3)\delta(y - 4) * \frac{1}{j\lambda z} e^{jkz\left[1 + \frac{x^2 + y^2}{2z^2}\right]}$$

$$\simeq \frac{1}{j\lambda z} e^{jkz\left[1 + \frac{(x-3)^2 + (y-4)^2}{2z^2}\right]}$$

EXAMPLE 5.3 A circular aperture with a radius of 2 mm is illuminated by a normally incident plane wave of wavelength equal to 0.5 μ. If the observation region is limited to a radius of 10 cm, find z in order to be in the Fresnel region.
Solution: With $L_1 = 0.2$ cm and $L_2 = 10$ cm, Eq. (5.2-4) gives

$$z \gg 10.2 \text{ cm}$$

If a factor of 10 is used to satisfy the inequality, we require

$$z \geq 1.02 \text{ m}$$

The second condition given by Eq. (5.2-7) is

$$z^3 \gg \frac{2\pi(0.102)^4}{4 \times 10^{-6}} \text{m}^3$$

$$z \gg 5.4 \text{ m}$$

Using a factor of 10, we require

$$z \geq 54 \text{ m}$$

It is observed that the second condition given by Eq. (5.2-7) yields an excessive value of z. However, this condition does not have to be true for the Fresnel approximation to be valid [Goodman]. In general, kr_{01} is very large, causing the quadratic phase factor $e^{jkr_{10}}$ to oscillate rapidly. Then, the major contribution to the diffraction integral comes from points near $x \sim x_0$ and $y \sim y_0$, called *points of stationary phase* [Stamnes, 1986]. Consequently, the Fresnel approximation can be expected to be valid in a region whose minimum z value is somewhere between the two values given by Eqs. (5.2-4) and (5.2-7). As a matter of fact, the Fresnel approximation has been used in the near field to analyze many optical phenomena. The major reasons for this apparent validity are the subject matter of Chapter 7.

EXAMPLE 5.4 Determine the Fresnel diffraction pattern if a square aperture of width D is illuminated by a plane wave with amplitude 1.
Solution: The input wave can be written as

$$U(x, y, 0) = \text{rect}\left(\frac{x}{D}\right)\text{rect}\left(\frac{y}{D}\right)$$

The Fresnel diffraction is written in the convolution form as

$$U(x_0, y_0, z) = \frac{e^{jkz}}{j\lambda z} \int\int_{-D/2}^{D/2} e^{j\frac{\pi}{\lambda z}[(x-x_0)^2+(y-y_0)^2]}dxdy$$

$$= \frac{e^{jkz}}{j}\left[\frac{1}{\sqrt{\lambda z}}\int_{-D/2}^{D/2} e^{j\frac{\pi}{\lambda z}(x-x_0)^2}dx\right]\left[\frac{1}{\sqrt{\lambda z}}\int_{-D/2}^{D/2} e^{j\frac{\pi}{\lambda z}(y-y_0)^2}dy\right] \quad (5.2\text{-}15)$$

Consider the integral

$$I_x = \frac{1}{\sqrt{\lambda z}}\int_{-D/2}^{D/2} e^{j\frac{\pi}{\lambda z}(x-x_0)^2}dx$$

Let

$$v = \sqrt{\frac{2}{\lambda z}}(x - x_0)$$

Then, I_x becomes

$$I_x = \frac{1}{\sqrt{2}}\int_{x_b}^{x_e} e^{j\frac{\pi}{2}v^2}dv \quad (5.2\text{-}16)$$

where x_b and x_e are given by

$$x_{\mathrm{b}} = -\sqrt{\frac{2}{\lambda z}}\left(\frac{D}{2} + x_0\right)$$

$$x_{\mathrm{e}} = \sqrt{\frac{2}{\lambda z}}\left(\frac{D}{2} + x_0\right)$$

Eq. (5.2-15) can be written as

$$I_x = \frac{1}{\sqrt{2}}\int_0^{x_{\mathrm{e}}} e^{j\frac{\pi}{2}v^2}\,\mathrm{d}v - \frac{1}{\sqrt{2}}\int_0^{x_{\mathrm{b}}} e^{j\frac{\pi}{2}v^2}\,\mathrm{d}v \qquad (5.2\text{-}17)$$

The Fresnel integrals $C(z)$ and $S(z)$ are defined by

$$C(z) = \int_0^z \cos\left(\frac{\pi v^2}{2}\right)\mathrm{d}v \qquad (5.2\text{-}18)$$

$$S(z) = \int_0^z \sin\left(\frac{\pi v^2}{2}\right)\mathrm{d}v \qquad (5.2\text{-}19)$$

In terms of $C(z)$ and $S(z)$, Eq. (5.2-17) is written as

$$I_x = \frac{1}{\sqrt{2}}\{[C(x_{\mathrm{e}}) - C(x_{\mathrm{b}})] + j[S(x_{\mathrm{e}}) - S(x_{\mathrm{b}})]\} \qquad (5.2\text{-}20)$$

The integral above also occurs with respect to y in Eq. (5.2-15). Let it be denoted by I_y, which can be written as

$$\begin{aligned} I_y &= \frac{1}{\sqrt{2}}\int_{y_{\mathrm{b}}}^{y_{\mathrm{e}}} e^{j\frac{\pi}{2}v^2}\,\mathrm{d}v \\ &= \frac{1}{\sqrt{2}}\{[C(y_{\mathrm{e}}) - C(y_{\mathrm{b}})] + j[S(y_{\mathrm{e}}) - S(y_{\mathrm{b}})]\} \end{aligned} \qquad (5.2\text{-}21)$$

where

$$y_{\mathrm{b}} = -\sqrt{\frac{2}{\lambda z}}\left(\frac{D}{2} + y_0\right)$$

$$Y_{\mathrm{e}} = \sqrt{\frac{2}{\lambda z}}\left(\frac{D}{2} - y_0\right)$$

Equation (5.2-15) for $U(x_0, y_0, z)$ can now be written as

$$U(x_0, y_0, z) = \frac{e^{jkz}}{2j} |x|y \tag{5.2-22}$$

The intensity $I(x_0, y_0, z) = |U(x_0, y_0, z)|^2$ can be written as

$$I(x_0, y_0, z) = \frac{1}{4}\{[C(x_e) - C(x_b)]^2 + [S(x_e) - S(x_b)]^2\}$$

$$\times \{[C(y_e) - C(y_b)]^2 + [S(y_e) - S(y_b)]^2\} \tag{5.2-23}$$

The Fresnel number N_F is defined by

$$N_F = \frac{D^2}{4\lambda z} \tag{5.2-24}$$

For constant D and λ, N_F decreases as z increases.

The Fresnel diffraction intensity from a square aperture is visualized in Figures 5.2–5.5. As z gets less, the intensity starts resembling the shape of the input square aperture. This is further justified in Example 5.5. Figure 5.2 shows the image of the square aperture used. Figure 5.3 is the Fresnel intensity diffraction pattern from the square aperture. Figure 5.4 is the same intensity diffraction pattern as a 3-D plot.

Figure 5.2.   The input square aperture for diffraction studies.

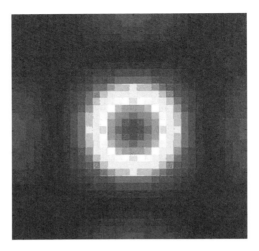

Figure 5.3. The intensity diffraction pattern from a square aperture in the Fresnel region.

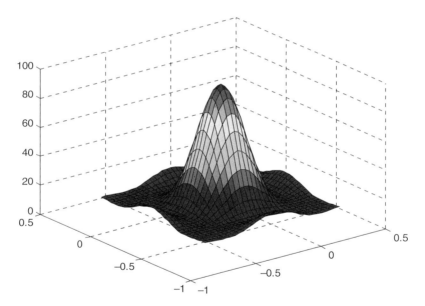

Figure 5.4. The same intensity diffraction pattern from the square aperture in the Fresnel region as a 3-D plot.

EXAMPLE 5.5 Using the Fresnel approximation, show that the intensity of the diffracted beam from a rectangular aperture looks like the rectangular aperture for small z.

Solution: The intensity pattern from the rectangular aperture is given by Eq. (5.2-22) where D_x and D_y are used instead of D along the x- and y-directions, respectively.

The variables for very small z can be written as

$$x_b = -\sqrt{\frac{2}{\lambda z}}\left(\frac{D}{2} + x_0\right) \rightarrow \begin{cases} -\infty & x_0 > -D_x/2 \\ \infty & x_0 < -D_x/2 \end{cases}$$

$$x_e = \sqrt{\frac{2}{\lambda z}}\left(\frac{D}{2} - x_0\right) \rightarrow \begin{cases} \infty & x_0 < D_x/2 \\ -\infty & x_0 > D_x/2 \end{cases}$$

$$y_b = -\sqrt{\frac{2}{\lambda z}}\left(\frac{D}{2} + y_0\right) \rightarrow \begin{cases} -\infty & y_0 > -D_y/2 \\ \infty & y_0 < -D_y/2 \end{cases}$$

$$y_e = \sqrt{\frac{2}{\lambda z}}\left(\frac{D}{2} - y_0\right) \rightarrow \begin{cases} \infty & y_0 < D_y/2 \\ -\infty & y_0 > D_y/2 \end{cases}$$

The asymptotic limits of cosine and sine integrals are given by

$$C(\infty) = S(\infty) = 0.5, \quad C(-\infty) = S(-\infty) = -0.5$$

Consequently, Eq. (5.2-22) becomes

$$I(x_0, y_0, z) \simeq \text{rect}\left(\frac{x_0}{D}\right)\text{rect}\left(\frac{y_0}{D}\right)$$

for small z.

EXAMPLE 5.6 Show under what conditions the angular spectrum method reduces to the Fresnel approximation.
Solution: The transfer function in the angular spectrum method is given by

$$H_A(f_x, f_y) = \begin{cases} e^{jz\sqrt{k^2 - 4\pi^2(f_x^2 + f_y^2)}} & \sqrt{f_x^2 + f_y^2} < \frac{1}{\lambda} \\ 0 & \text{otherwise} \end{cases}$$

The transfer function in the Fresnel approximation is given by

$$H_F(f_x, f_y) = e^{jkz}e^{-j\pi\lambda z(f_x^2 + f_y^2)}$$

If

$$\sqrt{k^2 - 4\pi^2(f_x^2 + f_y^2)} \cong k - \frac{4\pi^2(f_x^2 + f_y^2)}{2k}$$

that corresponds to keeping the first two terms of the Taylor expansion $H_A(f_x, f_y)$ equals $H_F(f_x, f_y)$. The approximation above is valid when $|\lambda f_x|, |\lambda f_y| \ll 1$. In conclusion, the Fresnel approximation and the angular spectrum method are approximately equivalent for small angles of diffraction.

5.3   FFT IMPLEMENTATION OF FRESNEL DIFFRACTION

Fresnel diffraction can be implemented either as convolution given by Eq. (5.2-10) or by directly using the DFT after discretizing Eq. (5.2-13). The convolution implementation is similar to the implementation of the method of the angular spectrum of plane waves. The second approach is discussed below.

Let $U(m_1, m_2, 0) = U(\Delta x m_1, \Delta y m_2, 0)$ be the discretized $u(x,y,0)$. Thus, x and y are discretized as $\Delta x m_1$ and $\Delta x m_2$, respectively. Similarly, x_0 and y_0 are discretized as $\Delta x_0 n_1$ and $\Delta x_0 n_2$, respectively. We write

$$U'(m_1, m_2, 0) = U(m_1, m_2, 0)e^{j\frac{k}{2z}[(\Delta x m_1)^2 + (\Delta y m_2)^2]} \tag{5.3-1}$$

In order to use the FFT, the Fourier kernel in Eq. (5.2-13) must be expressed as

$$e^{-j\frac{2\pi}{\lambda z}(x_0 x + y_0 y)} = e^{-j\frac{2\pi}{\lambda z}(\Delta x \Delta x_0 n_1 m_1 + \Delta y \Delta y_0 n_2 m_2)} = e^{-j2\pi\left(\frac{n_1 m_1}{N_1} + \frac{n_2 m_2}{N_2}\right)} \tag{5.3-2}$$

where $N_1 \times N_2$ is the size of the FFT to be used. Hence,

$$\frac{2\pi}{\lambda z} x_0 x = \frac{2\pi}{\lambda z} \Delta x \Delta x_0 n_1 m_1 = 2\pi \frac{n_1 m_1}{N_1}$$

This yields

$$\Delta x \Delta x_0 = \frac{\lambda z}{N_1} \tag{5.3-3}$$

Similarly, we have

$$\Delta y \Delta y_0 = \frac{\lambda z}{N_2} \tag{5.3-4}$$

In other words, small Δx or Δy gives large Δx_0 or Δy_0 and vice versa, respectively. Equations (5.3-3) and (5.3-4) can be compared to Eq. (4.4-12) for the angular spectrum method. For $f_x = x_0/\lambda z$ and $f_y = y_0/\lambda z$, these equations are the same. However, now the output is directly given by Eq. (5.2-3), there is no inverse transform to be computed, and hence the input and output windows can be of different sizes.

In order to use the FFT, $U'(m_1, m_2, 0)$ is shifted by $\left(\frac{N_1}{2}, \frac{N_2}{2}\right)$, as discussed previously in Section 4.4. Suppose the result is $U''(m_1, m_2, 0)$.

The discretized output wave $U(n_1, n_2, z) = U(\Delta x_0 n_1, \Delta y_0 n_2, z)$ is expressed as

$$U(n_1, n_2, z) = \frac{e^{jkz}}{j\lambda z} e^{j\frac{k}{2z}[(\Delta x_0 n_1)^2 + (\Delta y_0 n_2)^2]} \Delta x \Delta y \sum_{m_1=0}^{N_1-1} \sum_{m_2=0}^{N_2-1} U''(m_1, m_2, 0)e^{-j2\pi\left(\frac{n_1 m_1}{N_1} + \frac{n_2 m_2}{N_2}\right)}$$

$$\tag{5.3-5}$$

$U(n_1, n_2, z)$ is shifted again by $\left(\frac{N_1}{2}, \frac{N_2}{2}\right)$ in order to correspond to the real wave.

EXAMPLE 5.7 Consider Example 4.4. With the same definitions used in that example, the pseudo program for Fresnel diffraction can be written as

$$u1 = u * \exp\left[j\frac{k}{2z}(x^2 + y^2)\right]$$

$$\text{u1p} = \text{fftshift}(\text{u1})$$

$$u3 = \frac{e^{jkz}}{j\lambda z}e^{j\frac{k}{2z}(x_0^2 + y_0^2)} * u2$$

$$\text{U} = \text{fftshift}(\text{u3})$$

5.4 PARAXIAL WAVE EQUATION

The Fresnel approximation is actually a solution of the paraxial wave equation derived below. Consider the Helmholtz equation . If the field is assumed to be propagating mainly along the z-direction, it can be expressed as

$$U(x, y, z) = U'(x, y, z)e^{jkz} \tag{5.4-1}$$

where $U'(x, y, z)$ is assumed to be a slowly varying function of z. Substituting Eq. (5.4-1) in the Helmholtz equation yields

$$\nabla^2 U' + 2jk\frac{\delta}{\delta z}U' = 0 \tag{5.4-2}$$

As $U'(x, y, z)$ is a slowly varying function of z, Eq. (5.4-2) can be approximated as

$$\frac{\delta^2 U'}{\delta x^2} + \frac{\delta^2 U'}{\delta y^2} + 2jk\frac{\delta}{\delta z}U' = 0 \tag{5.4-3}$$

Equation (5.4-3) is called the *paraxial wave equation.*
   Consider

$$U_1(x, y) = \frac{1}{j\lambda z}e^{\frac{jk(x^2 + y^2)}{2z}} \tag{5.4-4}$$

It is easy to show that U_1 is a solution of Eq. (5.4-3). As Eq. (5.4-3) is linear and shift invariant, $U_1(x - x_1, y - y_1)$ for arbitrary x_1 and y_1 is also a solution. Superimposing such solutions, a general solution is found as

$$U(x, y, z) = \frac{1}{j\lambda z} \iint g(x_1, y_1) e^{\frac{jk((x-x_1)^2 + (y-y_1)^2)}{2z}} dx_1 dy_1 \qquad (5.4\text{-}5)$$

Consider the Fresnel approximation given by

$$U(x_0, y_0, z) = \frac{1}{j\lambda z} \iint U(x, y, 0) e^{\frac{jk((x_0-x)^2 + (y_0-y)^2)}{2z}} dx dy \qquad (5.4\text{-}6)$$

Choosing $g(x, y) = U(x, y, 0)$, it is observed that Eqs. (5.4-5) and (5.4-6) are the same. Hence, the Fresnel approximation is a solution of the paraxial wave equation.

5.5   DIFFRACTION IN THE FRAUNHOFER REGION

Consider Eq. (5.2-14). If

$$z \gg \frac{k}{2}(x^2 + y^2)_{\max} = \frac{kL_1^2}{2}, \qquad (5.5\text{-}1)$$

then, $U'(x, y, 0)$ is approximately equal to $U(x,y,0)$. When this happens, the *Fraunhofer region* is valid, and Eq. (5.2-13) becomes

$$U(x_0, y_0, z) = \frac{e^{jkz}}{j\lambda z} e^{\frac{jk}{2z}(x_0^2 + y_0^2)} \int\!\!\!\int_{-\infty}^{\infty} U(x, y, 0) e^{-j\frac{2\pi}{\lambda z}(x_0 x + y_0 y)} dx dy \qquad (5.5\text{-}2)$$

Aside from multiplicative amplitude and phase factors, $U(x_0, y_0, z)$ is observed to be the 2-D Fourier transform of $U(x,y,0)$ at frequencies $f_x = x_0/\lambda z$ and $f_y = y_0/\lambda z$. It is also observed that diffraction in the Fraunhofer region does not have the form of a convolution.

EXAMPLE 5.8 Find the wave field in the Fraunhofer region due to a wave emanating from a circular aperture of diameter D at $z = 0$.
Solution: The circular aperture of diameter D means

$$U(x, y, 0) = \text{cyl}(r/D)$$

where $r = \sqrt{x^2 + y^2}$.
   From Example 2.6, the 2-D Fourier transform of $U(x,y,0)$ is given by

$$U(x, y, 0) \leftrightarrow \frac{D^2 \pi}{4} \text{somb}\left(D\sqrt{f_x^2 + f_y^2}\right)$$

Figure 5.5.   The intensity diffraction pattern from a square aperture in the Fraunhofer region.

where $f_x = x_0/\lambda z$ and $f_y = y_0/\lambda z$. Using Eq. (5.5-2) gives

$$U(x_0, y_0, z) = \frac{e^{jkz}}{j\lambda z} e^{j\frac{k}{2z}(x_0^2 + y_0^2)} \frac{D^2 \pi}{4} \operatorname{somb}\left(D\sqrt{f_x^2 + f_y^2}\right)^2$$

The intensity distribution equal to $U^2(x_0, y_0, z)$ is known as the *Airy pattern*. It is given by

$$I(x_0, y_0, z) = \left(\frac{kD^2}{8z}\right)^2 \left[\operatorname{somb}\left(D\sqrt{f_x^2 + f_y^2}\right)\right]^2$$

The Fraunhofer diffraction intensity from the same square aperture of Figures 5.2 is visualized in Figures 5.5 and 5.6. Figure 5.5 is the Fraunhofer intensity diffraction pattern from the square aperture. Figure 5.6 is the same intensity diffraction pattern as a 3-D plot.

EXAMPLE 5.9   Determine when the Fraunhofer region is valid if the wavelength is 0.6 μ (red light) and an aperture width of 2.5 cm is used.
Solution:

$$z \gg \frac{kL_1^2}{2} = \frac{\pi L_1^2}{\lambda}$$

$$\frac{\pi L_1^2}{\lambda} = \frac{\pi \cdot (1.25 \cdot 10^{-2})^2}{0.6 \cdot 10^{-6}} = 830\,\mathrm{m}$$

This large distance will be much reduced when the lenses are discussed in Chapter 8.

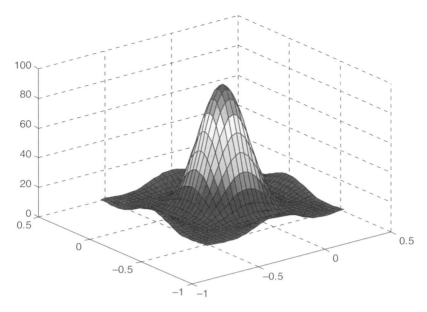

Figure 5.6. The same intensity diffraction pattern from the square aperture in the Fraunhofer region as a 3-D plot.

EXAMPLE 5.10 Consider Example 5.7. With the same definitions used in that example, the pseudo program for Fresnel diffraction can be written as

$$u1 = \text{fftshift}(u1)$$
$$u2 = \text{fft}(u1)$$
$$u3 = \frac{e^{jkz}}{j\lambda z} e^{j\frac{k}{2z}(x_0^2 + y_0^2)} * u2$$
$$U = \text{fftshift}(u3)$$

5.6   DIFFRACTION GRATINGS

Diffraction gratings are often analyzed and synthesized by using Fresnel and Fraunhofer approximations. Some major application areas for diffraction gratings are spectroscopy, spectroscopic imaging, optical communications, and networking.

A diffraction grating is an optical device that periodically modulates the amplitude or the phase of an incident wave. Both the amplitude and the phase may also be modulated. An array of alternating opaque and transparent regions is called a transmission amplitude grating. Such a grating is shown in Figures 5.7 and 5.8.

Figure 5.7.   Planar view of a transmission amplitude grating.

If the entire grating is transparent, but varies periodically in optical thickness, it is called a transmission phase grating. A reflective material with periodic surface relief also produces distinct phase relationships upon reflection of a wave. These are known as reflection phase gratings. Reflecting diffraction gratings can also be generated by fabricating periodically ruled thin films of aluminum evaporated on a glass substrate.

Diffraction gratings are very effective to separate a wave into different wavelengths with high resolution. In the simplest case, a grating can be considered as a large number of parallel, closely spaced slits. Regardless of the number of slits, the peak intensity occurs at diffraction angles governed by the following grating equation:

$$d \sin \phi_m = m\lambda \tag{5.6-1}$$

where d is the spacing between the slits, m is an integer called *diffraction order*, and ϕ is the diffraction angle. Equation (5.6-1) shows that waves at different

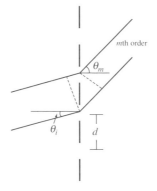

Figure 5.8.   Periodic amplitude grating.

wavelengths travel in different directions. The peak intensity at each wavelength depends on the number of slits in the grating.

The *transmission function* of a grating is defined by

$$t(x, y) = \frac{U(x, y, 0_+)}{U(x, y, 0_-)} \tag{5.6-2}$$

where $U(x, y, 0_-)$ and $U(x, y, 0_+)$ are the wave functions immediately before and after the grating, respectively. If the grating is of reflection type, the transmission function becomes the *reflection function*.

$t(x,y)$ is, in general, complex. If $\arg(t(x,y))$ equals zero, the grating modifies the amplitude only. If $|t(x, y)|$ equals a constant, the grating modifies the phase only.

Gratings are sometimes manufactured to work with the reflected light, for example, by metalizing the surface of the grating. Then, the transmission function becomes the *reflection function*.

The phase of the incident wave can be modulated by periodically varying the thickness or the refractive index of a transparent plate. Reflecting diffraction gratings can be generated by fabricating periodically ruled thin films of aluminum evaporated on a glass substrate.

5.7   FRAUNHOFER DIFFRACTION BY A SINUSOIDAL AMPLITUDE GRATING

Assuming a plane wave incident perpendicularly on the grating with unity amplitude, the transmission function is the same as $U(x,y,0)$ given by

$$U(x, y, 0) = \left[\frac{1}{2} + \frac{m}{2}\cos(2\pi f_0 x)\right]\text{rect}\left(\frac{x}{D}\right)\text{rect}\left(\frac{y}{D}\right) \tag{5.7-3}$$

where the diffraction grating is assumed to be limited to a square aperture of width D. $U(x,y,0)$ can be written as

$$U(x, y, 0) = f(x, y)g(x, y) \tag{5.7-4}$$

where

$$f(x, y) = \frac{1}{2} + \frac{m}{2}\cos(2\pi f_0 x) \tag{5.7-5}$$

$$g(x, y) = \text{rect}\left(\frac{x}{D}\right)\text{rect}\left(\frac{y}{D}\right) \tag{5.7-6}$$

The FTs of $f(x,y)$ and $g(x,y)$ are given by

$$F(f_x,f_y) = \frac{1}{2}\delta(f_x,f_y) + \frac{m}{4}[\delta(f_x + f_0,f_y) + \delta(f_x - f_0,f_y)] \qquad (5.7\text{-}7)$$

$$G(f_x,f_y) = D^2 \sin c(Df_x) \sin c(Df_y) \qquad (5.7\text{-}8)$$

The FT of $U(x,y,0)$ is the convolution of $F(f_x,f_y)$ with $G(f_x,f_y)$:

$$U(x,y,0) \longleftrightarrow F(f_x,f_y) * G(f_x,f_y) \qquad (5.7\text{-}9)$$

where

$$F(f_x,f_y) * G(f_x,f_y)$$

$$= \frac{D^2}{2}\sin c(Df_y)\left\{\left[\sin c(Df_x) + \frac{m}{2}(\sin c[D(f_x + f_0)] + \sin c[D(f_x - f_0)])\right]\right\}$$

$$(5.7\text{-}10)$$

Then, the wave field at z is given by

$$U(x_0,y_0,z) = \frac{e^{jkz}}{j\lambda z} e^{j\frac{k}{2z}(x_0^2 + y_0^2)} F(f_x,f_y)^* G(f_x,f_y) \qquad (5.7\text{-}11)$$

where $f_x = x_0/\lambda z$ and $f_y = y_0/\lambda z$.

5.8 FRESNEL DIFFRACTION BY A SINUSOIDAL AMPLITUDE GRATING

In order to compute Fresnel diffraction by a sinusoidal amplitude grating, we can use the convolution form of the diffraction integral given by Eq. (5.2-8). The transfer function of the system is rewritten here as

$$H(f_x,f_y) = e^{jkz}e^{-j\pi\lambda z(f_x^2 + f_y^2)} \qquad (5.8\text{-}1)$$

where e^{jkz} can be suppressed as it is a constant phase factor.

The Fourier transform of the transmittance function is given by Eq. (5.7-9). In order to simplify the computations, we will assume that the diffraction grating aperture is large so that the Fourier transform of the transmittance function of the grating can be approximated by Eq. (5.7-7). Then, the frequencies that contribute are given by

$$(f_x,f_y) = (0,0), \quad (-f_0,0), \quad (f_0,0)$$

The propagated wave has the following Fourier transform:

$$U(x_0,y_0,z) \longleftrightarrow \frac{1}{2}\delta(f_x,f_y) + \frac{m}{4}[e^{-j\pi\lambda zf_0^2}\delta(f_x - f_0,f_y) + e^{-j\pi\lambda zf_0^2}\delta(f_x + f_0,f_y)] \qquad (5.8\text{-}2)$$

The inverse Fourier transform yields

$$U(x_0, y_0, z) = \frac{1}{2} + \frac{m}{4} e^{-j\pi\lambda z f_0^2} [e^{j2\pi f_0 x_0} + e^{-j2\pi f_0 x_0}]$$

$$= \frac{1}{2} [1 + m\, e^{-j\pi\lambda z f_0^2 \cos(2\pi f_0 x_0)}]$$

(5.8-3)

The intensity of the wave is given by

$$I(x_0, y_0, z) = \frac{1}{4} \left[1 + 2m\cos\left(\pi\lambda z f_0^2\right)\cos(2\pi f_0 x_0) + m^2\cos^2(2\pi f_0 x_0)\right]$$ (5.8-4)

EXAMPLE 5.11 Determine the intensity of the diffraction pattern at a distance z that satisfies

$$z = \frac{2n}{\lambda f_0^2} \quad n \text{ integers}$$

Solution: $I(x_0, y_0, z)$ is given by

$$I(x_0, y_0, z) = \frac{1}{4}[1 + m\cos(2\pi f_0 x_0)]^2$$

This is the image of the grating. Such images are called *Talbot images* or *self-images*.

EXAMPLE 5.12 Repeat the previous example if

$$z = \frac{(2n + 1)}{\lambda f_0^2} \quad n \text{ integer}$$

Solution: $I(x_0, y_0, z)$ is given by

$$I(x_0, y_0, z) = \frac{1}{4}[1 - m\cos(2\pi f_0 x_0)]^2$$

This is the image of the grating with a 180° spatial phase shift, referred to as *contrast reversal*. Such images are also called *Talbot images*.

EXAMPLE 5.13 Repeat the previous example if

$$z = \frac{2n-1}{2\lambda f_0^2} \quad n \text{ integer}$$

Solution: $I(x_0, y_0, z)$ is given by

$$I(x_0, y_0, z) = \frac{1}{4}\left[1 + \frac{m^2}{2} + \frac{m^2}{2}\cos(4\pi f_0 x_0)\right]$$

This is the image of the grating at twice the frequency, namely, $2f_0$. Such images are called *Talbot subimages*.

5.9 FRAUNHOFER DIFFRACTION WITH A SINUSOIDAL PHASE GRATING

As with the sinusoidal amplitude grating, with an incident plane wave on the grating, the transmission function for a *sinusodial phase grating* is given by

$$U(x, y, 0) = e^{j\frac{m}{2}\sin(2\pi f_0 x)}\text{rect}\left(\frac{x}{D}\right)\text{rect}\left(\frac{y}{D}\right) \tag{5.9-1}$$

$U(x,y,0)$ can be written as

$$U(x, y, 0) = f(x, y)g(x, y) \tag{5.9-2}$$

where

$$f(x, y) = e^{j\frac{m}{2}\sin(2\pi f_0 x)} \tag{5.9-3}$$

$$g(x, y) = \text{rect}\left(\frac{x}{D}\right)\text{rect}\left(\frac{y}{D}\right) \tag{5.9-4}$$

The analysis can be simplified by use of the following identity:

$$e^{j\frac{m}{2}\sin(2\pi f_0 x)} = \sum_{k=-\infty}^{\infty} J_k\left(\frac{m}{2}\right)e^{j2\pi f_0 kx} \tag{5.9-5}$$

where $J_k(\cdot)$ is the Bessel function of the first kind of order k. Using the above identity, the FT of $f(x,y)$ is given by

$$F(f_x, f_y) = \sum_{k=-\infty}^{\infty} J_k\left(\frac{m}{2}\right)\delta(f_x - kf_0, f_y) \tag{5.9-6}$$

The FT of $U(x,y,0)$ is given by

$$U(x, y, 0) \leftrightarrow F(f_x, f_y) * G(f_x, f_y) \tag{5.9-7}$$

where

$$F(fx_1, f_y) * G(f_x, f_y) = D^2 \sum_{k=-\infty}^{\infty} J_k\left(\frac{m}{2}\right) \text{sinc}[D(f_x - kf_0)] \text{sinc}(Df_y) \tag{5.9-8}$$

Then,

$$U(x_0, y_0, z) = \frac{e^{jkz}}{j\lambda z} e^{j\frac{k}{2z}(x_0^2 + y_0^2)} F(f_x, f_y) * G(f_x, f_y) \tag{5.9-9}$$

It is observed that sinusoidal amplitude grating in Section 5.7 has three orders of energy concentration due to three sin c functions, which do not significantly overlap if the grating frequency is much greater than $2/D$. In contrast, the sinusoidal phase grating has many orders of energy concentration. Whereas the central order is dominant in the amplitude grating, the central order may vanish in the phase grating when $J_0(m/2)$ equals 0.

5.10   DIFFRACTION GRATINGS MADE OF SLITS

Sometimes diffraction gratings are made of slits. A two-slit example is shown in Figure 5.9, where b is the slit width, and ϕ is the diffraction angle as shown in this figure.

The intensity of the wave field coming from the grating in the Fraunhofer region can be shown to be

$$I(\phi) = 2I_0 \left[\frac{\sin(kb\phi/2)}{kb\phi/2}\right]^2 [1 + \cos(k\Delta + kd\phi)] \tag{5.10-1}$$

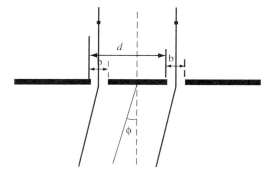

Figure 5.9.   A two-slit diffraction grating.

where Δ is the difference of the optical path lengths between the rays of two adjacent slits and I_0 is the initial intensity of the beam.

When there are N parallel slits, the intensity in the Fraunhofer region can be shown to be

$$I(\phi) = 2I_0 \left[\frac{\sin(kb\phi/2)}{kb\phi/2}\right]^2 \left[\frac{\sin(Nkd\phi/2)}{\sin(kd\phi/2)}\right]^2 \qquad (5.10\text{-}2)$$

The second factor above equals $N\cos(\pi Nm)/\cos(\pi m)$ when the grating equation $d\sin\phi = m\lambda$ is satisfied.

6

Inverse Diffraction

6.1 INTRODUCTION

Inverse diffraction involves recovery of the image of an object whose diffraction pattern is measured, for example, on a plane. In the case of the Fresnel and Fraunhofer approximations, the inversion is straightforward. In the very near field, the angular spectrum representation can be used, but then there are some technical issues that need to be addressed.

The geometry to be used in the following sections is shown in Figure 6.1. The observation plane and the measurement plane are assumed to be at $z = z_0$ and $z = z_r$, respectively. Previously, z_r was chosen equal to 0. The distance between the two planes is denoted by z_{0r}. The medium is assumed to be homogeneous. The problem is to determine the field at $z = z_0$, assuming it is known at $z = z_r$.

Like all inverse problems, the inverse diffraction problem is actually difficult if evanescent waves are to be incorporated into the solution [Vesperinas, 1991]. Then, the inverse diffraction problem involves a singular kernel [Shewell, Wolf, 1968]. When the evanescent waves are avoided as discussed in the succeeding sections, the problem is well behaved.

This chapter consists of four sections. Section 2 is on inversion of the Fresnel and Fraunhofer approximations. Section 6.3 describes the inversion of the angular spectrum representation. Section 6.4 discusses further analysis of the results of Section 6.3.

6.2 INVERSION OF THE FRESNEL AND FRAUNHOFER REPRESENTATIONS

The Fresnel diffraction is governed by Eq. (5.2-13). As the integral is a Fourier transform, its inversion is given by

$$U(x, y, z_r) = \frac{je^{-jkz_{0r}}}{\lambda z} e^{-j\frac{k}{2z_{0r}}(x^2+y^2)} \int_{-\infty}^{\infty} \int_{-\infty}^{\infty} U(x_0, y_0, z_0) e^{-j\frac{k}{2z_{0r}}(x_0^2+y_0^2)} e^{j\frac{2\pi}{\lambda z_{0r}}(x_0 x + y_0 y)} dx_0 dy_0$$

$$(6.2\text{-}1)$$

Diffraction, Fourier Optics and Imaging, by Okan K. Ersoy
Copyright © 2007 John Wiley & Sons, Inc.

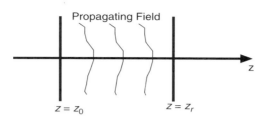

Figure 6.1. Geometry for inverse diffraction.

where z_r and z_0 are the input and output z variables, and $z_{0r} = z_0 - z_r$.

The Fraunhofer diffraction is governed by Eq. (5.4-2). Its inversion is given by

$$U(x, y, z_r) = \frac{je^{-jkz_{0r}}}{\lambda z_{0r}} \int\limits_{-\infty}^{\infty} \int\limits_{-\infty}^{\infty} U(x_0, y_0, z_0) e^{-j\frac{k}{2z_{0r}}(x_0^2 + y_0^2)} e^{j\frac{2\pi}{\lambda z_{0r}}(x_0 x + y_0 y)} dx_0 dy_0 \qquad (6.2\text{-}2)$$

6.3 INVERSION OF THE ANGULAR SPECTRUM REPRESENTATION

Below the method is first discussed in 2-D. Then, it is generalized to 3-D.

Equation (4.3-12) in 2-D can be written as

$$U(x_0, z_0) = \int\limits_{-\infty}^{\infty} A(f_x, z_r) e^{jz_{0r}\sqrt{k^2 - 4\pi^2 f_x^2}} e^{j2\pi f_x x_0} df_x \qquad (6.3\text{-}1)$$

As this is a Fourier integral, recovery of $A(f_x, z_r)$ involves the computation of the Fourier transform of $U(x_0, z_0)$:

$$A(f_x, z_r) = e^{-jz_{0r}\sqrt{k^2 - 4\pi^2 f_x^2}} \int\limits_{-\infty}^{\infty} U(x_0, z_0) e^{-j2\pi f_x x_0} dx_0 \qquad (6.3\text{-}2)$$

$U(x, z_r)$ is the inverse Fourier transform of $A(f_x, z_r)$:

$$U(x, z_r) = \int\limits_{-\infty}^{\infty} \left[\int\limits_{-\infty}^{\infty} U(x_0, z_0) e^{-j2\pi f_x x_0} dx_0 \right] e^{-jz_{0r}\sqrt{k^2 - 4\pi^2 f_x^2}} e^{j2\pi f_x x} df_x \qquad (6.3\text{-}3)$$

With $F[\bullet]$ and $F^{-1}[\bullet]$ indicating the forward and inverse Fourier transforms, Eq. (6.3-3) can be expressed as

$$U(x, z_r) = F^{-1} \left[F[U(x_0, z_0)] e^{-jz_{0r}\sqrt{k^2 - 4\pi^2 f_x^2}} \right] \qquad (6.3\text{-}4)$$

In 3-D, the corresponding equation is given by

$$U(x, y, z_r) = F^{-1}\left[F[U(x_0, y_0, z_0)]e^{-jz_{0r}\sqrt{k^2 - 4\pi^2(f_x^2 + f_y^2)}}\right] \qquad (6.3\text{-}5)$$

Note that z_r can be varied above to reconstruct the wave field at different depths. In this way, a 3-D wave field reconstruction is obtained.

If measurements are made with nonmonochromatic waves, resulting in a wave field $U(x_0, y_0, z_0, t)$, where t indicates the time dependence, a single time frequency component $U(x_0, y_0, z_0, f)$ can be chosen by computing

$$U(x_0, y_0, z_0, f) = \int_{-\infty}^{\infty} U(x_0, y_0, z_0, t)e^{-j2\pi ft}\,dt \qquad (6.3\text{-}6)$$

Then, the equations discussed above can be used with k given by

$$k = \frac{2\pi}{\lambda} = 2\pi\frac{f}{v} \qquad (6.3\text{-}7)$$

where v is the phase velocity. The technique discussed above was used with ultrasonic and seismic image reconstruction [Boyer, 1971; Boyer et al., 1970; Ljunggren, 1980]. It is especially useful when z_{0r} is small so that other approximations to the diffraction integral, such as Fresnel and Fraunhofer approximations discussed in Chapter 5, cannot be used.

6.4   ANALYSIS

Interchanging orders of integration in Eq. (6.4-3) results in

$$U(x, z_r) = \int_{-\infty}^{\infty} U(x_0, z_0)B(x, x_0)\,dx_0 \qquad (6.4\text{-}1)$$

where

$$B(x, x_0) = \int_{-\infty}^{\infty} e^{-jz_{0r}\sqrt{k^2 - 4\pi^2 f_x^2}}\,e^{-j2\pi f_x(x - x_0)}\,df_x \qquad (6.4\text{-}2)$$

When $2\pi f_x > k$, the propagator in the last integral becomes $e^{kz_{0r}\sqrt{4\pi^2 f_x^2 - k^2}}$, and the integral in Eq. (6.4-2) diverges. This is due to the inclusion of evanescent waves. In practice, it can be shown that the results are sufficiently accurate when the limits of integration are restricted such that $k > 2\pi f_x$ [Lalor, 1968]. Then, f_x is restricted to the range

$$|f_x| \leq \frac{1}{\lambda} \tag{6.4-3}$$

Suppose that the exact wave field at $z = z_r$ is $T(x,z_r)$. How does $U(x,z_r)$ as computed above compare to $T(x,z_r)$? In order to answer this question, we can first determine $U(x_0, z_0)$ in terms of $T(x,z_r)$ by forward diffraction. This is given by

$$U(x_0, z_0) = \int_{-M}^{M} \left[\int_{-\infty}^{\infty} T(x, z_r) e^{-j2\pi f_x x} dx \right] e^{jz_{0r}\sqrt{k^2 - 4\pi^2 f_x^2}} e^{j2\pi f_x x_0} df_x \tag{6.4-4}$$

where $|f_x| \leq M$ is used. The upper bound for M is $1/\lambda$.

Substituting this result in Eq. (6.4-3) and allowing f_x to the range $f_x \leq Q$ results in

$$U(x, z_r) = \int_{-Q}^{Q} \left\{ \int_{-M}^{M} \left[\int_{-\infty}^{\infty} T(x, z_r) e^{-j2\pi f_x' x} dx \right] e^{jz_{0r}\sqrt{k^2 - 4\pi^2 f_x'^2}} e^{j2\pi f_x' x_0} df_x' \right\}$$
$$\times e^{-jz_{0r}\sqrt{k^2 - 4\pi^2 f_x^2}} e^{j2\pi f_x x} df_x$$

Interchanging orders of integration, this can be written as [Van Rooy, 1971]

$$U(x, z_r) = \int_{-\infty}^{\infty} \left[\int_{-M}^{M} df_x' e^{-j2\pi f_x' x} e^{jz_{0r}\sqrt{k^2 - 4\pi^2 f_x'^2}} \right.$$
$$\left. \times \left(\int_{-Q}^{Q} df_x e^{-j2\pi f_x x} e^{-jz_{0r}\sqrt{k^2 - 4\pi^2 f_x^2}} \left[\int_{-\infty}^{\infty} dx' e^{j2\pi x'(f_x' - f_x)} \right] \right) \right] T(x, z_r) dx$$

As

$$\delta(f_x - f_x') = \int_{-\infty}^{\infty} e^{j2\pi x(f_x' - f_x)} dx$$

Eq. (6.4-4) becomes

$$U(x, z_r) = \int_{-\infty}^{\infty} \left[\int_{-M}^{M} e^{-j2\pi f_x' x'} e^{jz_{0r}\sqrt{k^2 - 4\pi^2 f_x'^2}} df_x' \int_{-Q}^{Q} \delta(f_x - f_x') e^{j2\pi f_x x} e^{-jz_{0r}\sqrt{k^2 - 4\pi^2 f_x^2}} df_x \right] T(x, z_r) dx$$

Depending on the values of M and Q, there are two possible cases that are discussed below.

Case I: $Q \geq M$
In this case, Eq. (6.4-5) reduces to

$$U(x, z_r) = \int_{-\infty}^{\infty} \left[\int_{-M}^{M} e^{j2\pi f_x'(x-x_0)} df_x' \right] T(x, z_r) dx$$
$$= T(x, z_r) * \sin c \left(\frac{2Mx}{\lambda} \right) \qquad (6.4\text{-}5)$$
$$= T(x, z_r)$$

because $T(x,z_r)$ is bandlimited.

Case II: $Q < M$
In this case, Eq. (6.4-12) reduces to

$$U(x, z_r) = T(x, z_r) * \sin c \left(\frac{2Qx}{\lambda} \right) \qquad (6.4\text{-}6)$$

This means lowpass filtering of $T(x,z_r)$ because the Fourier transform of the $\sin c$ function is the rectangular function. In other words, frequencies above Q/λ are filtered out. The reconstructed image may have smoothed edges and loss of detail as a consequence.

As Q is restricted to the values $Q \leq 1/\lambda$ in practice, the resolution achievable in the image is λ. This is in agreement with other classical results that indicate that linear imaging systems cannot resolve distances less than a wavelength.

EXAMPLE 6.1 Determine $U(x,z_r)$ when $U(x', z_0)$ is measured in an aperture $|x'| \leq R$.
Solution: In this case, Eq. (6.4-3) becomes

$$U(x, z_r) = \int_{-Q}^{Q} \left\{ e^{-jkz_{0r}\sqrt{k^2-4\pi^2 f_x^2}} \int_{-R}^{R} dx_0 e^{-j2\pi f_x x_0} \right.$$
$$\times \left. \left(\left[\int_{-M}^{M} df_x' e^{j2\pi f_x' x_0} \left[\int_{-\infty}^{\infty} dx T(x, z_r) e^{-j2\pi f_x' x} \right] \right] \right) \right\}^{j2\pi f_x x} df_x$$

Interchanging orders of integration, this can be written as

$$U(x, z_r) = \int\limits_{-\infty}^{\infty} \left[\int\limits_{-M}^{M} df_x'(e^{-j2\pi f_x' x} e^{jkz_{0r}\sqrt{k^2 - 4\pi^2 f_x'^2}} \right.$$

$$\left. \times \int\limits_{-Q}^{Q} df_x[e^{j2\pi f_x x} e^{-jz_{0r}\sqrt{k^2 - 4\pi^2 f_x^2}} 2R \sin c(2Rf_x)]) \right] T(x, z_r) dx$$

It is observed that the sin c function above mixes the plane wave components in the original object wave. This means that each plane wave component in the original object wave is truncated by the aperture (-R,R) and therefore "spreads" by diffraction.

EXAMPLE 6.2 When $M = Q = 1/\lambda$, show that $T^*(x, z_r)$ can be exactly recovered from $U(x_0, z_0)$ by forward propagation of $U^*(x_0, z_0)$ instead of inverse propagation.
Solution: Suppose that forward propagation from the plane $z = z_0$ to $z = 2z_0 - z_r$ is carried out. As

$$2z_0 - z_r - z_0 = z_{0r}$$

the forward propagation equation is given by

$$U(x, 2z_0 - z_r) = \int\limits_{-Q}^{Q} df_x \int\limits_{-\infty}^{\infty} U^*(x_0, z_0)e^{-j2\pi f_x x'} e^{jz_{0r}\sqrt{k^2 - 4\pi^2 f_x^2}} e^{j2\pi f_x x} dx_0$$

Rearranging orders of integration results in

$$U(x, 2z_0 - z_r) = \left[\int\limits_{-Q}^{Q} \int\limits_{-\infty}^{\infty} U(x_0, z_0)e^{j2\pi f_x x_0} e^{-jz_{0r}\sqrt{k^2 - 4\pi^2 f_x^2}} e^{-j2\pi f_x x} dx_0 \right]^* df_x = T^*(x, 0)$$

In other words, the original image can be reconstructed by back space propagation of the recorded wave or forward propagation of the complex conjugate of the recorded wave.

7

Wide-Angle Near and Far Field Approximations for Scalar Diffraction

7.1 INTRODUCTION

Approximations for computing forward and inverse diffraction integrals are of vital significance in many areas involving wave propagation. As discussed in Chapters 4 and 5, approximations such as the Fresnel approximation, the Fraunhofer approximation, and the more rigorous angular spectrum method (ASM) all involve the Fourier transform, its discrete counterpart, the discrete Fourier transform, and its fast computational routine, the fast Fourier transform (FFT).

The Fresnel approximation is valid at reasonable distances from the input plane whereas the Fraunhofer approximation is valid in the far field. The ASM is a rigorous solution of the Helmholtz equation; its numerical implementation is usually done with the FFT, and possibly other digital signal processing algorithms, with their related approximations [Mellin and Nordin, 2001; Shen and Wang, 2006].

When the sizes of the diffracting apertures are less than the wavelength, scalar diffraction theory yields nonnegligible errors, and other numerical methods such as the finite difference time domain (FDTD) method and the finite element method (FEM) may become necessary to use [Kunz, 1993; Taflove and Hagness, 2005]. However, these methods are not practical with large scale simulations as compared with methods utilizing the FFT. With diffracting aperture sizes of the order of the wavelength used and in the near field, the ASM has been found to give satisfactory results [Mellin and Nordin, 2001]. With the ASM, one disadvantage is that the input and output plane sizes are the same. The output plane size is usually desired to be considerably larger than the input plane size in most applications, and this can be done with the ASM only by repeatedly using the ASM in short distances and additionally using filtering schemes to make the output size progressively larger.

The Fresnel, Fraunhofer approximations, and the ASM can be considered to be valid in practice at small angles of diffraction. It is desirable to have approximate methods that are valid at wide angles of diffraction in the near field as well as the far field and also based on the Fourier integral to be implemented with the FFT so that

Diffraction, Fourier Optics and Imaging, by Okan K. Ersoy
Copyright © 2007 John Wiley & Sons, Inc.

large scale computations can be carried out in a reasonable amount of time and storage. Even if the spatial frequencies are sampled such that the FFT cannot be used, the Fourier integral representation is still desirable as it is separable in the two variables of integration, reducing 4-D tensor operations to 2-D matrix operations.

In this chapter, a new set of approximations having such features is discussed. The approximations are based on the Taylor expansion around the radial distance of a point on the output plane from the origin. With these approximations, what makes possible to utilize the FFT is semi-irregular sampling at the output plane. Semi-irregular sampling means sampling is first done in a regular array, and the array points are subsequently perturbed to satisfy the FFT conditions.

Some interesting results will be related to the similarity of the proposed approximation to the Fresnel approximation even though the new approximation is valid in the near field and at wide angles whereas the Fresnel approximation is not. This is believed to be one reason why the Fresnel approximation has often been used in the near field in many applications. The results obtained in such studies are actually valid at the output sampling points perturbed in position, as discussed in this paper.

This chapter consists of eight sections. Section 7.2 is a review of Fresnel and Fraunhofer approximations previously discussed in Chapter 5. Section 7.3 introduces the new methods with the radial set of approximations. Section 7.4 provides further higher order improvements and error analysis. Section 7.5 shows how the method can be used in inverse diffraction and iterative optimization applications. Section 7.6 provides numerical simulation examples in 2-D and 3-D geometries. Section 7.7 is about how to increase accuracy by centering input and output plane apertures around some center coordinates and possibly also using smaller subareas in large scale simulations. Section 7.8 covers conclusions.

7.2 A REVIEW OF FRESNEL AND FRAUNHOFER APPROXIMATIONS

The first Rayleigh–Sommerfeld diffraction integral can be written as

$$U(x,y,z) = \frac{1}{j\lambda} \iint\limits_{-\infty}^{\infty} U(x,y,0) \frac{z}{r^2} e^{jkr} dxdy \qquad (7.2\text{-}1)$$

where $U(x,y,0)$ is the input field, $U(x_0,y_0,z)$ is the output field, k is the wave number equal to $2\pi/\lambda$, λ is the wavelength, and r is the radius vector length from the point $(x,y,0)$ to the point (x_0,y_0,z). In general, z is often large enough to approximate z/r^2 by $1/z$. Then, Eq. (7.2-1) becomes

$$U(x,y,z) = \frac{1}{j\lambda z} \iint\limits_{-\infty}^{\infty} U(x,y,0)e^{jkr} dxdy \qquad (7.2\text{-}2)$$

The constant term $\frac{1}{j\lambda z}$ will be neglected in the rest of the chapter.

In the Fresnel approximation commonly used in diffraction problems, the radius vector length is approximated by the Taylor's series expansion with two terms [Mezouari and Harvery, 2003; Southwell, 1981; Steane and Rutt, 1989]:

$$r = \sqrt{z^2 + (x_0 - x)^2 + (y_0 - y)^2} \cong z\left(1 + \frac{g}{2}\right) \tag{7.2-3}$$

where

$$g = \frac{(x_0 - x)^2 + (y_0 - y)^2}{z^2} \tag{7.2-4}$$

The magnitude of the maximum phase error E_{max} in radians is estimated by the next term in the Taylor series as

$$E_{max} \leq \frac{1}{8} z g^2 k \tag{7.2-5}$$

The Fresnel approximation is not sufficiently good in problems where E_{max} is larger than, say, 1 radian. However, this condition has often been relaxed in various applications because good results have been generally obtained. This was partially explained by using the method of stationary phase [Goodman]. In this chapter, another explanation for these results is provided.

The advantage in the Fresnel approximation is that the variables are separable so that the integral in Eq. (7.2-2) can be written as a Fourier integral. For example, Eq. (7.2-4) is rewritten as

$$g = \frac{v}{z^2} + \frac{w}{z^2} - \frac{2(x_0 x + y_0 y)}{z^2} \tag{7.2-6}$$

where

$$v = x_0^2 + y_0^2 \tag{7.2-7}$$
$$w = x^2 + y^2 \tag{7.2-8}$$

Then, with the help of Eq. (7.2-3), Eq. (7.2-2) can be approximated as a Fourier integral by

$$U(x_0, y_0, z) = e^{jk\frac{v}{2z}} \iint u'(x, y) e^{-\frac{jk}{z}(x_0 x + y_0 y)} dx dy \tag{7.2-9}$$

where

$$u'(x, y) = u(x, y) e^{jk\frac{w}{2z}} \tag{7.2-10}$$

Equation (7.2-9) becomes a Fourier integral by writing it as

$$U(x_0, y_0, z) = e^{jk\frac{v}{2z}} \iint u'(x, y) e^{-j2\pi(f_x x + f_y y)} \mathrm{d}x \mathrm{d}y \qquad (7.2\text{-}11)$$

where the spatial frequencies f_x and f_y are defined by

$$f_x = \frac{x_s}{\lambda_z} \qquad (7.2\text{-}12)$$

$$f_y = \frac{y_s}{\lambda_z} \qquad (7.2\text{-}13)$$

where x_s equals x_0, and y_s equals y_0 at this point.

In the Fraunhofer approximation, the condition $z \gg w$ allows $u'(x, y) \approx u(x, y)$ so that Eq. (7.2-9) becomes

$$U(x_0, y_0, z) = e^{jk\frac{v}{2z}} \iint u(x, y) e^{-\frac{jk}{z}(x_0 x + y_0 y)} \mathrm{d}x \mathrm{d}y \qquad (7.2\text{-}14)$$

Equation (7.2-9) can be evaluated with the fast Fourier transform [Brigham, 1974] by letting

$$x_s = m_x \Delta_{0x}, \quad m_x = \text{integer} \qquad (7.2\text{-}15)$$

$$y_s = m_y \Delta_{0y}, \quad m_y = \text{integer} \qquad (7.2\text{-}16)$$

$$x = n_x \Delta_x, \quad n_x = \text{integer} \qquad (7.2\text{-}17)$$

$$y = n_y \Delta_y, \quad n_y = \text{integer} \qquad (7.2\text{-}18)$$

$$\Delta_{0x} \Delta_x = \frac{\lambda z}{M_x} \qquad (7.2\text{-}19)$$

$$\Delta_{0y} \Delta_y = \frac{\lambda z}{M_y} \qquad (7.2\text{-}20)$$

where M_x and M_y are the number of samples in the x- and y-directions, respectively. In any other approximation, it is necessary to keep Eqs. (7.2-19) and (7.2-20) valid if FFT is to be used.

7.3 THE RADIAL SET OF APPROXIMATIONS

In many applications such as those involving digital holography and diffractive optical elements, the coordinates $(x, y, 0)$ are usually much smaller than the object coordinates (x_0, y_0, z). Then, Eq. (7.2-3) will be written as

$$r = r_0 \sqrt{1 + h} \qquad (7.3\text{-}1)$$

where

$$r_0 = \sqrt{z^2 + x_0^2 + y_0^2} \tag{7.3-2}$$

$$h = \frac{w}{r_0^2} - \frac{2(x_0 x + y_0 y)}{r_0^2} \tag{7.3-3}$$

In order to be able to use the Fourier transform, Eq. (7.3-3) is to be approximated by

$$h = \frac{w}{z^2} - \frac{2(x_0 x + y_0 y)}{r_0^2} \tag{7.3-4}$$

The simplest radial set of approximations is obtained by approximating the radial distance r as

$$r \approx r_0 + \frac{w^2}{2z} - \frac{xx_0 + yy_0}{r_0} \tag{7.3-5}$$

Equations (7.3-4) and (7.3-5) are also valid when x, y, x_0, and y_0 are sufficiently smaller than z (sufficiency conditions will be more clear later). Equation (7.3-5) is further improved in the next section. In order to obtain a Fourier transform relationship with the diffraction integral given by Eq. (7.2-9), the following is defined:

$$x_0 = \frac{x_s}{z} r_0 \tag{7.3-6}$$

$$y_0 = \frac{y_s}{z} r_0 \tag{7.3-7}$$

where (x_s, y_s) is the regular output sampling point. The simultaneous solution of Eqs. (7.3-6) and (7.3-7) is given by

$$x_0 = C(x_s, y_s) x_s \tag{7.3-8}$$
$$y_0 = C(x_s, y_s) y_s \tag{7.3-9}$$

where

$$C(x_s, y_s) = \left[1 - \frac{x_s^2 + y_s^2}{z^2} \right]^{-\frac{1}{2}} \tag{7.3-10}$$

When $z \gg w$ as in the Fraunhofer approximation, the term involving w^2 in Eq. (7.3-5) is skipped. The resulting approximations will be referred to as the *near field approximation* (NFA) and the *far field approximation* (FFA) below. The general case covering both NFA and FFA will be referred to as NFFA.

With the new approximations, Eq. (7.2-11) is replaced by

$$U(x_0, y_0, z) = e^{jkr_0} \iint u'(x, y) e^{-j2\pi(f_x x + f_y y)} dx dy \qquad (7.3\text{-}11)$$

where the spatial frequencies are still given by Eqs. (7.2-12) and (7.2-13) due to Eqs. (7.3-6) and (7.3-7). With FFA, $u'(x, y)$ equals $u(x, y)$.

We note that Eq. (7.3-11) looks like Eq. (7.2-11) for the Fresnel approximation, but the phase factor outside the integral and the output sampling points are different. Equation (31) is straightforward to compute with the FFT, and the results are valid at the sampling points defined by Eqs. (7.3-8) and (7.3-9). The spatial frequencies can also be written as

$$f_x = \frac{x_0}{\lambda r_0} \qquad (7.3\text{-}12)$$

$$f_y = \frac{y_0}{\lambda r_0} \qquad (7.3\text{-}13)$$

It is observed that f_x and f_y are still given by Eqs. (7.2-12) and (7.2-13) with respect to the regular sampling points (x_s, y_s) or by Eqs. (7.3-12) and (7.3-13) with respect to the actual sampling points (x_0, y_0).

7.4 HIGHER ORDER IMPROVEMENTS AND ANALYSIS

We write r in Eq. (7.3-1) in a Taylor series as

$$r = r_0 \left(1 + \frac{h}{2} - \frac{h^2}{8} + \frac{h^3}{16} \bullet \bullet \bullet \bullet \bullet + (-1)^{n-1} \frac{1 * 3 * 5 \bullet \bullet \bullet *(2n-1)}{2 * 4 * 6 \bullet \bullet \bullet *2n} h^{2m-1} \bullet \bullet \bullet \bullet \bullet \right)$$
$$(7.4\text{-}1)$$

where it is desirable to attain the terms shown in order to obtain high accuracy. In order to be able to use the Fourier transform, we only keep the terms in the Taylor series that do not have factors $(x_0 x)^n$ other than $x_0 x$.

We also define

$$a_n = (-1)^{n-1} \frac{1 * 3 * 5 \bullet \bullet \bullet *(2n-1)}{2 * 4 * 6 \bullet \bullet \bullet *2n} \frac{1}{z^{2n}} \qquad (7.4\text{-}2)$$

Then, r is approximated by

$$r \approx r_0 - \frac{x x_0 + y y_0}{r_0} + p(w) \qquad (7.4\text{-}3)$$

where

$$p(w) = p(x, y) = \sum_{n=1}^{\infty} \frac{a_n}{z^{2n-1}} w^n \tag{7.4-4}$$

Equations (7.3-8) and (7.3-9) remain the same because the term involving xx_0 and yy_0 is not modified. When $z \gg w$ as in the Fraunhofer approximation, the term $p(w)$ in Eq. (7.4-3) is skipped. With the new approximations, Eq. (7.3-11) is replaced by

$$U(x_0, y_0, z) = e^{jkr_0} \iint u''(x, y) e^{-j2\pi(f_x x + f_y y)} \mathrm{d}x \mathrm{d}y \tag{7.4-5}$$

where

$$u''(x, y) = u(x, y) e^{jkp(w)} \tag{7.4-6}$$

For a given z, the maximum error in the Fresnel approximation is proportional to $(x_0^2 + y_0^2)^2$ whereas, in the NFFA, it is approximately given by

$$E = -\frac{v(x^2 + y^2)}{4z^3} - \frac{(x_0 x + y_0 y)^2}{2z^3} \tag{7.4-7}$$

This error corresponds to the neglected terms in the second component $-\frac{u^2}{2}$ of the Taylor series expansion.

We note that changes are made only in the object coordinates with Eqs. (7.3-8) and (7.3-9). Otherwise, the integral in Eq. (7.4-5) is the same as the integral in Eq. (7.2-9) or Eq. (7.2-14). The output points (x_0, y_0) are determined by Eqs. (7.3-8) and (7.3-9) in terms of (x_s, y_s) given by Eqs. (7.2-15) and (7.2-16).

An interesting observation is that the Fresnel approximation examples cited in the literature are approximately correct in the near field at the points (x_0, y_0) given by Eqs. (7.3-8) and (7.3-9), and not at the points (x_s, y_s). As such, these results are really the NFFA results.

In order to use the FFT, it is necessary to satisfy Eqs. (7.2-19) and (7.2-20). $m_x \Delta_{0x}$ and $m_y \Delta_{0y}$ are the output sampling positions that are modified to do so. It is possible to consider other sampling strategies in which $m_x \Delta_{0x}$ and $m_y \Delta_{0y}$ are replaced by some other output sampling positions, say, x_s and y_s. The results obtained above are valid in all such cases.

7.5 INVERSE DIFFRACTION AND ITERATIVE OPTIMIZATION

Inverse diffraction involves recovery of $u(x, y)$ from $U(x_0, y_0, z)$ as discussed in Chapter 6. In the case of the Fresnel approximation, Eq. (7.2-9) is a Fourier transform, and its inversion yields

$$u'(x, y) = \iint \hat{U}(x_0, y_0, z) e^{j2\pi(f_x x + f_y y)} \mathrm{d}f_x \mathrm{d}f_y \tag{7.5-1}$$

where $u'(x, y)$ is given by Eq. (7.2-10), the spatial frequencies f_x, f_y are given by Eqs. (7.2-12), (7.2-13), and

$$\hat{U}(x_0, y_0, z) = U(x_0, y_0, z)e^{-jk\frac{y}{2z}} \tag{7.5-2}$$

Similarly, the integrals in Eqs. (7.3-11) and (7.4-5) are Fourier transforms. Inverse Fourier transformation of Eq. (7.3-11) yields

$$u'(x, y) = \iint U'(x_0, y_0, z)e^{j2\pi(f_x x + f_y y)}\mathrm{d}x\mathrm{d}y \tag{7.5-3}$$

where

$$U'(x_0, y_0, z) = U(x_0, y_0, z)e^{-jkr_0} \tag{7.5-4}$$

The inverse Fourier transformation of Eq. (7.4-5) similarly yields

$$u''(x, y) = \iint U'(x_0, y_0, z)e^{j2\pi(f_x x + f_y y)}\mathrm{d}x\mathrm{d}y \tag{7.5-5}$$

Iterative optimization techniques are often used to design optical devices such as DOE's [Lee, 1970, 1975, 1979; Zhuang and Ersoy, 1995]. These techniques are discussed in Chapters 15 and 16. For this purpose, many iterations are carried out between the input and output planes. Equations (7.3-11) and (7.5-3) as well as Eqs. (7.4-5) and (7.5-5) are Fourier transform pairs. Hence, iterative optimization can be carried out exactly as before with these equations.

7.6 NUMERICAL EXAMPLES

Below the 2-D case is considered first with y variable neglected without loss of generality. In the initial computer simulations, the relevant parameters are chosen as follows:

$$\lambda = 0.6328\,\mu, \ x = 1\,\text{mm}, \ N = 256, \ \Delta x/\lambda = 7$$

The minimum z distance for the approximations to be valid as a function of offset x_0 at the output plane can be computed by specifying $k(r - r(\text{approximate}))$ to be less than 1 radian. The result with input offset x equal to 1 mm is shown in Figure 7.1. It is observed that for very small $(x_0 - x)$, the two approximations behave almost the same, but this quickly changes as $(x_0 - x)$ increases.

Figure 7.2 shows the ratio z/x_0 as a function of the output plane offset x_0. This ratio approaches 2 with the new approximation whereas it approaches 36 with the Fresnel approximation as the offset x_0 increases. We note that with the FFT, the maximum output offset equals $z/(\Delta x/\lambda)$, Δx being the input plane sampling

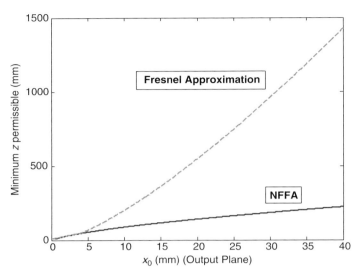

Figure 7.1. Minimum z distance for the approximations to be valid as a function of x_0 at the output plane with x equal to 1 mm.

interval. Hence, the new approximation is sufficiently accurate at all the output sampling points for $\Delta x/\lambda$ greater than 2. Figure 7.3 shows the growth of the difference between x_0 and x_s calculated with Eq. (28) as x_0 increases when $z = 10$ cm.

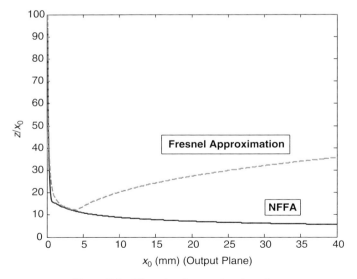

Figure 7.2. The ratio z/x_0 as a function of x_0.

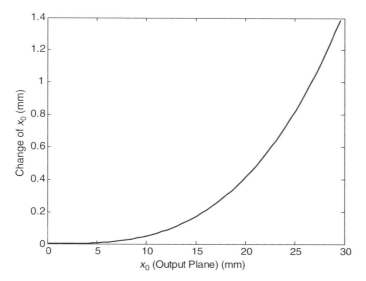

Figure 7.3. Change of x_0 in mm as a function of x_0 at the output plane when $z = 20$ cm and $x = 1$ mm.

Equations (7.3-8) and (7.3-9) show the relationship with the regularly sampled output points as in the Fresnel or Fraunhofer approximations, and the semi-irregularly sampled output points as in the NFFA. Examples of these relationships are given in Tables 7.1 thru 7.3. The (x,y) sampling coordinate values in these tables are shown as complex numbers $x + iy$. Table 7.1 shows the example of regularly sampled values. Table 7.2 shows the corresponding example of semi-irregularly sampled values. Table 7.3 is the difference between Tables 7.1 and 7.2.

In the next set of experiments in 2-D, a comparison of the exact, Fresnel and NFFA approximation integrals was carried out for diffraction of a converging spherical wave from a diffracting aperture. The parameters of the experiments are as follows:

Diffracting aperture size $(D_I) = 30\,\mu$, z *(focusing distance)* $= 5\,\text{mm}$,
$N = 512$, $\lambda = 0.6328\,\mu$

The numerical integration was carried out with the adaptive Simpson quadrature algorithm [Garner]. The output aperture sampling was chosen according to the FFT conditions. Then, the full output aperture size is given by

$$D_O = \frac{N\lambda z}{D_I} \qquad (7.6\text{-}1)$$

However, only the central portion of the output aperture corresponding to 64 sampling points and an extent of 6.7 mm was actually computed. As $z = 5$ mm, this represents a very wide-angle output field.

Table 7.1. Output plane position coordinates in mm with FFT and the Fresnel approximation.

−8.0998 − 8.0998i	−8.0682 − 8.0998i	−8.0366 − 8.0998i	−8.0049 − 8.0998i	−7.9733 − 8.0998i	−7.9416 − 8.0998i	−7.9100 − 8.0998i	−7.8784
−8.0998 − 8.0682i	−8.0682 − 8.0682i	−8.0366 − 8.0682i	−8.0049 − 8.0682i	−7.9733 − 8.0682i	−7.9416 − 8.0682i	−7.9100 − 8.0682i	−7.8784
−8.0998 − 8.0366i	−8.0682 − 8.0366i	−8.0366 − 8.0366i	−8.0049 − 8.0366i	−7.9733 − 8.0366i	−7.9416 − 8.0366i	−7.9100 − 8.0366i	−7.8784
−8.0998 − 8.0049i	−8.0682 − 8.0049i	−8.0366 − 8.0049i	−8.0049 − 8.0049i	−7.9733 − 8.0049i	−7.9416 − 8.0049i	−7.9100 − 8.0049i	−7.8784
−8.0998 − 7.9733i	−8.0682 − 7.9733i	−8.0366 − 7.9733i	−8.0049 − 7.9733i	−7.9733 − 7.9733i	−7.9416 − 7.9733i	−7.9100 − 7.9733i	−7.8784
−8.0998 − 7.9416i	−8.0682 − 7.9416i	−8.0366 − 7.9416i	−8.0049 − 7.9416i	−7.9733 − 7.9416i	−7.9416 − 7.9416i	−7.9100 − 7.9416i	−7.8784
−8.0998 − 7.9100i	−8.0682 − 7.9100i	−8.0366 − 7.9100i	−8.0049 − 7.9100i	−7.9733 − 7.9100i	−7.9416 − 7.9100i	−7.9100 − 7.9100i	−7.8784

Table 7.2. Output plane position coordinates in mm with FFT and the NFFA.

$-8.1535 - 8.1535i$	$-8.1214 - 8.1533i$	$-8.0894 - 8.1531i$	$-8.0573 - 8.1529i$	$-8.0253 - 8.1527i$	$-7.9932 - 8.1525i$	$-7.9612 - 8.1523i$
$-8.1533 - 8.1214i$	$-8.1212 - 8.1212i$	$-8.0892 - 8.1210i$	$-8.0571 - 8.1208i$	$-8.0251 - 8.1206i$	$-7.9930 - 8.1204i$	$-7.9610 - 8.1202i$
$-8.1531 - 8.0894i$	$-8.1210 - 8.0892i$	$-8.0890 - 8.0890i$	$-8.0569 - 8.0888i$	$-8.0249 - 8.0886i$	$-7.9928 - 8.0884i$	$-7.9608 - 8.0881i$
$-8.1529 - 8.0573i$	$-8.1208 - 8.0571i$	$-8.0888 - 8.0569i$	$-8.0567 - 8.0567i$	$-8.0247 - 8.0565i$	$-7.9926 - 8.0563i$	$-7.9606 - 8.0561i$
$-8.1527 - 8.0253i$	$-8.1206 - 8.0251i$	$-8.0886 - 8.0249i$	$-8.0565 - 8.0247i$	$-8.0245 - 8.0245i$	$-7.9924 - 8.0243i$	$-7.9604 - 8.0240i$
$-8.1525 - 7.9932i$	$-8.1204 - 7.9930i$	$-8.0884 - 7.9928i$	$-8.0563 - 7.9926i$	$-8.0243 - 7.9924i$	$-7.9922 - 7.9922i$	$-7.9602 - 7.9920i$
$-8.1523 - 7.9612i$	$-8.1202 - 7.9610i$	$-8.0881 - 7.9608i$	$-8.0561 - 7.9606i$	$-8.0240 - 7.9604i$	$-7.9920 - 7.9602i$	$-7.9600 - 7.9600i$

Table 7.3. Error in output position in mm (difference between Tables 7.1 and 7.2).

$-0.0537 - 0.0537i$	$-0.0532 - 0.0535i$	$-0.0528 - 0.0532i$	$-0.0524 - 0.0530i$	$-0.0520 - 0.0528i$	$-0.0516 - 0.0526i$	$-0.0512 - 0.0524i$
$-0.0535 - 0.0532i$	$-0.0530 - 0.0530i$	$-0.0526 - 0.0528i$	$-0.0522 - 0.0526i$	$-0.0518 - 0.0524i$	$-0.0514 - 0.0522i$	$-0.0510 - 0.0520i$
$-0.0532 - 0.0528i$	$-0.0528 - 0.0526i$	$-0.0524 - 0.0524i$	$-0.0520 - 0.0522i$	$-0.0516 - 0.0520i$	$-0.0512 - 0.0518i$	$-0.0508 - 0.0516i$
$-0.0530 - 0.0524i$	$-0.0526 - 0.0522i$	$-0.0522 - 0.0520i$	$-0.0518 - 0.0518i$	$-0.0514 - 0.0516i$	$-0.0510 - 0.0514i$	$-0.0506 - 0.0512i$
$-0.0528 - 0.0520i$	$-0.0524 - 0.0518i$	$-0.0520 - 0.0516i$	$-0.0516 - 0.0514i$	$-0.0512 - 0.0512i$	$-0.0508 - 0.0510i$	$-0.0504 - 0.0508i$
$-0.0526 - 0.0516i$	$-0.0522 - 0.0514i$	$-0.0518 - 0.0512i$	$-0.0514 - 0.0510i$	$-0.0510 - 0.0508i$	$-0.0506 - 0.0506i$	$-0.0502 - 0.0504i$
$-0.0524 - 0.0512i$	$-0.0520 - 0.0510i$	$-0.0516 - 0.0508i$	$-0.0512 - 0.0506i$	$-0.0508 - 0.0504i$	$-0.0504 - 0.0502i$	$-0.0500 - 0.0500i$

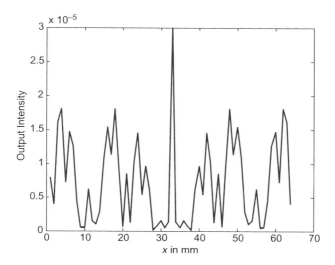

Figure 7.4. The output intensity from an aperture with the exactly computed and NFFA approximated diffraction integral.

Figure 7.4 shows the output diffraction pattern intensities for both the exact method and the NFFA method almost perfectly overlapping each other. Figure 7.5 shows the normalized intensity error defined by

$$\text{Error} = \frac{I_{\text{Exact}} - I_{\text{NFFA}}}{I_{\text{Exact}}} \qquad (7.6\text{-}2)$$

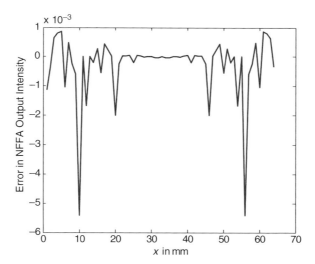

Figure 7.5. The output intensity error between the exactly computed and NFFA approximated diffraction integrals.

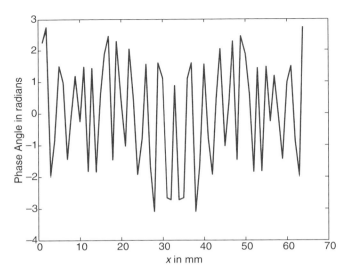

Figure 7.6. The output phase angle with the exactly computed and NFFA approximated diffraction integral.

where I_{Exact} and I_{NFFA} are the exact and NFFA intensities. Figure 7.6 shows the output phase angles for the exact method and the NFFA method almost perfectly overlapping each other.

Figure 7.7 shows the output diffraction pattern intensity computed with the Fresnel approximation in comparison with the exact method. Figure 7.8 shows the

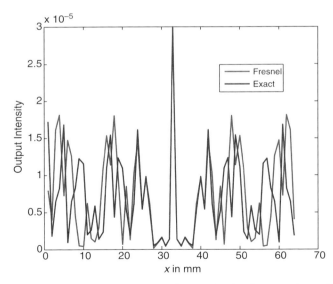

Figure 7.7. The output intensity with the Fresnel approximation in comparison with the exactly computed diffraction integral.

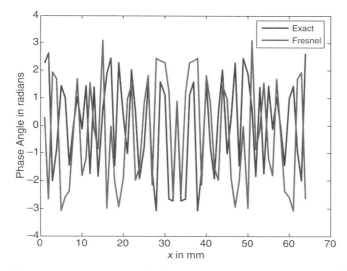

Figure 7.8. The output phase angle with the Fresnel approximation in comparison with the output phase angle with the exactly computed diffraction integral.

output phase angle computed with the Fresnel approximation in comparison with the exact method. It is observed that the errors with the Fresnel approximation in the geometry considered are nonnegligible.

In the next set of experiments, 3-D results are discussed. Figure 7.9 shows the input square aperture used in the simulations. Two sets of parameters were used. In the first set, the parameters are as follows:

$$Z = 50 \, \text{mm}, \quad D_I \, (\text{input square aperture side}) = 0.2 \, \text{mm}, \quad N = 512, \quad \lambda = 0.6328 \, \mu$$

Figure 7.9. The input square aperture used in simulations.

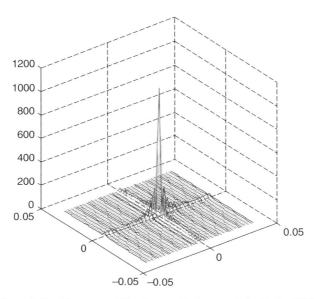

Figure 7.10. The output diffraction amplitude computed with the NFFA.

With the FFT conditions, the output plane aperture size becomes 0.081 mm on a side with these parameters. Figure 7.10 shows the 3-D plot of the output diffraction amplitude computed with the NFFA. Figure 7.11 shows the corresponding 3-D plot of the output diffraction amplitude computed with the Fresnel approximation.

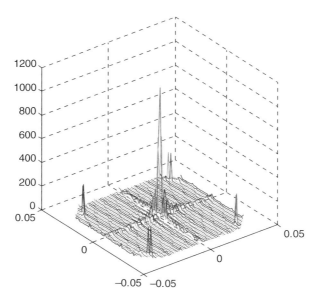

Figure 7.11. The output diffraction amplitude computed with the Fresnel approximation.

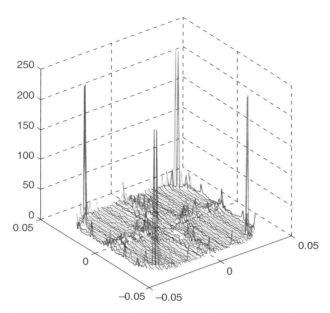

Figure 7.12. The error in diffraction amplitude (absolute difference between Figs. 7.11 and 7.12).

Figure 7.12 shows the error in diffraction amplitude (absolute difference between Figures 7.10 and 7.11). It is observed that the Fresnel approximation makes nonnegligible error in amplitude in certain parts of the observation plane in this geometry.

The second set of parameters was chosen as follows:

$$z = 100\,\text{mm}, \quad D_I \text{ (input square aperture side)} = 2\,\text{mm}, \quad N = 512, \quad \lambda = 0.6328\,\mu$$

With the FFT conditions, the output plane aperture size becomes 16.2 mm on a side with these parameters. Figure 7.13 shows the 3-D plot of the output diffraction amplitude computed with the NFFA. Figure 7.14 shows the corresponding 3-D plot of the output diffraction amplitude computed with the Fresnel approximation. Figure 7.15 shows the error in diffraction amplitude (absolute difference between Figures 7.13 and 7.14). It is observed that the nonnegligible error in amplitude due to the Fresnel approximation in certain parts of the observation plane is less serious in this geometry because z is larger and the output size is smaller.

The final set of experiments was done with the checkerboard image of Figure 16 as the desired output amplitude diffraction pattern. The parameters of the experiments were as follows:

$$z = 80\,\text{mm}, \quad D_I \text{ (input square aperture side)} = 0.8\,\text{mm}, \quad N = 256, \quad \lambda = 0.6328\,\mu$$

With the FFT conditions, the output plane aperture size becomes 16.2 mm on a side with these parameters. The input diffraction pattern was computed in two ways. The

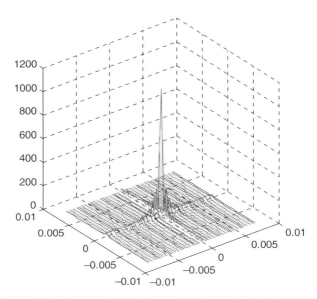

Figure 7.13.   The output diffraction amplitude computed with the NFFA.

first one is by inverse Fresnel approximation given by Eqs. (7.5-1) and (7.5-2). The second one is by inverse NFFA approximation given by Eqs. (7.5-3) and (7.5-4). Figure 16 also shows the NFFA reconstruction of the output diffraction amplitude at the semi-irregularly sampled output points, meaning that it is the same as the

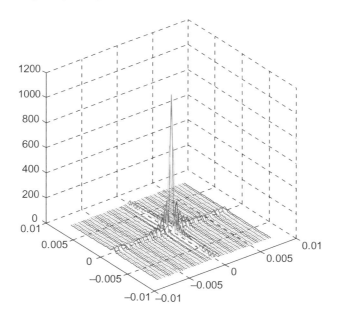

Figure 7.14.   The output diffraction amplitude computed with the Fresnel approximation.

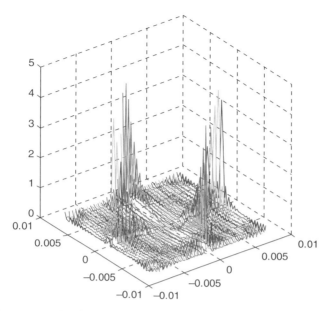

Figure 7.15. The error in diffraction amplitude (absolute difference between Figs. 7.14 and 7.15).

original image. Figure 17 shows the output diffraction amplitude computed with the NFFA at the regularly sampled output points, starting with the input diffraction pattern obtained with the inverse Fresnel approximation. As the NFFA is believed to be accurate, the output diffraction amplitude in an actual physical implementation would look like Figure 17 at the regularly sampled output points.

Figure 7.16. The checkerboard image and its NFFA reconstruction from the field computed with the inverse NFFA ($z = 80$ mm, input size $= 0.8 \times 0.8$ mm, output size $= 16.2 \times 16.2$ mm).

Figure 7.17. The NFFA reconstruction of the checkerboard image from the field computed with the inverse Fresnel approximation ($z = 80$ mm, input size $= 0.8 \times 0.8$ mm, output size $= 16.2 \times 16.2$ mm).

7.7 MORE ACCURATE APPROXIMATIONS

If the output space coordinates are centered around (x_c, y_c), and the input space coordinates are centered around (x_I, y_I), better accuracy can be obtained by defining

$$x_0 = x_0' + x_c \tag{7.7-1}$$

$$y_0 = y_0' + y_c \tag{7.7-2}$$

$$x = x' + x_I \tag{7.7-3}$$

$$y = y' + y_I \tag{7.7-4}$$

$$q = (x_c - x_I)^2 + (y_c - y_I)^2 \tag{7.7-5}$$

$$r_c = [z^2 + q]^{1/2} \tag{7.7-6}$$

The following is defined:

$$x_{cI} = x_c - x_I \tag{7.7-7}$$

$$y_{cI} = y_c - y_I \tag{7.7-8}$$

$$v' = x_0'^2 + y_0'^2 + 2[x_{cI}x_0' + y_{cI}y_0'] \tag{7.7-9}$$

$$w' = x'^2 + y'^2 - 2[x_{cI}x' + y_{cI}y'] \tag{7.7-10}$$

$$g' = \frac{v' + w'}{2r_c^2} - \frac{x_0'x + y_0'y}{r_c^2} \tag{7.7-11}$$

Then, the analysis remains valid with z replaced by r_c and g, v, w replaced by g', v', w', respectively. In this case, the error is approximately given by

$$E \simeq -\frac{v'(x^2 + y^2)}{4r_c^3} - \frac{(x_0 x + y_0 y)^2}{2r_c^3} \qquad (7.7\text{-}12)$$

These approximations would be especially useful in large scale simulations by dividing the input and/or output areas into smaller subareas and applying the NFFA between such subareas. With a parallel processing system, large-scale simulations can then be carried out in near real time.

7.8 CONCLUSIONS

We conclude that the Fresnel and Fraunhofer approximations can be considerably improved with the NFFA, especially for wide-angle computations while still utilizing the FFT for computation by more accurately sampling the output points.

The major difference between the Fresnel/Fraunhofer approximations and the NFFA is indeed the different sampling points. The different phase factors outside the integrals in Eqs. (7.2-11) and (7.3-11) are comparable except for very wide angles of diffraction and do not affect the diffraction intensity pattern at the output plane. Hence, when the Fresnel approximation is used as it has always been, the results are still sufficiently accurate at the new sampling points determined by Eqs. (7.3-8) and (7.3-9), but now these results are really the NFFA results at the new sampling points. The new permissible z-distance in the NFFA is smaller than the permissible z-distance with the Fresnel approximation by a considerable factor such as 18.

As long as semi-irregular output point is acceptable, the NFFA can be considered to be as accurate as the angular spectrum method, but also with wide angles of diffraction, at most geometries considered in practice.

8

Geometrical Optics

8.1 INTRODUCTION

Geometrical optics (ray optics) involves approximate treatment of wave propagation in which the wavelength λ is considered to be infinitesimally small. In practice, this means λ is much smaller than the spatial sizes of all disturbances to the amplitude or phase of a wave field. For example, if a phase shift of 2π radians occurs over a distance of many wavelengths due to a phase-shifting aperture, geometrical optics can be reliably used. For example, ray optics can be used in a large core multimode optical fiber.

In geometrical optics, a wave field is usually described by *rays* which travel in different optical media according to a set of rules. The *Eikonal equation* forms the basis of geometrical optics. It can be derived from the Helmholtz equation as the wavelength approaches zero.

This chapter consists of eight sections. Section 8.2 introduces the physical fundamentals of the propagation of rays. This is followed by the ray equation in Section 8.3. For rays traveling close to the optical axis, it is shown that the simplification of the ray equation results in the paraxial ray equation.

The eikonal equation is derived in Section 8.4. Local spatial frequencies and rays are illustrated in Section 8.5 with examples. The *meridional rays*, which are rays traveling in a single plane containing the optical axis and another orthogonal axis such as the y-axis, are described in Section 8.6 in terms of 2×2 matrix algebra. The theory is generalized to thick lenses in Section 8.7. The entrance and exit pupils of a complex optical system are described in Section 8.8.

8.2 PROPAGATION OF RAYS

In geometrical optics, the location and direction of rays are of primary concern. For example, in image formation by an optical system, the collection of rays from each point of an object are redirected towards a corresponding point of an image.

Diffraction, Fourier Optics and Imaging, by Okan K. Ersoy
Copyright © 2007 John Wiley & Sons, Inc.

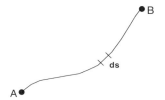

Figure 8.1. Optical path length between A and B.

Optical components centered about an optical axis are usually small such that the rays travel at small angles. Such rays are called *paraxial rays* and are the basis of *paraxial optics*.

The basic postulates of geometrical optics are as follows:

1. A wave field travels in the form of rays.
2. A ray travels with different velocities in different optical media: the *refractive index*, n of an optical medium is a number greater than or equal to unity. If the velocity equals c when n equals 1, the velocity equals c/n when $n \neq 1$. n is often a function of position, say, $n(\mathbf{r})$ in a general optical medium.
3. *Fermat's principle*: Rays travel along the path of least-time. In order to understand this statement better, consider Figure 8.1. The infinitesimal time of travel in the differential length element ds along the path as shown in Figure 8.1 is given by

$$\frac{ds}{c/n(\mathbf{r})} = \frac{n(\mathbf{r})ds}{c} \tag{8.2-1}$$

We define *optical path length* L_p between the points A and B as

$$L_p = \int_A^B n(\mathbf{r})ds \tag{8.2-2}$$

Then, the time of travel between A and B is given by L_p/c.

Fermat's principle states that L_p is an extremum with respect to neighboring paths:

$$\delta L_p = 0 \tag{8.2-3}$$

Dividing this equation by v, the variation of travel time also satisfies

$$\frac{L_p}{v} = 0 \tag{8.2-4}$$

This extremum is typically a minimum such that rays follow the path of least time.

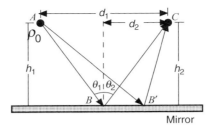

Figure 8.2. Minimum path between A and B after reflection from a mirror.

In a *homogeneous medium*, n is constant. Then, the path of minimum optical path length between two points is the straight line between these two points. This is often stated as "rays travel in straight lines."

Fermat's principle can be used to prove what happens when a ray is reflected from a mirror, or is reflected and refracted at the boundary between two optical media, as illustrated in the following examples.

EXAMPLE 8.1 Show that when a ray reflects from a mirror, the angle of reflection should be equal to the angle of incidence in a medium of constant n.

Solution: The geometry involved is shown in Figure 8.2. θ_1 and θ_2 are the angles of incidence and of reflection, respectively. The path between A and C after reflection from the mirror should be a minimum.

The path length is given by

$$L_p = n(AB + BC)$$

which can be written as

$$L_p = n([h_1^2 + (d_1 - d_2)^2]^{1/2} + [h_2^2 + d_2^2]^{1/2})$$

L_p depends on d_2 as the independent variable. Setting $\delta L_p / \delta d_2 = 0$ yields

$$\frac{d_1 - d_2}{[h_1^2 + (d_1 - d_2)^2]^{1/2}} = \frac{d_2}{[h_2^2 + d_2^2]^{1/2}}$$

which is the same as

$$\sin(\theta_1) = \sin(\theta_2)$$

or

$$\theta_1 = \theta_2$$

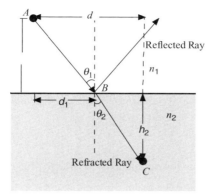

Figure 8.3. Reflection and refraction at the boundary between two optical media.

EXAMPLE 8.2 (a) An incident ray is split into two rays at the boundary between two optical media with indices of refraction equal to n_1 and n_2, respectively. This is visualized in Figure 8.3.

Show that

$$n_1 \sin \theta_1 = n_2 \sin \theta_2 \qquad (8.2\text{-}5)$$

This is called *Snell's law.*

(b) For rays traveling close to the optical axis, show that

$$n_1 \theta_1 \cong n_2 \theta_2 \qquad (8.2\text{-}6)$$

This is called *paraxial approximation* in geometrical optics.

Solution: (a) The optical path length between A and C is given by

$$L_p = n_1 \overline{AB} + n_2 \overline{BC}$$

By Fermat's principle, L_p is to be minimized. L_p can be written as

$$L_p = n_1 [h_1^2 + d_1^2]^{1/2} + n_2 [h_2^2 + (d - d_1)^2]^{1/2}$$

The independent variable is d_1. Setting $\delta L_p / \delta d_1 = 0$ yields

$$\frac{n_1 d_1}{[h_1^2 + d_1^2]^{1/2}} = \frac{n_2 (d - d_1)}{[h_2^2 + (d - d_1)^2]^{1/2}}$$

which is the same as

$$n_1 \sin \theta_1 = n_2 \sin \theta_2$$

(b) When the rays travel close to the optical axis, $\sin \theta_1 \cong \theta_1$ and $\sin \theta_2 \cong \theta_2$ so that

$$n_1 \theta_1 \cong n_1 \theta_2$$

EXAMPLE 8.3 A thin lens has the property shown in Figure 8.3.
A ray incident at an angle θ_1 leaves the lens at an angle θ_2 given by

$$\theta_2 = \theta_1 - \frac{y}{f} \tag{8.2-7}$$

where y is the distance from the optical axis, and f is the focal length of the lens given by

$$\frac{1}{f} = (n - 1)\left(\frac{1}{R_1} - \frac{1}{R_2}\right) \tag{8.2-8}$$

n is the index of refraction of the lens material, R_1 and R_2 are the radii of curvature of the two surfaces of the lens.

   Using the information given above, show the lens law of image formation.
Solution: Consider Figure 8.4(b). As a consequence of Eq. (8.2-7), and paraxial approximation such that $\sin \theta \cong \theta$, the ray starting at P_1 and parallel to the optical axis passes through the focal point F since θ_1 equals zero; the ray starting at P_1 and passing through the center point of the lens on the optical axis goes through the lens without change in direction since y equals 0. The two rays intersect at P_2, which determines the image.

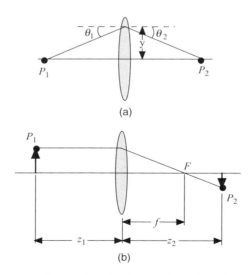

(a)

(b)

Figure 8.4.   (a) Ray bending by a thin lens, (b) Image formation by a thin lens.

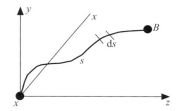

Figure 8.5. The trajectory of a ray in a GRIN medium.

8.3 THE RAY EQUATIONS

In a *graded-index (GRIN) material*, the refractive index is a function $n(r)$ of position
r. In such a medium, the rays follow curvilinear paths in accordance with Fermat's
principle.

The trajectory of a ray can be represented in terms of the coordinate functions
$x(s)$, $y(s)$ and $z(s)$, s being the length of the trajectory. This is shown in Figure 8.5.

The trajectory is such that Fermat's principle is satisfied along the trajectory
between the points A and B:

$$\delta \int_A^B n(\mathbf{r}) \mathrm{d}s = 0 \qquad (8.3\text{-}1)$$

r being the position vector. Using calculus of variations, Eq. (8.3-1) can be converted
to the *ray equation* given by

$$\frac{\mathrm{d}}{\mathrm{d}s}\left(n(\mathbf{r})\frac{\mathrm{d}\mathbf{r}}{\mathrm{d}s}\right) = \nabla n(\mathbf{r}) \qquad (8.3\text{-}2)$$

where ∇ is the gradient operator. The ray equation is derived rigorously in Section
8.4 by using the Eikonal equation.

The ray equation given above is expressed in vector form. In Cartesian
coordinates, the component equations can be written as

$$\frac{\mathrm{d}}{\mathrm{d}s}\left(n\frac{\mathrm{d}x}{\mathrm{d}s}\right) = \frac{\partial n}{\partial x}$$

$$\frac{\mathrm{d}}{\mathrm{d}s}\left(n\frac{\mathrm{d}x}{\mathrm{d}s}\right) = \frac{\partial n}{\partial y}$$

$$\frac{\mathrm{d}}{\mathrm{d}s}\left(n\frac{\mathrm{d}x}{\mathrm{d}s}\right) = \frac{\partial n}{\partial z}$$

Solution of Eq. (8.3-2) is not trivial. It can be considerably simplified by using
the paraxial approximation in which the rays are assumed to be traveling close to
the optical axis so that $\mathrm{d}s \cong \mathrm{d}z$. Equation (8.3-2) can then be written in terms of two

equations given by

$$\frac{d}{dz}\left(n\frac{dx}{dz}\right) = \frac{\partial n}{\partial x} \tag{8.3-3}$$

$$\frac{d}{dz}\left(n\frac{dy}{dz}\right) = \frac{\partial n}{\partial y} \tag{8.3-4}$$

Equations (8.3-3) and (8.3-4) are called the *paraxial ray equations*.

EXAMPLE 8.4 Show that in a homogeneous medium, rays follow straight lines.
Solution: When n is constant, Eqs. (8.3-3) and (8.3-4) become

$$\frac{d^2x}{dz^2} = \frac{d^2y}{dz^2} = 0$$

Hence,

$$x = Az + B$$
$$y = Cz + D \tag{8.3-5}$$

which represent straight lines.

EXAMPLE 8.5 If $n = n(y)$, show how to utilize the paraxial ray equations.
Solution: Since n is not a function of x, the variation along the x direction is still given by Eq. (8.3-5). For the y-direction, Eq. (8.3-4) can be written as

$$\frac{d^2y}{dz^2} = \frac{1}{n}\frac{dn}{dy} \tag{8.3-6}$$

Given $n(y)$ and the initial conditions $y(0)$ and $\frac{dy}{dz}\big|_{z=0}$, Eq. (8.3-6) can be solved for $y(z)$.

8.4   THE EIKONAL EQUATION

A monochromatic wave field traveling in a medium in which the refractive index changes slowly with respect to the wavelength has a complex amplitude, which can be approximated as

$$U(\mathbf{r}) = A(\mathbf{r})e^{-jk_0S(\mathbf{r})} \tag{8.4-1}$$

where k_0 equals the free space wave number $2\pi/\lambda_0$, λ_0 is the free space wavelength, and $k_0S(\mathbf{r})$ is the phase of the wave. $S(\mathbf{r})$ is a function of the refractive index $n(\mathbf{r})$ and is called the *Eikonal (function)*.

Constant phase surfaces are defined by

$$S(\mathbf{r}) = \text{constant} \tag{8.4-2}$$

Such surfaces are called *wavefronts*. Power flows in the direction of **k** which is perpendicular to the wavefront at a point **r**. The trajectory of a ray can be shown to be always perpendicular to the wavefront at every point. In other words, a ray follows the direction of power flow.

Equation (8.4-1) can be substituted in the Helmholtz equation

$$(\nabla^2 + k^2)U(\mathbf{r}) = 0 \tag{8.4-3}$$

to yield

$$k_0^2[n^2 - |\nabla S|^2]A + \nabla^2 A = 0 \tag{8.4-4}$$

and

$$k_0[2\nabla S \bullet \nabla A + A\nabla^2 S] = 0 \tag{8.4-5}$$

Equation (8.4-4) is rewritten as

$$|\nabla S|^2 = n^2 + \frac{\lambda_0^2}{2\pi^2}\frac{\nabla^2 A}{A} \tag{8.4-6}$$

As $\lambda_0 \to 0$, the second term on the right-hand side above can be neglected so that

$$|\nabla S(\mathbf{r})|^2 \cong n^2(\mathbf{r}) \tag{8.4-7}$$

Equation (8.4-7) is known as the *Eikonal equation*. It can be used to determine wavefronts. Once the wavefronts are known, the ray trajectories are determined as being perpendicular to the wavefronts. Three types of wavefronts are shown in Figure 8.6.

The Eikonal equation can be interpreted as the limit of the Helmholtz equation as the wavelength approaches zero.

Once $S(\mathbf{r})$ is obtained by using Eq. (8.4-7), $A(\mathbf{r})$ can be determined by using Eq. (8.4-5). Thereby, the wave field $U(\mathbf{r})$ is known.

Equation (8.4-7) can also be written as

$$\left[\frac{\partial S}{\partial x}\right]^2 + \left[\frac{\partial S}{\partial y}\right]^2 + \left[\frac{\partial S}{\partial z}\right]^2 = n^2(\mathbf{r}) \tag{8.4-8}$$

Then, the index of refraction is given by

$$n(\mathbf{r}) = |\nabla S(\mathbf{r})| \tag{8.4-9}$$

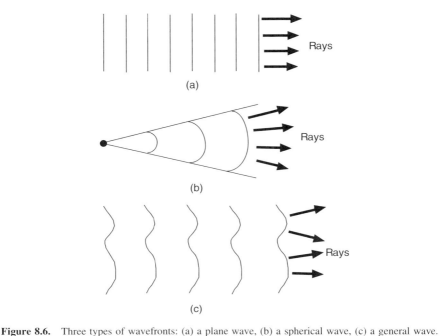

Figure 8.6. Three types of wavefronts: (a) a plane wave, (b) a spherical wave, (c) a general wave.

EXAMPLE 8.6 Show that the optical path length between two points A and B equals $S(\mathbf{r}_B) - S(\mathbf{r}_A)$.

Solution: Since $n(\mathbf{r}) = |\nabla S(\mathbf{r})|$, the optical path length is given by

$$L_p = \int_A^B n(\mathbf{r})d\mathbf{r} = \int_A^B |\nabla S(\mathbf{r})|d\mathbf{r} = S(\mathbf{r}_B) - S(\mathbf{r}_A)$$

Note that the optical path length is similar to potential difference in electrostatics. Then, the phase $S(\mathbf{r})$ is analogous to potential at a point.

8.5 LOCAL SPATIAL FREQUENCIES AND RAYS

In previous chapters, an arbitrary wave was analyzed in terms of plane wave components with direction cosines. Optical rays can also be associated with direction cosines. In order to do so, it is necessary to consider the nonstationary characteristics of a wave field.

In general, a nonstationary signal $h(t)$ is a signal whose frequency content changes as a function of t. Such signals are not well characterized by a single Fourier transform representation. Instead a short-time Fourier transform can be used by computing the Fourier transform of $h(t)\omega(t)$, $\omega(t)$ being a window function of limited extent.

The 1-D case outlined above can be generalized to the 2-D case in terms of nonstationary 2-D signals, which is relevant to wave fields. Thus, let us write a wave field at constant z as

$$h(x, y) = A(x, y)e^{j\phi(x,y)} \tag{8.5-1}$$

where z is suppressed. $A(x,y)$ will be assumed to be slowly varying.

The *local spatial frequencies* are defined by

$$f'_x = \frac{1}{2\pi}\frac{\partial}{\partial x}\phi(x, y) \tag{8.5-2}$$

$$f'_y = \frac{1}{2\pi}\frac{\partial}{\partial y}\phi(x, y) \tag{8.5-3}$$

f'_x and f'_y are also defined to be zero in regions where $h(x,y)$ is zero.

In general, f'_x and f'_y are not the same as the Fourier transform spatial frequencies f_x and f_y discussed in previous chapters. Good agreement between the two sets occurs only when the phase $\phi(x, y)$ varies slowly in the x–y plane so that $\phi(x, y)$ can be approximated by three terms of its Taylor series expansion near any point (x_0, y_0). The terms involved are $\phi(x_0, y_0)$,

$$\frac{\partial}{\partial x}\phi(x, y)\Big|_{\substack{x=x_0 \\ y=y_0}} \quad \text{and} \quad \frac{\partial}{\partial y}\phi(x, y)\Big|_{\substack{x=x_0 \\ y=y_0}}.$$

The *local direction cosines* are defined by

$$\alpha' = \lambda f'_x \tag{8.5-4}$$

$$\beta' = \lambda f'_y \tag{8.5-5}$$

$$\gamma' = \sqrt{1 - \alpha'^2 - \beta'^2} \tag{8.5-6}$$

It can be shown that these direction cosines are exactly the same as the direction cosines of a ray at (x,y) on a plane defined by constant z.

EXAMPLE 8.7 Determine the local spatial frequencies of a plane wave field.
Solution: The spatial part of a plane wave field can be written as

$$h(x, y) = e^{j2\pi(f_x x + f_y y)}$$

Then, with $\phi(x, y) = 2\pi(f_x x + f_y y)$

$$f'_x = \frac{1}{2\pi}\frac{\partial}{\partial x}\phi(x, y) = f_x$$

$$f'_y = \frac{1}{2\pi}\frac{\partial}{\partial y}\phi(x, y) = f_y$$

In this case, the local spatial frequencies and the Fourier transform spatial frequencies are the same.

EXAMPLE 8.8 A finite chirp function is defined by

$$h(x, y) = e^{j\pi\alpha(x^2+y^2)} \text{rect}\left(\frac{x}{2D_x}\right) \text{rect}\left(\frac{y}{2D_y}\right)$$

(a) Determine the local spatial frequencies.
(b) Compare the local frequencies to the properties of the Fourier transform of $g(x,y)$.

Solution: (a) The phase is given by

$$\phi(x, y) = \pi\alpha(x^2 + y^2)$$

in a rectangular region around the origin of dimensions $2D_x$ by $2D_y$. In this region, the local spatial frequencies are given by

$$f_x' = \frac{1}{2\pi} \frac{\partial}{\partial x} \phi(x, y) = \alpha x$$

$$f_y' = \frac{1}{2\pi} \frac{\partial}{\partial y} \phi(x, y) = \alpha y$$

Outside the rectangular region, f_x' and f_y' are zero by definition. Hence, they can be written as

$$f_x' = \alpha x \, \text{rect}\left(\frac{x}{2D_x}\right)$$

$$f_y' = \alpha x \, \text{rect}\left(\frac{y}{2D_y}\right)$$

Note that f_x' and f_y' vary linearly with x and y, respectively, inside the given rectangular region.

(b) $h(x,y)$ is a separable function:

$$h(x, y) = h_1(x)h_2(y) = \left[e^{j\pi\alpha x^2} \text{rect}\left(\frac{x}{2D_X}\right)\right]\left[e^{j\pi\alpha y^2} \text{rect}\left(\frac{y}{2D_y}\right)\right]$$

Hence, the Fourier transform of $h(x,y)$ equals $H_1(f_x)H_2(f_y)$.
The Fourier transform of $h_1(x)$ is given by

$$H_1(f_x) = \int_{-D_x}^{D_x} e^{j\pi\alpha x^2} e^{-j2\pi f_x x} dx \qquad (8.5\text{-}7)$$

Let $v = \sqrt{2\alpha}(x - (f_x/\alpha))$. Then, Eq. (8.5-7) can be written by change of variables as

$$H_1(f_x) = \frac{1}{\sqrt{2\alpha}} e^{-j\frac{\pi f_x^2}{\alpha}} \int_{-L_1}^{L_2} e^{j\frac{\pi v^2}{2}} dv = \frac{1}{\sqrt{2\alpha}} e^{-j\frac{\pi f_x^2}{\alpha}} [C(L_2) - C(L_1) + jS(L_2) - jS(L_1)]$$

(8.5-8)

where $C(\bullet)$ and $S(\bullet)$ are the Fresnel cosine and sine integrals given by Eqs. (5.2-18) and (5.2-19), and L_1, L_2 are defined by

$$L_1 = \sqrt{2\alpha}\left(D_x + \frac{f_x}{\alpha}\right)$$

(8.5-9)

$$L_2 = \sqrt{2\alpha}\left(D_x - \frac{f_x}{\alpha}\right)$$

(8.5-10)

$H_2(f_y)$ is similarly computed. It can be shown that the amplitude of $H_1(f_x)$ fast approaches zero outside the rectangular region [Goodman, 2004]. In other words, the frequency components are negligible away from the rectangular region. The fact that local frequencies are zero outside the rectangular region also suggests the same result. However, similarity between local frequencies and Fourier frequencies is not easy to show in more complicated problems.

8.6   MATRIX REPRESENTATION OF MERIDIONAL RAYS

Meridional rays are rays traveling in a single plane containing the z-axis called the optical axis. The transverse axis can be called the y-axis. Under the paraxial approximation $(\sin\theta \cong \theta)$, rays traveling in such a plane can be characterized in terms of a 2×2 matrix called the *ray-transfer matrix*.

A meridional ray at $z = z_i$ in the paraxial approximation is characterized by its position y_i and angle θ_i, as shown in Figure 8.7. This will be denoted as a pair (y_i, θ_i). Consider the traveling of the ray within an optical system from $z = z_1$ to $z = z_2$ as shown in Figure 8.8.

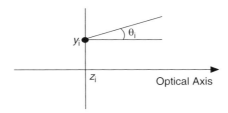

Figure 8.7.   A ray at $z = z_i$ characterized by (y_i, θ_i).

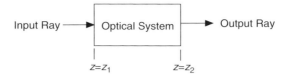

Figure 8.8. The travel of a ray through an optical system between $z = z_1$ and $z = z_2$.

In the paraxial approximation, (y_1, θ_1) and (y_2, θ_2) are related by

$$\begin{bmatrix} y_2 \\ \theta_2 \end{bmatrix} = \begin{bmatrix} A & B \\ C & D \end{bmatrix} \begin{bmatrix} y_1 \\ \theta_1 \end{bmatrix} \tag{8.6-1}$$

where the matrix **M** with elements A, B, C, D is called the *ray-transfer matrix*. The ray transfer matrices of elementary optical systems are discussed below.

1. **Propagation in a medium of constant index of refraction n:** This is as shown in Figure 8.9.

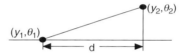

Figure 8.9. Propagation in a medium of constant index of refraction n.

M is given by

$$\mathbf{M} = \begin{bmatrix} 1 & d \\ 0 & 1 \end{bmatrix} \tag{8.6-2}$$

2. **Reflection from a planar mirror:** This is as shown in Figure 8.10.

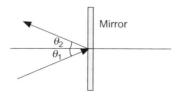

Figure 8.10. Reflection from a planar mirror.

M is given by

$$\mathbf{M} = \begin{bmatrix} 1 & 0 \\ 0 & 1 \end{bmatrix} \tag{8.6-3}$$

3. **Refraction at a planar boundary:** This is as shown in Figure 8.11.

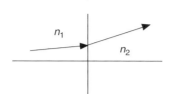

Figure 8.11. Refraction at a plane boundary.

M is given by

$$\mathbf{M} = \begin{bmatrix} 1 & 0 \\ 0 & \dfrac{n_1}{n_2} \end{bmatrix} \tag{8.6-4}$$

4. **Reflection from a spherical mirror:** This is as shown in Figure 8.12.

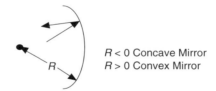

Figure 8.12. Reflection from a spherical mirror.

M is given by

$$\mathbf{M} = \begin{bmatrix} 1 & 0 \\ \dfrac{2}{R} & 1 \end{bmatrix} \tag{8.6-5}$$

5. **Refraction at a spherical boundary:** This is as shown in Figure 8.13.

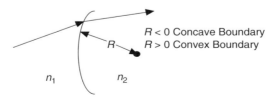

Figure 8.13. Refraction at a spherical boundary.

$\mathbf{M}$ is given by

$$\mathbf{M} = \begin{bmatrix} 1 & 0 \\ \dfrac{n_1 - n_2}{n_2 R} & \dfrac{n_1}{n_2} \end{bmatrix} \qquad (8.6\text{-}6)$$

6. **Transmission through a thin lens:** This is as shown in Figure 8.14.

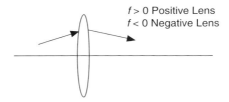

Figure 8.14.   Transmission through a thin lens.

$\mathbf{M}$ is given by

$$\mathbf{M} = \begin{bmatrix} 1 & 0 \\ -\dfrac{1}{f} & 1 \end{bmatrix} \qquad (8.6\text{-}7)$$

7. **Cascaded optical components:** Suppose N optical components are cascaded, as shown in Figure 8.15.

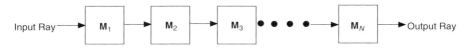

Figure 8.15.   Transmission through N optical components.

The ray-transfer matrix for the total system is given by

$$\mathbf{M} = \mathbf{M}_N \mathbf{M}_{N-1} \cdots \mathbf{M}_1 \qquad (8.6\text{-}8)$$

EXAMPLE 8.9 Show that the ray-transfer matrix for a thin lens is given by Eq. (8.6-7).
Solution: A thin lens with an index of refraction n_2 has two spherical surfaces with radii R_1 and R_2, respectively. Suppose that the index of refraction outside the lens

equals n_1. Transmission through the lens involves refraction at the two spherical surfaces. Hence, the total ray-transfer matrix is given by

$$
\mathbf{M} = \mathbf{M}_2\mathbf{M}_1 = \begin{bmatrix} 1 & 0 \\ \dfrac{n_2 - n_1}{n_1 R_2} & \dfrac{n_2}{n_1} \end{bmatrix} \begin{bmatrix} 1 & 0 \\ \dfrac{n_1 - n_2}{n_2 R_1} & \dfrac{n_1}{n_2} \end{bmatrix} = \begin{bmatrix} 1 & 0 \\ -\dfrac{1}{f} & 1 \end{bmatrix}
$$

where the focal length f is determined by

$$
\frac{1}{f} = \frac{n_2 - n_1}{n_1} \left(\frac{1}{R_1} - \frac{1}{R_2} \right)
$$

EXAMPLE 8.10 Determine the ray-transfer matrix for the following system:

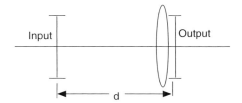

Solution: The system is a cascade of free-space propagation at a distance d and transmission through a thin lens of focal length f. Hence, the ray-transfer matrix is given by

$$
\mathbf{M} = \mathbf{M}_2\mathbf{M}_1 = \begin{bmatrix} 1 & 0 \\ -\dfrac{1}{f} & 1 \end{bmatrix} \begin{bmatrix} 1 & d \\ 0 & 1 \end{bmatrix} \begin{bmatrix} 1 & d \\ -\dfrac{1}{f} & 1 - \dfrac{d}{f} \end{bmatrix}
$$

EXAMPLE 8.11 (a) Derive the imaging equation for a single thin lens of focal length f, (b) determine the magnification.
Solution: The object is assumed to be at a distance d_1 from the lens; the image forms at a distance d_2 from the lens as shown in Figure 8.16.

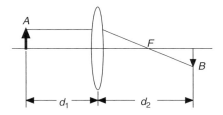

Figure 8.16. Imaging by a single lens.

For imaging to occur, the rays starting at a point A must converge to a point B regardless of angle. The total ray-transfer matrix is given by

$$\mathbf{M} = \begin{bmatrix} 1 & d_2 \\ 0 & 1 \end{bmatrix} \begin{bmatrix} 1 & 0 \\ -\dfrac{1}{f} & 1 \end{bmatrix} \begin{bmatrix} 1 & d_1 \\ 0 & 1 \end{bmatrix} = \begin{bmatrix} 1 - \dfrac{d_2}{f} & d_1 + d_2 - \dfrac{d_1 d_2}{f} \\ -\dfrac{1}{f} & 1 - \dfrac{d_1}{f} \end{bmatrix}$$

y_2 and y_1 are related by

$$y_2 = \left(1 - \frac{d_2}{f}\right) y_1 + \left(d_1 + d_2 - \frac{d_1 d_2}{f}\right) \theta_1$$

y_2 cannot depend on θ_1. Hence,

$$d_1 + d_2 - \frac{d_1 d_2}{f} = 0$$

or

$$\frac{1}{f} = \frac{1}{d_1} + \frac{1}{d_2}$$

(b) When imaging occurs, we have

$$y_2 = \left(1 - \frac{d_2}{f}\right) y_1$$

Hence, the magnification A is given by

$$A = \frac{y_2}{y_1} = 1 - \frac{d_2}{f}.$$

EXAMPLE 8.12 Two planes in an optical system are referred to as *conjugate planes* if the intensity distribution on one plane is the image of the intensity distribution on the other plane (imaging means the two distributions are the same except for magnification or demagnification). (a) What is the ray-transfer matrix between two conjugate planes? (b) How are the angles θ_1 and θ_2 related on two conjugate planes?

Solution: (a) In the previous problem, the ray transfer matrix $\mathbf{M}$ was calculated in general. Under the imaging condition, $\mathbf{M}$ becomes

$$\mathbf{M} = \begin{bmatrix} m_y & 0 \\ -\dfrac{1}{f} & m_\theta \end{bmatrix}$$

where m_y and m_θ equal $(1 - d_2/f)$, $(1 - d_1/f)$, respectively.

(b) θ_2 and θ_1 are related by

$$\theta_2 = -\frac{y^1}{f} + m_\theta \theta_1$$

When y_1 equals zero, we get

$$\theta_2 = m_\theta \theta_1$$

Hence, m_θ is the magnification factor of angles of rays starting from and converging to the optical axis.

EXAMPLE 8.13 Consider a positive lens. A parallel ray bundle that passes through the lens converges to a focused point behind the lens. This point is called the *rear focal point* or the *second focal point* of the lens. The plane passing through the rear focal point and perpendicular to the optical axis is called the *rear focal plane* or the *second focal plane*.

Consider the point in front of a positive lens such that the rays emerging from the point become a parallel bundle once they pass through the lens. This point is called the *front focal point* or the *first focal point* of the lens. The focal points and focal planes are visualized in Figures 8.17 and 8.18.

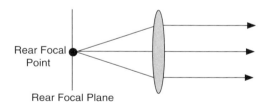

Figure 8.17.   The rear focal point and the rear focal plane of a positive lens.

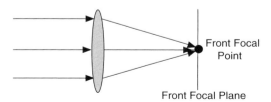

Figure 8.18.   The front focal point and the front focal plane of a positive lens. Determine the ray-transfer matrix between the front and rear focal planes of a positive lens.

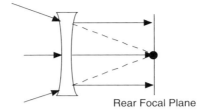

Figure 8.19.   The rear focal point and the rear focal plane of a negative lens.

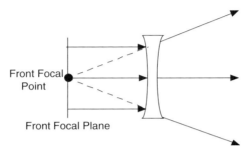

Figure 8.20.   The front focal point and the front focal plane of a negative lens.

Solution: In this case, d_1 and d_2 equal f. Hence, **M** becomes

$$\mathbf{M} = \begin{bmatrix} 1 - \dfrac{d_2}{f} & d_1 + d_2 - \dfrac{d_1 d_2}{f} \\[2mm] -\dfrac{1}{f} & 1 - \dfrac{d_1}{f} \end{bmatrix} = \begin{bmatrix} 0 & f \\[2mm] -\dfrac{1}{f} & 0 \end{bmatrix} \qquad (8.6\text{-}9)$$

EXAMPLE 8.14 The focal points and focal planes of a negative lens can be similarly defined. This is shown in Figures 8.19 and 8.20.

Note that the front and rear focal points and focal planes are the reverse of the front and rear focal points and focal planes of a positive lens. The ray-transfer matrix between the two planes can be shown to be also given by Eq. (8.6-9).

8.7   THICK LENSES

A *thick lens* can be analyzed just like a thin lens by defining two *principal planes*. Consider the front focal point of a thick lens as shown in Figure 8.21. By definition, the rays emerging from this point exit the lens as a parallel bundle. An incident ray on the lens intersect with the existing ray from the lens projected backwards at a

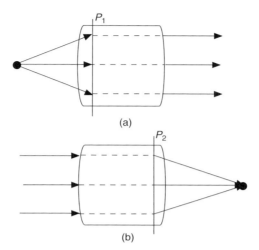

(a)

(b)

Figure 8.21.   (a) P_1: first principal plane, (b) P_2: second principal plane.

certain point. The plane passing through this point and perpendicular to the optical axis is called the *first principal plane*.

By definition, a parallel bundle of rays passing through a lens converge to the second focal point of the lens. An incident ray on the lens projected forward into the lens intersect the ray emerging towards the second focal point of the lens at a certain point. The plane passing through this point and perpendicular to the optical axis is called the *second principle plane*.

The ray-transfer matrix between the two principle planes of a thick lens can be shown to be the same as the regular ray-transfer matrix, namely,

$$\mathbf{M} = \begin{bmatrix} 1 & 0 \\ -\dfrac{1}{f} & 1 \end{bmatrix}$$

Consequently, the focal points, the focal length, and the imaging distances are as shown in Figure 8.22. The focal length is defined as the distance of a principle plane from the corresponding focal point. The two focal lengths are the same.

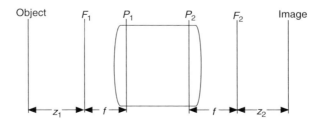

Figure 8.22.   Object, image, principal and focal planes, focal lengths, and object and image distances in a thick lens.

EXAMPLE 8.15 Show that the distances d_1 and d_2 between two conjugate planes satisfy the lens law.

Solution: The ray-transfer matrix between the two planes is given by

$$\mathbf{M} = \begin{bmatrix} 1 & d_2 \\ 0 & 1 \end{bmatrix} \begin{bmatrix} 1 & 0 \\ -\dfrac{1}{f} & 1 \end{bmatrix} \begin{bmatrix} 1 & d_1 \\ 0 & 1 \end{bmatrix}$$

This is the same ray-transfer matrix discussed in Example 8.11. Hence, the imaging condition is given by

$$\frac{1}{f} = \frac{1}{d_1} + \frac{1}{d_2}$$

8.8   ENTRANCE AND EXIT PUPILS OF AN OPTICAL SYSTEM

An optical system typically contains a number of physical apertures. One of these apertures causes the most severe limitation for the system. It will be called the *effective physical aperture*.

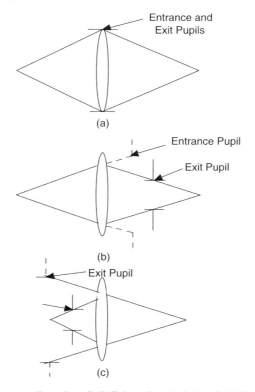

Figure 8.23.   (a) Entrance pupil = exit pupil, (b) Exit pupil = physical pupil, (c) Entrance pupil = exit pupil.

The *entrance pupil* of the system is the image of the effective physical aperture when looking from the object space. The *exit pupil* of the system is the image of the effective physical aperture when looking from the image space.

In order to clarify these concepts, consider Figure 8.23. In part (a), the effective physical aperture is on the plane of the lens. In this case, the entrance and exit pupils coincide, and are the same as the effective physical aperture. In part (b), the effective physical aperture is the same as the exit pupil. Its image seen from the object side is a virtual image, which is the entrance pupil. In part (c), the effective physical aperture is the same as the entrance pupil. Its image seen from the image space is a virtual image, which is the exit pupil.

In considering diffraction effects, it is the exit pupil that effectively limits the wave field passing through the optical system. As far as diffraction is concerned, the lens system has the exit pupil as its aperture, and Fraunhofer diffraction can be used to study diffraction effects on the image plane. For this purpose, the distance on the optical axis is the distance from the exit pupil to the image plane.

9

Fourier Transforms and Imaging with Coherent Optical Systems

9.1 INTRODUCTION

In this chapter, the uses of lenses to form the Fourier transform or the image of an incoming coherent wave are discussed. Both issues are discussed from the point of view of diffraction.

Imaging can also be discussed from the point of view of geometrical optics using rays as discussed in Chapter 8. The results obtained with diffraction are totally consistent with those of geometrical optics.

This chapter consists of seven sections. Section 9.2 discusses phase transformation with a thin lens, using its physical geometry. The phase function obtained is used in Section 9.3 to show how lenses form an output field, which is related to the Fourier transform of the input field in various geometries. This leads to the linear filtering interpretation of imaging in Section 9.4.

The theory discussed up to this point is illustrated with phase contrast microscopy in Section 9.5 and scanning confocal microscopy in Section 9.6. The last section highlights operator algebra for complex optical systems.

9.2 PHASE TRANSFORMATION WITH A THIN LENS

A lens consists of an optically dense material in which the phase velocity is less than the velocity in air. The thickness of the lens is modulated so that a desired phase modulation is achieved at the aperture of the lens.

A *thin lens* has three important parameters: n, the material index of refraction by which factor the phase velocity is reduced, R_1 and R_2, which are the radii of the two circular faces of the lens, as shown in Figure 9.1. If the material is glass, n is approximately 1.5. By convention, as rays travel from left to right, each convex surface hit by rays has positive radius of curvature and each concave surface has

Diffraction, Fourier Optics and Imaging,   by Okan K. Ersoy
Copyright © 2007 John Wiley & Sons, Inc.

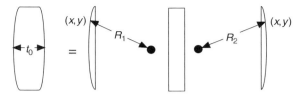

Figure 9.1. The three parts of a thin lens.

negative radius of curvature. A thin lens has the property that a ray enters and exits the lens at about the same (x,y) coordinates.

The three parameters discussed about can be combined in a single parameter called f, the focal length, by

$$f = \left[(n-1) \left(\frac{1}{R_1} - \frac{1}{R_2} \right) \right]^{-1} \tag{9.2-1}$$

f is positive or negative, depending on R_1 and R_2. For example, a double-convex lens has positive f whereas a double-concave lens has negative f. In the following discussion, f will be assumed to be positive.

The phase transformation of a perfect lens can be written as

$$t(x,y) = e^{jkt_0} e^{-j\frac{k}{2f}(x^2+y^2)} \tag{9.2-2}$$

where t_0 is the maximum thickness of the lens. Since e^{jkt_0} gives a constant phase change, it can be neglected. Equation (9.2-2) will be derived below. The finite extent of the lens aperture can be taken into account by defining a pupil function $P(x,y)$:

$$P(x,y) = \begin{cases} 1 & \text{inside the lens aperture} \\ 0 & \text{otherwise} \end{cases} \tag{9.2-3}$$

In order to drive Eq. (9.2-2), the lens is considered in three parts as shown in Figure 9.1. The thickness $\Delta(x,y)$ at a location (x,y) can be shown to be [Goodman, 2004]

$$\Delta(x,y) = t_0 - R_1 \left[1 - \sqrt{1 - \frac{x^2+y^2}{R_1^2}} \right] + R_2 \left[1 - \sqrt{1 - \frac{x^2+y^2}{R_2^2}} \right] \tag{9.2-4}$$

where t_0 is the total thickness of the lens. This expression is simplified by using paraxial approximation stated by

$$\sqrt{1 - \frac{x^2+y^2}{R_i}} \cong 1 - \frac{x^2+y^2}{2R_i^2} \tag{9.2-5}$$

where R_i is R_1 or R_2. Then, the thickness function becomes

$$\Delta(x,y) = t_0 - \frac{x^2+y^2}{2}\left(\frac{1}{R_1} - \frac{1}{R_2}\right) \tag{9.2-6}$$

The phase transformation $\theta(x,y)$ by the lens is expressed as

$$\begin{aligned} \theta(x,y) &= kn\Delta(x,y) + k(t_0 - \Delta(x,y)) \\ &= kt_0 + k(n-1)\Delta(x,y) \end{aligned} \tag{9.2-7}$$

Using Eqs. (9.2-1) and (9.2-6), Eq. (9.2-7) becomes

$$\theta(x,y) = kt_0 - \frac{k}{2f}(x^2+y^2) \tag{9.2-8}$$

Equation (9.2-8) leads to Eq. (9.2-2).

9.3 FOURIER TRANSFORMS WITH LENSES

In Section 5.4, the Fraunhofer region was observed to be very far from the initial plane for reasonable aperture sizes. This limitation is removed with the use of a lens. Three separate cases will be discussed depending upon whether the initial wave field is incident on the lens, in front of the lens or behind the lens.

9.3.1 Wave Field Incident on the Lens

This geometry is shown in Figure 9.2.

Suppose the wave field incident on the lens is $U(x, y, 0)$. The wave field behind the lens becomes

$$U'(x,y,0) = U(x,y,0)P(x,y)e^{-j\frac{k}{2f}(x^2+y^2)} \tag{9.3-1}$$

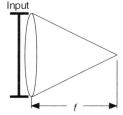

Figure 9.2. The geometry for input placed against the lens.

This complex amplitude can be substituted in Eq. (5.2-13), the Fresnel diffraction formula, to find the wave field at $z = f$, the focal plane. If the focal plane coordinates are (x_f, y_f, f), the result is given by

$$U(x_f, y_f, f) = e^{j\frac{k}{2f}\left(x_f^2 + y_f^2\right)} \int\!\!\!\int\limits_{-\infty}^{\infty} U'(x, y, 0)P(x, y)e^{-j2\pi(f_x x + f_y y)}\,dxdy \qquad (9.3\text{-}2)$$

where constant terms are neglected, and the spatial frequencies are $f_x = x_f/\lambda f$, $f_y = x_f/\lambda f$. Equation (9.3-2) shows that the output complex amplitude is the Fraunhofer diffraction pattern of the input complex amplitude within the pupil function of the lens.

9.3.2 Wave Field to the Left of the Lens

This geometry is shown in Figure 9.3.

Consider $U(x_1, y_1, -d_1)$, a wave field at the plane $z = -d_1$, $d_1 > 0$. Its angular spectrum $A(f_x, f_y, -d_1)$ and the angular spectrum $A(f_x, f_y, 0^-)$ of $U(x, y, 0)$ are related by

$$A(f_x, f_y, 0^-) = A(f_x, f_y, -d_1)e^{-j\pi\lambda d_1 (f_x^2 + f_y^2)} \qquad (9.3\text{-}3)$$

If $U(x, y, 0)$ has an extent less than the extent of $P(x,y)$, Eq. (9.3-2) can be written as

$$U(x_f, y_f, f) = A(f_x, f_y, 0)e^{j\frac{k}{2f}(x_f^2 + y_f^2)} \qquad (9.3\text{-}4)$$

or, using Eq. (9.3-3),

$$U(x_f, y_f, f) = A(f_x, f_y, -d_1)e^{j\frac{k}{2f}\left(1 - \frac{d_1}{f}\right)\left(x_f^2 + y_f^2\right)} \qquad (9.3\text{-}5)$$

When $d_1 = f > 0$, the phase factor becomes unity such that

$$U(x_f, y_f, f) = A(f_x, f_y, -f) \qquad (9.3\text{-}6)$$

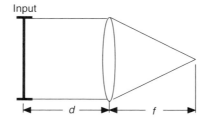

Figure 9.3. The geometry for input placed to the left of the lens.

Thus, the wave field at the focal plane of the lens is the 2-D Fourier transform of the wave field at $z = -f$ if the effect of $P(x,y)$ is neglected.

The limitation imposed by the finite lens aperture is called *vignetting* and can be avoided by choosing d_1 small. To include the effect of the finite lens aperture, geometrical optics approximation can be used [Goodman]. With this approach, the initial wave field is approximated such that the finite pupil function at the lens equals $P(x_1 + (d_1/f)x_f + (d_1/f)y_f)$. Then, $A(f_x,f_y,-d_1)$ with $f_x = x_f/\lambda_f$ and $f_y = y_f/\lambda_f$ is computed with respect to this pupil function. The wave field at $z = f$ becomes

$$U(x_f,y_f,f) = e^{j\frac{k}{2f}\left(1-\frac{d_1}{f}\right)(x_f^2+y_f^2)} \iint\limits_{-\infty}^{\infty} U(x,y,-d_1)P\left(x+\frac{d_1}{f}x_f,y+\frac{d_1}{y}y_f\right)e^{-j\frac{2\pi}{\lambda_f}(xx_f+yy_f)}\mathrm{d}x\mathrm{d}y$$

$$(9.3\text{-}7)$$

9.3.3   Wave Field to the Right of the Lens

The third possibility is to consider the initial wave field behind the lens, a distance d from the focal plane of the lens. This is the case when the lens is illuminated with a plane perpendicular wave field, and an object transparency is located at $z = f - d$ as shown in Figure 9.4.

By geometrical optics, the amplitude of the spherical wave at this plane is proportional to f/d and can be neglected as constant. If $(x_2, y_2, f - d)$ indicates a point on this plane, the equivalent pupil function on the object plane is $P(x_2(f/d), y_2(f/d))$ due to the converging spherical wave. If $U(x_2, y_2, f - d)$ is the wave field due to the object transparency, the total wave field at the object plane is given by

$$U'(x_2,y_2,f-d) = e^{-j\frac{k}{2d}(x_2^2+y_2^2)}P\left(x_2\frac{f}{d},y_2\frac{f}{d}\right)U(x_2,y_2,f-d)   (9.3\text{-}8)$$

Let $A'(f_x,f_y,f-d)$ be the angular spectrum of $U'(x_2,y_2,f-d)$. Provided that the Fresnel approximation is valid for the distance d, the angular spectrum of the wave field at the focal plane is given by

$$A(x_f,y_f,f) = A'(f_x,f_y,f-d)e^{j\frac{k}{2d}(x_f^2+y_f^2)}   (9.3\text{-}9)$$

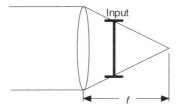

Figure 9.4.   The geometry for the input to the right of the lens.

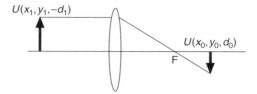

$U(x_1, y_1, -d_1)$

$U(x_0, y_0, d_0)$

F

Figure 9.5. The geometry used for image formation.

where constant terms are neglected f_x and f_y are $x_f/\lambda d$ and $y_f/\lambda d$. With this geometry, the scale of the Fourier transform can be varied by adjusting d. The size of the transform is made smaller by decreasing d. In optical spatial filtering, this is a useful feature because the size of the Fourier transform must be the same as the size of the spatial filter located at the focal plane [Goodman].

9.4 IMAGE FORMATION AS 2-D LINEAR FILTERING

Lenses are generally known as imaging devices. Imaging by a positive, aberration-free thin lens in the presence of monochromatic waves is discussed below.

Consider Figure 9.5 in which transparent object is placed on plane $z = -d_1$, and illuminated by a wave so that the wave field is $U(x_1, y_1, -d_1)$ on this plane.

The wave field $U(x_0, y_0, d_0)$ on the plane $z = d_0$ is to be determined. Then, under what conditions an image is formed is discussed.

Since wave propagation is a linear phenomenon, the fields at (x_0, y_0, d_0) and $(x, y, -d_1)$ can always be related by a superposition integral:

$$U(x_0, y_0, d_0) = \iint\limits_{-\infty}^{\infty} h(x_0, y_0, x_1, y_1)U(x_1, y_1, -d_1)\mathrm{d}x_1\mathrm{d}y_1 \qquad (9.4\text{-}1)$$

where $h(x_0, y_0, x_1, y_1)$ is the impulse response of the system. In order to find h, $U(x, y, -d_1)$ will be assumed to be a delta function at $(x_1, y_1, -d_1)$. This is physically equivalent to a spherical wave originating at this point.

We will assume that the lens at $z = 0$ has positive focal length f. All the constant terms due to wave propagation will be neglected. The field incident on the lens within the paraxial approximation is given by

$$U(x, y, 0) = \mathrm{e}^{j\left[\frac{k}{2d_1}((x-x_1)^2 + (y-y_1)^2)\right]} \qquad (9.4\text{-}2)$$

The field after the lens is given by

$$U'(x, y, 0) = U(x, y, 0)P(x, y)\mathrm{e}^{-j\frac{k}{2f}(x^2 + y^2)} \qquad (9.4\text{-}3)$$

The Fresnel diffraction formula, Eq (5.2-13), yields

$$U(x_0, y_0, d_0) = h(x_0, y_0, x_1, y_1)$$

$$= \iint\limits_{-\infty}^{\infty} U'(x, y, 0) e^{j\left[\frac{k}{2d_0}\left((x_0 - x)^2 + (y_0 - y)^2\right)\right]} dxdy \qquad (9.4\text{-}4)$$

In practical imaging applications, the final image is detected by a detector system that is sensitive to intensity only. Consequently, the phase terms of the form $e^{j\theta}$ can be neglected if they are independent of the integral in Eq. (9.4-4). Then, Eq. (9.4-4) can be simplified to

$$h(x_0, y_0, x_1, y_1) = \iint\limits_{-\infty}^{\infty} P(x, y) e^{j\frac{kD}{2}(x^2 + y^2)} e^{-jk\left[\left(\frac{x_0}{d_0} + \frac{x_1}{d_1}\right)x + \left(\frac{y_0}{d_0} + \frac{y_1}{d_1}\right)y\right]} dxdy \qquad (9.4\text{-}5)$$

where

$$D = \frac{1}{d_1} + \frac{1}{d_0} - \frac{1}{f} \qquad (9.4\text{-}6)$$

Consider the case when D is equal to zero. In this case, Eq. (8.4-5) becomes

$$h(x_0, y_0, x_1, y_1) = \iint\limits_{-\infty}^{\infty} P(x, y) e^{-j\frac{2\pi}{\lambda d_0}[(x_0 + Mx_1)x + (y_0 + My_1)y]} dxdy \qquad (9.4\text{-}7)$$

where

$$M = \frac{d_0}{d_1} \qquad (9.4\text{-}8)$$

If $P(x,y)$ is neglected, Eq. (8.4-7) is the same as

$$h(x_0, y_0, x_1, y_1) = \iint\limits_{-\infty}^{\infty} e^{-j2\pi\left[\left(x_1 + \frac{x_0}{M}\right)x' + \left(y_1 + \frac{y_0}{M}\right)y'\right]} dx'dy' \qquad (9.4\text{-}9)$$

where $x' = x/\lambda d_0$ and $y' = y/\lambda d_0$, and a constant term is dropped.
Equation (9.4-9) is the same as

$$h(x_0, y_0, x_1, y_1) = \delta\left(x_1 + \frac{x_0}{M}, y_1 + \frac{y_0}{M}\right) \qquad (9.4\text{-}10)$$

Using this result in Eq. (9.4-1) gives

$$U(x_0, y_0, d_0) = U\left(-\frac{x_0}{M}, -\frac{y_0}{M}, -d_1\right) \qquad (9.4\text{-}11)$$

We conclude that $U(x_0, y_0, d_0)$ is the magnified and inverted image of the field at $z = d_1$, and M is the magnification when D given by Eq. (9.4-6) equals zero. Then, d_0 and d_1 are related by

$$\frac{1}{d_0} = \frac{1}{f} - \frac{1}{d_1} \qquad (9.4\text{-}12)$$

This relation is known as the *lens law*.

9.4.1   The Effect of Finite Lens Aperture

Above the effect of the finite size pupil function $P(x,y)$ was neglected. Letting

$$x_1' = -Mx_1 \qquad (9.4\text{-}13)$$
$$y_1' = -My_1 \qquad (9.4\text{-}14)$$

Equation (9.4-7) can be written as

$$
\begin{aligned}
h(x_0, y_0, x_1', y_1') &= \iint\limits_{-\infty}^{\infty} P(\lambda d_0 x, \lambda d_0 y) e^{-j2\pi[(x_0 - x_1')x + (y_0 - y_1')y]} \mathrm{d}x\mathrm{d}y \\
&= h(x_0 - x_1', y_0 - y_1')
\end{aligned}
\qquad (9.4\text{-}15)
$$

Then, Eq. (8.4-1) can be written as

$$U(x_0, y_0, d_0) = \iint\limits_{-\infty}^{\infty} h(x_0 - x_1', y_0 - y_1') U(x_1', y_1', -d_1) \mathrm{d}x_1' \mathrm{d}y_1' \qquad (9.4\text{-}16)$$

where a constant term is again dropped. Equation (9.4-16) is a 2-D convolution:

$$U(x_0, y_0, d_0) = h(x_0, y_0) * U\left(-\frac{x_0}{M}, -\frac{y_0}{M}, -d_1\right) \qquad (9.4\text{-}17)$$

where $U((-x_0/M), (-y_0/M), -d_1)$ is the ideal image, and

$$h(x_0, y_0) = \iint\limits_{-\infty}^{\infty} P(\lambda d_0 x, \lambda d_0 y) e^{-j2\pi(x_0 x + y_0 y)} \mathrm{d}x\mathrm{d}y \qquad (9.4\text{-}18)$$

The impulse response is observed to be the 2-D FT of the scaled pupil function of the lens. The final image is the convolution of the perfect image with the system impulse response. This smoothing operation can strongly attenuate the fine details of the image.

In a more general imaging system with many lenses, Eqs. (9.4-17) and (9.4-18) remain valid provided that $P(\bullet,\bullet)$ denotes the finite equivalent exit pupil of the system, and the system is diffraction limited [Goodman]. An optical system is diffraction limited if a diverging spherical wave incident on the entrance pupil is mapped into a converging spherical wave at the exit pupil.

9.5   PHASE CONTRAST MICROSCOPY

In this and the next section, the theory discussed in the previous sections is illustrated with two advanced imaging techniques. Phase contrast microscopy is a technique to generate high contrast images of transparent objects, such as living cells in cultures, thin tissue slices, microorganisms, lithographic patterns, fibers and the like. This is achieved by converting small phase changes in to amplitude changes, which can then be viewed in high contrast. In this process, the specimen being viewed is not negatively perturbed.

The image of an industrial phase contrast microscope is shown in Figure 9.6.

When light passes through a transparent object, its phase is modified at each point. Hence, the transmission function of the specimen with coherent illumination can be written as

$$t(x, y) = e^{j(\theta_0 + \theta(x,y))} \qquad (9.5\text{-}1)$$

where θ_0 is the average phase, and the phase shift $\theta(x, y)$ is considerably less than

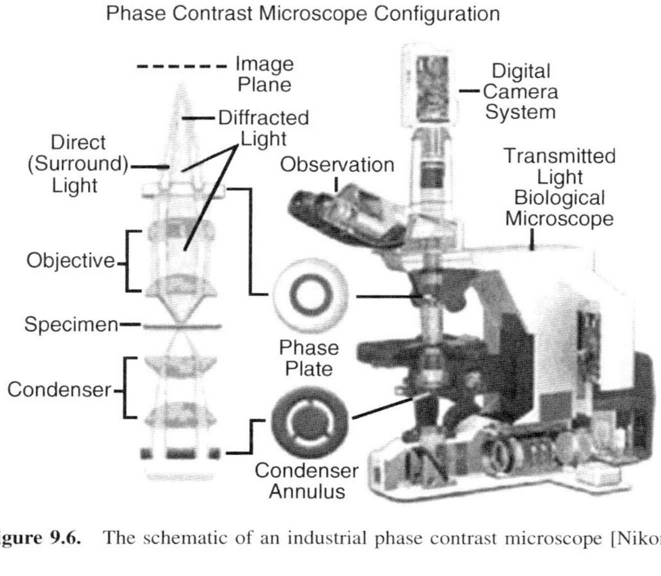

Figure 9.6.   The schematic of an industrial phase contrast microscope [Nikon].

2π. Hence, $t(x, y)$ can be approximated as

$$t(x, y) = e^{j\theta_0}[1 + j\theta(x, y)] \qquad (9.5\text{-}2)$$

where the last factor is the first two terms of the Taylor series expansion of $e^{j\theta(x,y)}$.

A microscope is sensitive to the intensity of light that can be written as

$$I(x, y) = |1 + j\theta(x, y)|^2 \approx 1 \qquad (9.5\text{-}3)$$

Hence, no image is observable. We note that Eq. (9.5-3) is true because the first term of unity due to undiffracted light is in phase quadrature with $j\theta(x, y)$ generated by the diffracted light. In order to circumvent this problem, a phase plate yielding $\pi/2$ or $3\pi/2$ phase shift with the undiffracted light can be used. This is usually achieved by using a glass substrate with a transparent dielectric dot giving $\pi/2$ or $3\pi/2$ phase shift, typically by controlling thickness at the focal point of the imaging system. The undiffracted light passes through the focal point whereas the diffracted light from the specimen is spread away from the focal point since it has high spatial frequencies. On the imaging plane, the intensity can now be written as

$$I(x, y) = \left| e^{j\pi/2} + j\theta(x, y) \right|^2 \approx 1 + 2\theta(x, y) \qquad (9.5\text{-}4)$$

with $\pi/2$ phase shift, and

$$I(x, y) = \left| e^{j3\pi/2} + j\theta(x, y) \right|^2 \approx 1 - 2\theta(x, y) \qquad (9.5\text{-}5)$$

with $3\pi/2$ phase shift. Equations (9.5-4) and (9.5-5) are referred to as positive and negative phase contrast, respectively. In either case, the phase variation $\theta(x, y)$ is observable as an image since it is converted into intensity. An example of imaging with phase contrast microscope in comparison to a regular microscope is shown in Figure 9.7.

Living Cells in Brightfield and Phase Contrast

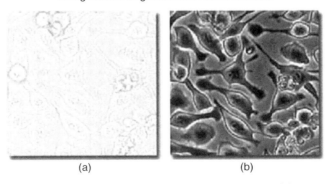

(a)   (b)

Figure 9.7. (a) Image of a specimen with a regular microscope, (b) image of the same specimen with a phase contrast microscope [Nikon].

It is possible to further improve the method by designing more complex phase plates. For example, contrast can be modulated by varying the properties of the phase plate such as absorption, refractive index, and thickness. Apodized phase contrast objectives have also been utilized.

9.6   SCANNING CONFOCAL MICROSCOPY

The conventional microscope is a device which images the entire object field simultaneously. In scanning microscopy, an image of only one object point is generated at a time. This requires scanning the object to generate an image of the entire field. A diagram of a reflection mode scanning optical microscope is shown in Figure 9.8. Such a system can also be operated in the transmission mode. Modern systems feed the image information to a computer system which allows digital postprocessing and image processing.

In scanning confocal microscopy, light from a point source probes a very small area of the object, and another point detector ensures that only light from the same object area is detected. A particular configuration of a scanning confocal microscope is shown in Figure 9.8 [Wilson, 1990]. In this system, the image is generated by scanning both the source and the detector in synchronism. Another configuration is shown in Figure 9.9. In this mode, the confocal microscope has excellent depth discrimination property (Figure 9.10). Using this property, parallel sections of a thick translucent object can be imaged with high resolution.

9.6.1   Image Formation

Equation (9.4-16) can be written as

$$U(x_0, y_0) = \int\limits_{-\infty}^{\infty} \int\limits_{-\infty}^{\infty} h\left(x_1 + \frac{x_0}{M}, y_1 + \frac{y_0}{M}\right) U(x_1, y_1) \mathrm{d}x_1 \mathrm{d}y_1 \qquad (9.6\text{-}1)$$

where d_0 and d_1 are dropped from the input output complex amplitudes, to be understood from the context. The impulse response function given by Eq. (9.4-18)

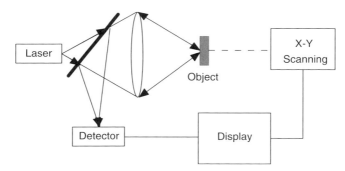

Figure 9.8.   Schematic diagram of a reflection mode scanning confocal microscope.

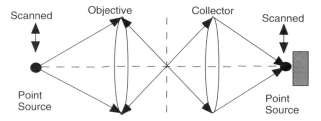

Figure 9.9. A confocal optical microscope system.

can be written as

$$h(x, y) = \int\limits_{-\infty}^{\infty} \int\limits_{-\infty}^{\infty} P(\lambda d_0 x_1, \lambda d_0 y_1) e^{-2\pi j(x_1 x + y_1 y)} dx_1 dy_1 \qquad (9.6\text{-}2)$$

where $P(x_1, y_1)$ is the pupil function of the lens system.

In confocal microscopy, since one object point is imaged at a time, $U(x_1, y_1)$ can be written as

$$U(x_1, y_1) = \delta(x_1)\delta y(x_1) \qquad (9.6\text{-}3)$$

Hence,

$$U(x_0, y_0) = \int\limits_{-\infty}^{\infty} \int\limits_{-\infty}^{\infty} h\left(x_1 + \frac{x_0}{M}, y_1 + \frac{y_0}{M}\right)\delta(x_1)\delta y(x_1) dx_1 dy_1$$

$$= h\left(\frac{x_0}{M}, \frac{y_0}{M}\right) \qquad (9.6\text{-}4)$$

The intensity of the point image becomes

$$I(x_0, y_0) = \left| h\left(\frac{x_0}{M}, \frac{y_0}{M}\right) \right|^2 \qquad (9.6\text{-}5)$$

Assuming a circularly symmetric pupil function of radius a, $P(x_1, y_1)$ can be written as

$$(x_1, y_1) = P(\rho) = \begin{cases} 1 & \rho \le 1/2 \\ 0 & \text{otherwise} \end{cases} \qquad (9.6\text{-}6)$$

where

$$\rho = \frac{\sqrt{x_1^2 + y_1^2}}{2a} \tag{9.6-7}$$

$P(\rho)$ is the cylinder function studied in Example 2.6. The impulse response function becomes the Hankel transform of $P(\rho)$, and is given by

$$h(x_0, y_0) = h(r_0) = \frac{\pi}{4} \text{somb}(r_0) \tag{9.6-8}$$

where

$$r_0 = \sqrt{x_0^2 + y_0^2} \tag{9.6-9}$$

$$\text{somb}(r_0) = \frac{2J_1(\pi r_0)}{\pi r_0} \tag{9.6-10}$$

The intensity of the point image becomes

$$I(x_0, y_0) = \left| \frac{MJ_1(\pi r_0/M)}{2r_0} \right|^2 \tag{9.6-11}$$

In practice, there are typically two lenses used, as seen in Figure 9.9. They are called objective and collector lenses. Suppose that the two lenses have impulse response functions h_1 and h_2, respectively. By repeated application of Eq. (9.7-1), it can be shown that the image intensity can be written as

$$I(x_0, y_0) = |h_t * U(x_1, y_1)|^2 \tag{9.6-12}$$

where

$$h_t = h_1 h_2 \tag{9.6-13}$$

If the lenses are equal, with impulse response h given by Eq. (9.7-7), the output intensity can be written as

$$I(x_0, y_0) = \left| \frac{MJ_1(\pi r_0/M)}{2r_0} \right|^4 \tag{9.6-14}$$

EXAMPLE 9.1 Determine the transfer function of the two-lens scanning confocal microscope.
Solution: Since $h_t = h_1 h_2$, we have

$$H_t(f_x, f_y) = H_1(f_x, f_y) * H_2(f_x, f_y)$$

where, for two equal lenses,

$$H_1(f_x, f_y) = H_2(f_x, f_y) = P(\lambda d_0 x_1, \lambda d_0 y_1)$$

If the pupil function P is circularly symmetric, $H_t(f_x, f_y)$ is also symmetric, and can be written as

$$H_t(f_x, f_y) = H_t(\rho)$$

where

$$\rho = \sqrt{f_x^2 + f_y^2}$$

Hence,

$$H_t(\rho) = H_1(\rho) * H_2(\rho)$$

It can be shown that this convolution is given by

$$H_t(\rho) = \frac{2}{\pi} \left[\cos^{-1}\left(\frac{\rho}{2}\right) - \frac{\rho}{2} \sqrt{1 - \left(\frac{\rho}{2}\right)^2} \right]$$

where

$$P(\rho) = \begin{cases} 1 & 0 \leq \rho \leq 1 \\ 0 & \text{otherwise} \end{cases}$$

9.7 OPERATOR ALGEBRA FOR COMPLEX OPTICAL SYSTEMS

The results derived in Sections 9.1–9.5 can be combined together in an operator algebra to analyze complex optical systems with coherent illumination [Goodman]. For simple results to be obtained, it is necessary to assume that the results are valid within the paraxial approximations involved in Fresnel diffraction and geometrical optics.

An operator algebra is discussed below in which the apertures are assumed to be separable in rectangular coordinates so that 2-D systems are discussed in terms of 1-D systems.

The operators involve fundamental operations that occur within a complex optical system. If $U(x)$ is the current field, $O(u)[U_1(x)]$ will denote the transformation of $U_1(x)$ by the operator $O(u)$ characterized by a variable u. Below the most important operators are described.

1. Fourier transform

$$U_2(v) = F(v)[U_1(x)] = \int_{-\infty}^{\infty} U_1(x)e^{-j2\pi vx}dx \qquad (9.7\text{-}1)$$

   where v denotes frequency.

2. Free-space propagation

$$U_2(x_2) = P(z)[U_1(x_1)] = \frac{1}{\sqrt{j\lambda z}} \int_{-\infty}^{\infty} U_1(x_1)e^{j\frac{k}{2z}(x_2-x_1)^2}dx_1 \qquad (9.7\text{-}2)$$

   where $k = 2\pi/\lambda$, λ is the wavelength, and z is the distance of propagation.
   For a distance of propagation closer than the Fresnel region, the angular spectrum method can be used instead of Fresnel diffraction.

3. Scaling by a constant

$$U_2(x) = S(a)[U_1(x)] = \sqrt{|a|}\, U_1(ax) \qquad (9.7\text{-}3)$$

   where a is a scaling constant.

4. Multiplication by a quadratic phase factor

$$U_2(x) = Q(b)[U_1(x)] = e^{j\frac{k}{2}bx^2}U_1(x) \qquad (9.7\text{-}4)$$

   where b is the parameter controlling the quadratic phase factor.

Each of these operators has an inverse. They are shown in Table 9.1.
In order to use the operator algebra effectively, it is necessary to show how two successive operations are related. These relationships are summarized in Table 9.2.
In this table, each element, say, E equals the successive operations with the row operator $R(v_1)$ followed by the column operator $C(v_2)$:

$$E = C(v_2)R(v_1) \qquad (9.7\text{-}5)$$

Table 9.1. Inverse operators.

	$F(v)$	$P(z)$	$S(a)$	$Q(b)$
Inverse Operator	$F(-v)$	$P(-z)$	$S\left(\dfrac{1}{a}\right)$	$Q(-b)$

Table 9.2. Operators and their algebra.

	$F(v)$	$P(z)$	$S(a)$	$Q(b)$
$F(v)$	$S(-1)$	$Q(-\lambda_z^2)F(v)$	$S\left(\dfrac{1}{a}\right)F(v)$	$P\left(-\dfrac{b}{\lambda^2}\right)F(v)$
$P(z)$	$F(v)Q(-\lambda^2 z)$	$P(z_2)P(z_1)$ $= P(z_1 + z_2)$	$S(a)P(a_z^2)$	$Q\left[\left(\dfrac{1}{b}+z\right)^{-1}\right]\bullet$ $S\left[(1+bz)^{-1}\right]\bullet$ $P\left[(z^{-1}+b)^{-1}\right]$
$S(a)$	$F(v)S\left(\dfrac{1}{a}\right)$	$P\left(\dfrac{z}{a^2}\right)S(a)$	$S(a_2)S(a_1)$ $= S(a_1 + a_2)$	$Q(a^2 b)S(a)$
$Q(b)$	$F(v)P\left(-\dfrac{b}{\lambda^2}\right)$	$P[(z^{-1}+b)^{-1}]\bullet$ $S(1+bz)\bullet$ $Q[(b^{-1}+z)^{-1}]$	$S(a)Q\left(\dfrac{b}{a^2}\right)$	$Q(b_2)Q(b_1)$ $= Q(b_1 + b_2)$

In addition, the following relations are often used to simplify results:

$$P(z) = Q\left(\frac{1}{z}\right)S\left(\frac{1}{\lambda z}\right)F(v)Q\left(\frac{1}{z}\right) \tag{9.7-6}$$

$$S\left(\frac{1}{\lambda f}\right)F(v) = P(f)Q\left(\frac{1}{f}\right)P(f) \tag{9.7-7}$$

Equation (9.7-6) corresponds to Fresnel diffraction. Equation (9.7-7) shows that the fields at the front and back focal planes of a lens are related by a scaled Fourier transform.

EXAMPLE 9.2 (a) Simplify the following operator equation:

$$O = Q\left(-\frac{1}{f}\right)P(f)Q\left(-\frac{1}{f}\right)P(f)Q\left(-\frac{1}{f}\right)P(f)$$

(b) Which optical system does this operator correspond to?

(c) Determine the final field U_{end} in terms of the initial field U_1.

Solution: (a) The first three operators on the right hand side correspond to $S\left(\frac{1}{\lambda f}\right)$ $F(v)$ by Eq. (9.7-7). Next $P(f)$ is replaced by $Q\left(\frac{1}{f}\right)S\left(\frac{1}{\lambda f}\right)F(v_2)Q\left(\frac{1}{f}\right)$ by using Eq. (9.7-6). We get

$$O = S\left(\frac{1}{\lambda f}\right)F(v_2)S\left(\frac{1}{\lambda f}\right)F(v_1)$$

(b) This operator corresponds to the optical system shown in Figure 9.10.

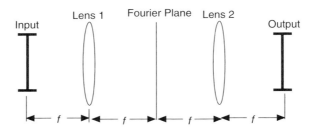

Figure 9.10. Final optical system for Example 9.2.

(c) Let us follow up each operator as an integral equation:

$$U_2(v_1) = F(v_1)[U_1(x)] = \int_{-\infty}^{\infty} U_1(x)e^{-j2\pi v_1 x}dx$$

$$U_3(v_1) = S\left(\frac{1}{\lambda f}\right)[U_2(v_1)] = \frac{1}{\sqrt{\lambda f}} \int_{-\infty}^{\infty} U_1(x)e^{-j2\pi \frac{v_1}{\lambda f} x}dx$$

$$U_4(v_2) = F(v_2)[U_3(v_1)] = \int_{-\infty}^{\infty} U_3(v_1)e^{-j2\pi v_2 v_1}dr_1$$

$$U_5(v_2) = S\left(\frac{1}{\lambda f}\right)[U_4(v_2)] = \frac{1}{\sqrt{\lambda f}} \int_{-\infty}^{\infty} U_3(r_1)e^{-j2\pi \frac{v_2}{\lambda f} v_1}dv_1$$

Using the integral equation for $U_3(v_1)$, the last equation is written as

$$U_5(v_2) = \frac{1}{\lambda f} \int\!\!\!\int_{-\infty}^{\infty} U_1(x)e^{-j2\pi \frac{v_1}{\lambda f} x}e^{-j2\pi \frac{v_2}{\lambda f} r_1}dxdv_1$$

$$= \frac{1}{\lambda f} \int_{-\infty}^{\infty} U_1(x)dx\left[\int_{-\infty}^{\infty} e^{-j2\pi(\frac{x+v_2}{\lambda f})v_1}dv_1\right]$$

$$= \frac{1}{\lambda f} \int_{-\infty}^{\infty} U_1(x)\delta\left(\frac{x+v_2}{\lambda f}\right)dx$$

$$= \int_{-\infty}^{\infty} U_1(x)\delta(x+v_2)dx$$

$$= U_1(-v_2)$$

EXAMPLE 9.3 Show that the optical system shown below results in a Fourier transform in which the spatial frequency can be adjusted by changing z.

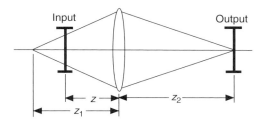

Solution: The operator O relating the output to the input is given by

$$O = P(z_2)Q\left(-\frac{1}{f}\right)P(z)Q\left(\frac{1}{z_1 - z}\right) \qquad (9.7\text{-}8)$$

By using the lens law, namely, $\frac{1}{f} = \frac{1}{z_1} + \frac{1}{z_2}$, $Q\left(-\frac{1}{f}\right)$ can be written as

$$Q\left(-\frac{1}{f}\right) = Q\left(-\frac{1}{z_1} - \frac{1}{z_2}\right) \qquad (9.7\text{-}9)$$

From Table 9.2, we use the following relations:

$$P(z_2)Q\left(-\frac{1}{z_1} - \frac{1}{z_2}\right) = Q\left(\frac{z_1 + z_2}{z_2^2}\right)S\left(-\frac{z_1}{z_2}\right)P(-z_1) \qquad (9.7\text{-}10)$$

$$P(z - z_1) = Q\left(\frac{1}{z - z_1}\right)S\left(-\frac{z_1}{z_2}\right)Q\left(\frac{1}{z - z_1}\right)S\left(\frac{1}{\lambda(z - z_1)}\right)F(r) \quad (9.7\text{-}11)$$

Substituting the results in Eqs. (9.7-9), (9.7-10), and (9.7-11) in Eq. (9.7-1) yields

$$O = Q\left(\frac{z_1 + z_2}{z_2^2}\right)S\left(-\frac{z_1}{z_2}\right)Q\left(\frac{1}{z - z_1}\right)S\left(\frac{1}{\lambda(z - z_1)}\right)F(r)$$

Using the fact that

$$Q(b)S(a) = S(a)Q\left(\frac{b}{a^2}\right)$$

from Table 9.2 results in

$$S\left(-\frac{z_1}{z_2}\right)Q\left(\frac{1}{z - z_1}\right) = Q\left(\frac{z_1 + z_2}{z_2^2}\right)S\left(-\frac{z_1}{z_2}\right)$$

This yields

$$O = Q\left(\frac{z_1 + z_2}{z_2^2}\right) Q\left(\frac{z_1 + z_2}{z_1^2}\right) S\left(-\frac{z_1}{z_2}\right) S\left(\frac{1}{\lambda(z - z_2)}\right) F(r)$$

Combining successive Q as well as S operations results in

$$O = Q\left[\frac{(z_1 + z_2)z - z_1 z_2}{z_2^2(z - z_1)}\right] S\left[\frac{z_1}{\lambda z_2(z_1 - z)}\right] F(r)$$

As an integral equation, O corresponds to

$$U_2(x_2) = e^{j\frac{k}{2}\frac{(z_1+z_2)z - z_1 z_2}{z_2^2(z_1 - z)}x_0^2} \int\limits_{-\infty}^{\infty} U_1(x) e^{-j\frac{2\pi z_1 x_0}{\lambda z_2(z_1 - z)}x} \mathrm{d}x$$

By varying z, the spatial frequency $\frac{z_1 x_0}{z_2(z_1 - z)}$ can be adjusted.

10

Imaging with Quasi-Monochromatic Waves

10.1 INTRODUCTION

In previous chapters, wave propagation and imaging with coherent and monochromatic wave fields were discussed. In this chapter, this is extended to wave propagation and imaging with quasi-monochromatic coherent or incoherent wave fields.

Monochromatic wave fields have a single temporal frequency f. *Nonmonochromatic wave fields* have many temporal frequencies. *Quasi-monochromatic wave fields* have a temporal frequency spread Δf, which is much less than the average temporal frequency f_c. In practical imaging applications, wave fields can usually be assumed to be quasi-monochromatic.

This chapter consists of 10 sections. The first few sections lay the groundwork for the theory that is pertinent for analyzing quasi-monochromatic waves. Section 10.2 introduces the *Hilbert transform* that is closely related to the Fourier transform. Its main property is swapping the cosine and sine frequency components. It is a tool that is needed to define the *analytic signal* described in Section 10.3. The analytic signal is complex, with its real part equal to a real signal and its imaginary part equal to the Hilbert transform of the same real signal. In addition to its use in analyzing quasi-monochromatic waves, it is commonly used in analyzing single-sideband modulation in communications.

Section 10.4 shows how to represent a quasi-monochromatic wave in terms of an analytic signal. The meanings of quasi-monochromatic, coherent, and incoherent waves are more closely examined in Section 10.5 with *spatial coherence* and *time coherence* concepts.

The theory developed up to this point for simple optical systems is generalized to more complex imaging systems in Section 10.6. Imaging with quasi-monochromatic waves is the subject of Section 10.7. The difference between *coherent imaging* and *incoherent imaging* becomes clear in this section. A diffraction-limited imaging system is considered as a linear system in Section 10.8, and its linear system properties are derived for coherent and incoherent imaging. One of these properties is the *optical transfer function*. How it can be computed with a computer is the topic

Diffraction, Fourier Optics and Imaging, by Okan K. Ersoy
Copyright © 2007 John Wiley & Sons, Inc.

of Section 10.9. All imaging systems have *aberrations*. They are described in Section 10.10, especially in terms of *Zernike polynomials*.

10.2 HILBERT TRANSFORM

The Hilbert transform and the analytic signal discussed in Section 10.3 are useful in characterizing quasi-monochromatic fields and image formation. They are also useful in a number of other applications such as single sideband modulation in communications [Ziemer and Tranter, 2002]. The Hilbert transform of a real signal $v(t)$ is defined as

$$v(t) = \frac{1}{\pi} \int_{-\infty}^{\infty} \frac{u(\tau)}{t - \tau} d\tau \qquad (10.2\text{-}1)$$

This is the convolution of $u(t)$ with the function $1/\pi t$.

Using the convolution theorem, Eq. (10.2-1) in the Fourier domain can be written as

$$V(f) = U(f)H(f) \qquad (10.2\text{-}2)$$

where $H(f)$ is the FT of $1/\pi t$, given by

$$H(f) = -j \operatorname{sgn}(f) \qquad (10.2\text{-}3)$$

The sgn function $\operatorname{sgn}(f)$ is defined by

$$\operatorname{sgn}(f) = \begin{cases} 1 & f > 0 \\ 0 & f = 0 \\ 1 & f < 0 \end{cases} \qquad (10.2\text{-}4)$$

Using Parseval's theorem and Eq. (10.2-2), it is observed that the energy of $v(t)$ is equal to the energy of $u(t)$:

$$\int_{-\infty}^{\infty} v^2(t) dt = \int_{-\infty}^{\infty} u^2(t) dt \qquad (10.2\text{-}5)$$

Equation (10.2-3) shows that the Hilbert transform shifts the phase of the spectral components by $\pi/2$. That is why it is also called the *quadrature filter*.

The inverse Hilbert transform can be shown to be

$$u(t) = -\frac{1}{\pi} \int\limits_{-\infty}^{\infty} \frac{v(\tau)}{t - \tau} d\tau \qquad (10.2\text{-}6)$$

The integral involved in the Hilbert transform is an improper integral, which is actually an abbreviation of the following:

$$\int\limits_{-\infty}^{\infty} \frac{u(\tau)}{t - \tau} d\tau = \lim_{\varepsilon \to 0} \left[\int\limits_{-\infty}^{t-\varepsilon} \frac{u(\tau)}{t - \tau} d\tau + \int\limits_{t+\varepsilon}^{\infty} \frac{u(\tau)}{t - \tau} d\tau \right] \qquad (10.2\text{-}7)$$

It can also be written as

$$\int\limits_{-\infty}^{\infty} \frac{u(\tau)}{t - \tau} d\tau = \int\limits_{-\infty}^{\infty} \frac{u(t + \tau)}{-\tau} d\tau \qquad (10.2\text{-}8)$$

EXAMPLE 10.1 Find the Hilbert transform of $u(t) = \cos(2\pi f_0 t + \phi(f_0))$.
Solution:

$$v(t) = \frac{1}{\pi} \int\limits_{-\infty}^{\infty} \frac{\cos(2\pi f_0(\tau))}{t - \tau} d\tau = \frac{1}{\pi} \int\limits_{-\infty}^{\infty} \frac{\cos(2\pi f_0(t + \tau))}{\tau} d\tau$$

$$= \frac{1}{\pi} \int\limits_{-\infty}^{\infty} \left[\frac{\cos(2\pi f_0 t + \phi(f)) \cos(2\pi f_0 \tau + \phi(f_0))}{-\tau} - \frac{\sin(2\pi f_0 t + \phi(f_0)) \sin(2 f_0 \tau)}{-\tau} \right] d\tau$$

$$(10.2\text{-}9)$$

The first term is odd in τ and integrates to zero. Then,

$$v(t) = \frac{\sin(2\pi f_0 t + \phi(f_0))}{\pi} \int\limits_{-\infty}^{\infty} \frac{\sin(2\pi f_0 \tau + \phi(f_0))}{\tau} d\tau = \sin(2\pi f_0 t + \phi(f_0))$$

$$(10.2\text{-}10)$$

EXAMPLE 10.2 Show that a real signal $u(t)$ and its Hilbert transform $v(t)$ are orthogonal to each other.

Solution: Using Parseval's theorem, we write

$$\int_{-\infty}^{\infty} u(t)v(t)\mathrm{d}t = \int_{-\infty}^{\infty} U(f)V^*(f)\mathrm{d}f$$

$$= j\int_{0}^{\infty} |U(f)|^2\mathrm{d}f - j\int_{-\infty}^{0} |U(f)|^2\mathrm{d}f = 0$$

$$U(f) = U_1(f) - jU_0(f) \tag{10.2-11}$$

$$u_1(t) = \frac{1}{2}\left[u(t) + u(-t)\right]$$

$$u_0(t) = \frac{1}{2}\left[u(t) - u(-t)\right]$$

$$(x_0, y_0, z)$$

$$t_2 - t_1$$

because $U(-f) = U^*(f)$.

EXAMPLE 10.3 The FT of $u(t)$ can be written as

$$U(f) = U_1(f) - jU_0(f)$$

where $U_1(f)$ and $U_0(f)$ are the cosine and sine parts of the FT of $u(t)$.

Show that $U_1(f)$ and $U_0(f)$ are a Hilbert transform pair when $u(t)$ is a causal signal ($u(t)$ equals zero for negative t).

Solution: The even and odd parts of $u(t)$ can be written as

$$u_1(t) = \frac{1}{2}[u(t) + u(-t)]$$

$$u_0(t) = \frac{1}{2}[u(t) - u(-t)]$$

When $u(t)$ is causal, it is straightforward to show that

$$u_1(t) = u_0(t)\mathrm{sgn}(t)$$
$$u_0(t) = u_1(t)\mathrm{sgn}(t) \tag{10.2-14}$$

Then,

$$U_1(f) = \int_{-\infty}^{\infty} u_1(t)\mathrm{e}^{-j2\pi ft}\mathrm{d}t$$

$$= \int_{-\infty}^{\infty} u_0(t)\mathrm{sgn}(t)\mathrm{e}^{-j2\pi ft}\mathrm{d}t \tag{10.2-15}$$

By the modulation property of the Fourier transform, $U_1(f)$ equals the convolution of the FT of $u_0(t)$ equal to $-jU_0(f)$ and the FT of sgn(t) equal to $\frac{1}{j\pi f}$:

$$U_1(f) = \int_{-\infty}^{\infty} -jU_0(v)\frac{1}{j\pi(f-v)}\,dv \qquad (10.2\text{-}16)$$

Thus,

$$U_1(f) = -\frac{1}{\pi}\int_{-\infty}^{\infty}\frac{U_0(v)}{f-v}\,dv \qquad (10.2\text{-}17)$$

It can be similarly shown that

$$U_0(f) = \frac{1}{\pi}\int_{-\infty}^{\infty}\frac{U_1(v)}{f-v}\,dv \qquad (10.2\text{-}18)$$

Equations (10.2-17) and (10.2-18) are also called *Kramers–Krönig relations* in electromagnetics.

10.3 ANALYTIC SIGNAL

The *analytic signal* is useful in understanding the properties of narrowband waveforms, wave propagation, and image formation. It is defined as

$$s(t) = u(t) + jv(t) \qquad (10.3\text{-}1)$$

where $u(t)$ is a real signal, and $v(t)$ is its Hilbert transform. Thus, the analytic signal is a means of converting a real signal to a complex signal. Taking the FT of both sides of Eq. (1.18.1) gives

$$S(f) = U(f) + jV(f) \qquad (10.3\text{-}2)$$

As $V(f) = -j\,\text{sgn}(f)U(f)$, we get

$$S(f) = \begin{cases} U(f) & f = 0 \\ 2U(f) & f > 0 \\ 0 & f < 0 \end{cases} \qquad (10.3\text{-}3)$$

Hence, the analytic signal is given by

$$s(t) = 2\int_{0}^{\infty} U(f)e^{j2\pi ft}\,df \qquad (10.3\text{-}4)$$

The analytic signal can be further written as

$$s(t) = |s(t)|e^{j\phi(t)} \tag{10.3-5}$$

where

$$|s(t)|^2 = u^2(t) + v^2(t) \tag{10.3-6}$$

$$\phi(t) = \tan^{-1}\frac{v(t)}{u(t)} \tag{10.3-7}$$

The analytic signal is often used with narrowband waveforms such as quasi-monochromatic wave fields with central frequency f_c. Then, $\phi(t)$ can be written as

$$\phi(t) = 2\pi f_c t + \phi'(t) \tag{10.3-8}$$

Equation (10.2-13) can be written as

$$s(t) = \mu(t)e^{j2\pi f_c t} \tag{10.3-9}$$

where

$$\mu(t) = |s(t)|e^{j\phi'(t)} \tag{10.3-10}$$

$\mu(t)$ is called the complex envelope. In optics, it is also referred to as the *phasor amplitude*.

EXAMPLE 10.4 Find the analytic signal corresponding to $u(t) = \cos(2\pi ft + \phi(f))$.
Solution: In Example 1.25, we found that

$$v(t) = \sin(2\pi ft + \phi(f)) \tag{10.3-11}$$

Hence,

$$s(t) = \cos(2\pi ft) + j\sin(2\pi ft) = e^{j2\pi ft} \tag{10.3-12}$$

EXAMPLE 10.5 Find the energy in the analytic signal.
Solution: The energy of the analytic signal is

$$\int_{-\infty}^{\infty} |s(t)|^2 dt = \int_{-\infty}^{\infty} u^2(t)dt + \int_{-\infty}^{\infty} v^2(t)dt = 2\int_{-\infty}^{\infty} u^2(t)dt = 2\int_{-\infty}^{\infty} |U(f)|^2 df$$

$$\tag{10.3-13}$$

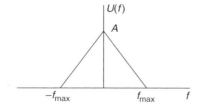

Figure 10.1. The spectrum of a lowpass signal.

It is observed that the analytic signal has twice the energy of the corresponding real signal.

EXAMPLE 10.6 The amplitude spectrum of a lowpass signal $u(t)$ is shown in Figure 10.1.

It is modulated by $\cos(2\pi f_0 t)$ to generate $g(t) = u(t)\cos(2\pi f_0 t)$. (a) Find and draw the amplitude spectrum of $g(t)$, and (b) Draw the amplitude spectrum of the analytic signal generated from $u(t)$, and (c) Find and draw the amplitude spectrum of $p(t) = u(t)\cos(2\pi f_0 t) - v(t)\sin(2\pi f_0 t)$.

Solution:

(a) $g(t)$ can be written as

$$g(t) = \frac{u(t)}{2}e^{j2\pi f_0 t} + \frac{u(t)}{2}e^{-j2\pi f_0 t} \qquad (10.3\text{-}14)$$

The FT of $g(t)$ is given by

$$G(f) = \frac{1}{2}U(f - f_0) + \frac{1}{2}U(f + f_0) \qquad (10.3\text{-}15)$$

The amplitude spectrum is given by

$$|G(f)| = \frac{1}{2}|U(f - f_0) + U(f + f_0)| \qquad (10.3\text{-}16)$$

$|G(f)|$ is shown in Figure 10.2.

The frequency components in the range $f_0 \leq f \leq f_0 + f_{max}$ and $-f_0 - f_{max} \leq f \leq f_0$ are known as the *upper sideband*. The frequency components in the range $-f_0 \leq f \leq -f_0 + f_{max}$ and $f_0 - f_{max} \leq f \leq f_0$ are known as the *lower sideband*.

(b) The analytic signal generated from $u(t)$ is given by

$$s(t) = u(t) + jv(t) \qquad (10.3\text{-}17)$$

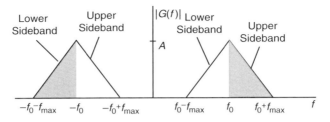

Figure 10.2. The amplitude spectrum of $G(f)$.

Its FT is

$$
\begin{aligned}
S(f) &= U(f) + jV(f) \\
&= U(f)[1 + jH(f)] \\
&= \begin{cases} 2U(f) & f \geq 0 \\ 0 & f < 0 \end{cases}
\end{aligned}
\tag{10.3-18}
$$

The amplitude spectrum of $s(t)$ is shown in Figure 10.3.
(c) $p(t) = u(t)\cos(2\pi f_0 t) - v(t)\sin(2\pi f_0 t)$ can be written as

$$
\begin{aligned}
p(t) &= \mathrm{Re}[(u(t) + jv(t))e^{j2\pi f_0 t}] \\
&= \mathrm{Re}[s(t)e^{j2\pi f_0 t}] \\
&= \frac{1}{2}[s(t)e^{j2\pi f_0 t} + s^*(t)e^{-j2\pi f_0 t}]
\end{aligned}
\tag{10.3-19}
$$

The FT of $p(t)$ can be written as

$$
\begin{aligned}
P(f) &= \frac{1}{2}S(f - f_0) + \frac{1}{2}S^*(-f - f_0) \\
&= \begin{cases} U(f - f_0) & f_0 \leq f \leq f_0 + f_{max} \\ U^*(-f - f_0) & -f_0 - f_{max} \leq f \leq -f_0 \\ 0 & \text{otherwise} \end{cases}
\end{aligned}
\tag{10.3-20}
$$

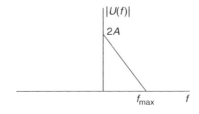

Figure 10.3. The amplitude spectrum of the analytic signal $s(t)$.

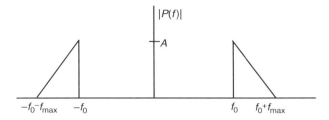

Figure 10.4. The amplitude spectrum of the single sideband signal $p(t)$.

The amplitude spectrum of $p(t)$ is shown in Figure 10.4.

$p(t)$ is known as the *single sideband signal* because it contains only the upper sideband of $u(t)$. If it was chosen instead as $u(t)\cos(2\pi f_0 t) + v(t)\sin(2\pi f_0 t)$, it would contain the lower sideband only. In this way, the frequency bandwidth required to transmit the signal over a channel is reduced by a factor of 2.

10.4   ANALYTIC SIGNAL REPRESENTATION OF A NONMONOCHROMATIC WAVE FIELD

Let $u(\mathbf{r}, t)$ represent the real representation of a nonmonochromatic wave field. As discussed in Section 2.7, $u(\mathbf{r}, t)$ can be written in Fourier representation as

$$u(\mathbf{r}, t) = 2 \int_0^\infty U(\mathbf{r}, f)\cos(2\pi ft + \phi(f)\mathrm{d}f \tag{10.4-1}$$

where $U(\mathbf{r}, f)$ and $\phi(f)$ are the amplitude and phase spectra of $u(\mathbf{r}, t)$ with respect to t, respectively.

The analytic signal corresponding to $u(\mathbf{r}, t)$ is given by

$$u_A(\mathbf{r}, t) = u(\mathbf{r}, t) + jv(\mathbf{r}, t) \tag{10.4-2}$$

where $v(\mathbf{r}, t)$ is the Hilbert transform of $u(\mathbf{r}, t)$. As the Hilbert transform of $\cos(2\pi ft + \phi(f))$ equals $\sin(2\pi ft + \phi(f))$ by Example 10.1, $v(\mathbf{r}, t)$ can be written as

$$v(\mathbf{r}, t) = 2 \int_0^\infty U(\mathbf{r}, f)\sin(2\pi ft + \phi(f))\mathrm{d}f \tag{10.4-3}$$

Hence, the analytic signal $u_A(\mathbf{r}, t)$ is given by

$$u_A(\mathbf{r}, t) = 2 \int_0^\infty U(\mathbf{r}, f)e^{j(2\pi ft + \phi(f))}\mathrm{d}f \tag{10.4-4}$$

Note that $U_c(\mathbf{r}, f) = \frac{1}{2} U(\mathbf{r}, f) e^{j\phi(f)}$ is the Fourier transform of $u(\mathbf{r}, t)$ with respect to t. Equation (10.4-4) shows that the analytic signal is obtained by integrating the spectrum of the original signal over positive frequencies only. $u(\mathbf{r}, t)$ can also be written as

$$u(\mathbf{r}, t) = \frac{1}{2}[s(\mathbf{r}, t) + s^*(\mathbf{r}, t)] \tag{10.4-5}$$

In the context of waves, the analytic signal is often referred to as the complex wave function [Saleh and Teich, 1991].

10.5   QUASI-MONOCHROMATIC, COHERENT, AND INCOHERENT WAVES

A wave field is called quasi-monochromatic if its temporal frequency spread Δf satisfies

$$\frac{\Delta f}{f_c} \ll 1 \tag{10.5-1}$$

where f_c is the mean temporal frequency of the wave field. In this case, $u_A(\mathbf{r}, t)$ can be written as

$$u_A(\mathbf{r}, t) = U_A(\mathbf{r}, t) e^{j2\pi f_c t} \tag{10.5-2}$$

where $U_A(\mathbf{r}, t)$ is the complex envelope of the analytic signal.

A wave field that is monochromatic is perfectly coherent. The subject of coherence of wave fields is very broad. For example, see references [Born and Wolf, 1969] and [Marathay]. A coherent wave field is characterized by its ability to generate constructive and destructive interference on a time-average basis when its different portions are combined at the same point.

Coherence is usually discussed in terms of *spatial coherence* and *temporal coherence*. Let $u_A(\mathbf{r}, t)$ be the analytic signal representation of a wave field. If $u_A(\mathbf{r}_1, t)$ and $u_A(\mathbf{r}_2, t)$ are related by a complex constant factor independent of time for all points $\mathbf{r}_1$ and $\mathbf{r}_2$, the wave field is called *spatially coherent*. In the following sections, spatially coherent waves are simply referred to as *coherent waves*.

If the relationship between $u_A(\mathbf{r}, t_1)$ and $u_A(\mathbf{r}, t_2)$ depends only on the time difference $t_2 - t_1$, the wave field is *temporally coherent*.

10.6   DIFFRACTION EFFECTS IN A GENERAL IMAGING SYSTEM

In Section 9.4, image formation with a single lens was discussed as 2-D linear filtering. It is possible to generalize the results obtained to more general imaging

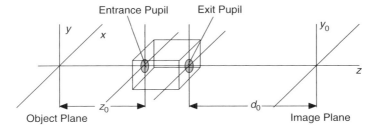

Figure 10.5. Model of an optical imaging system.

systems as well as to quasi-monochromatic sources with spatially coherent and spatially incoherent illumination.

A general imaging system typically consists of a number of lenses. Such an imaging system can be characterized in terms of an entrance pupil and an exit pupil, which are actually both images of an effective system aperture [Goodman]. This is shown in Figure 10.5. Diffraction effects can be expressed in terms of the exit pupil and the distance d_0 between the exit pupil and the image plane.

An optical system is called *diffraction limited* if a diverging spherical wave incident on the entrance pupil is mapped into a converging spherical wave at the exit pupil. For a real imaging system, this property is at best limited to finite areas of object and image planes. Aberrations are distortions modifying the spherical property. They are discussed in Section 10.9.

In a general imaging system with many lenses, Eqs. (9.4-17) and (9.4-18) remain valid provided that $P(\bullet,\bullet)$ denotes the finite equivalent exit pupil of the system, the equivalent focal length of the system and the distance from the exit pupil to the image plane is used, and the system is diffraction limited [Goodman].

We define the ideal image appearing in Eq. (8.4-17) as

$$U_g(x,y) = U\left(-\frac{x}{M}, -\frac{y}{M}\right) \qquad (10.6\text{-}1)$$

where $U(x,y)$ is the original image (wave field). Equation (9.4-16) is rewritten as

$$U(x_0, y_0) = \int\limits_{-\infty}^{\infty} \int\limits_{-\infty}^{\infty} h(x_0 - x, y_0 - y)U_g(x,y)\,\mathrm{d}x\mathrm{d}y \qquad (10.6\text{-}2)$$

where

$$h(x,y) = \int\limits_{-\infty}^{\infty} \int\limits_{-\infty}^{\infty} P(x_1, y_1)\mathrm{e}^{-j\frac{2\pi}{\lambda d_0}(xx_0 + yy_0)}\,\mathrm{d}x\mathrm{d}y \qquad (10.6\text{-}3)$$

10.7   IMAGING WITH QUASI-MONOCHROMATIC WAVES

In practical imaging systems, the conditions for quasi-monochromatic wave fields are usually satisfied. For example, in ordinary photography, films are sensitive to the visible range of the electromagnetic temporal spectrum. Then, it can be shown that the complex envelope of the analytic signal satisfies the imaging equations derived previously. Thus, Eq. (10.6-2) can be written as

$$U_A(x_0, y_0, d_0, t) = h(x_0, y_0) * U_G(x_0, y_0, t - \tau)$$

$$= \iint\limits_{-\infty}^{\infty} h(x_0 - x, y_0 - y)U_G(x, y, t - \tau)\mathrm{d}x\mathrm{d}y \qquad (10.7\text{-}1)$$

where $h(x_0, y_0)$ is given by Eq. (10.6-3) except that λ is replaced by λ_c, and $U_G(x_0, y_0, t)$ is the complex envelope of the analytic signal corresponding to the ideal image $U(-\frac{x_0}{M}, -\frac{y_0}{M}, t)$ predicted by geometrical optics; τ is a time delay associated with propagation to the image plane.

The amplitude transfer function $H(f_x, y_y)$ is the FT of $h(x, y)$:

$$H(f_x, f_y) = P(\lambda d_0 f_x, \lambda d_0 f_y) \qquad (10.7\text{-}2)$$

Detector systems are sensitive to intensity of the wave field, which can be written as

$$I(x_0, y_0, d_0) = \langle |U_A(x_0, y_0, d_0, t)|^2 \rangle \qquad (10.7\text{-}3)$$

where $\langle \bullet \rangle$ indicates an infinite time average. Using Eq. (10.7.1), $I(x_0, y_0, d_0)$ can be written as

$$I(x_0, y_0, d_0)$$

$$= \iint\limits_{-\infty}^{\infty} h(x_0 - x, y_0 - y)\left[\iint\limits_{-\infty}^{\infty} h^*(x_0 - x', y_0 - y')I_G(x, y; x', y')\mathrm{d}x'\mathrm{d}y' \right]\mathrm{d}x\mathrm{d}y$$

$$(10.7\text{-}4)$$

where

$$I_G(x, y; x', y') = \langle U_G(x, y, t - \tau_1)U_G^*(x', y', t - \tau_2) \rangle \qquad (10.7\text{-}5)$$

I_G is known as the *mutual intensity*. In Eq. (10.7-5), τ_1 and τ_2 are approximately equal because the impulse response h is limited to a small region around the image point and hence can be neglected.

10.7.1   Coherent Imaging

For perfectly coherent wave fields, we can write

$$U_G(x, y, t) = U_G(x, y)U_G(0, 0, t) \qquad (10.7.1\text{-}1)$$

where $U_G(x, y)$ is the phasor amplitude of $U_G(x, y, t)$ relative to the wave field at the origin.

We write

$$\langle U_G(x, y, t)U_G(x, y, t)\rangle = KU_G(x, y)U_G^*(x', y') \qquad (10.7.1\text{-}2)$$

where the constant K is given by

$$K = \langle |U_G(0, 0, t)|^2 \rangle \qquad (10.7.1\text{-}3)$$

Neglecting K, the intensity is written as

$$I(x_0, y_0, d_0) = |U_1(x_0, y_0)|^2 \qquad (10.7.1\text{-}4)$$

where

$$U_1(x_0, y_0) = h(x_0, y_0) * U_G(x_0, y_0) = \iint\limits_{-\infty}^{\infty} h(x_0 - x, y_0 - y)U_G(x, y)\mathrm{d}x\mathrm{d}y$$

$$(10.7.1\text{-}5)$$

Thus, the coherent imaging system is linear in complex amplitude of the analytic signal relative to the origin.

EXAMPLE 10.7 Determine the cutoff frequency of a diffraction-limited coherent imaging system with a circular effective pupil of radius R, assuming that the image forms at a distance d_0 from the system.
Solution: The pupil function in this case is given by

$$P(x, y) = \mathrm{circ}\left(\frac{\sqrt{x^2 + y^2}}{R}\right)$$

The amplitude transfer function is given by

$$H(f_x, f_y) = P(\lambda d_0 f_x, \lambda d_0 f_y) = \mathrm{circ}\left(\frac{\sqrt{f_x^2 + f_y^2}}{R/\lambda d_0}\right)$$

Hence, the cutoff frequency f_c is given by

$$f_c = \frac{R}{\lambda d_0}$$

10.7.2 Incoherent Imaging

When the wave field is perfectly incoherent $U_G(x, y, t)$ is spatially uncorrelated (equivalent to white noise). Hence, the time average can be written as

$$\langle U_G(x, y, t) U_G(x', y', t) \rangle = K I_G(x, y) \delta(x - x', y - y') \qquad (10.7.2\text{-}1)$$

where $I_G(x, y) = \langle |U_G(x, y, t)|^2 \rangle$. Neglecting K, Eq. (10.7-4) becomes

$$I(x_0, y_0, d_0) = \iint\limits_{-\infty}^{\infty} |h(x_0 - x, y_0 - y)|^2 I_G(x, y) \mathrm{d}x \mathrm{d}y \qquad (10.7.2\text{-}2)$$

We conclude that when the wave field is incoherent, the imaging system is linear in intensity. The ideal image intensity is filtered with the impulse response function $|h(x, y)|^2$.

10.8 FREQUENCY RESPONSE OF A DIFFRACTION-LIMITED IMAGING SYSTEM

In the previous section, coherent and incoherent imaging was considered as linear systems whose input is the ideal image in amplitude or intensity and whose output is the actual image, which is the convolution of the input with a system impulse response. This is shown in Figure 10.6. In this section, the system properties of coherent and incoherent imaging are studied in more detail.

10.8.1 Coherent Imaging System

A coherent imaging system is linear in amplitude mapping. Let us denote the 2-D FT of the output image U_1 and the ideal image U_G as V_1 and V_G, respectively. Then, by convolution theorem,

$$V_1(f_x, f_y) = H(f_x, f_y) V_G(f_x, f_y) \qquad (10.8.1\text{-}1)$$

Figure 10.6.   Linear system interpretation of coherent and incoherent imaging.

where $H(f_x, f_y)$ is the 2-D FT of the impulse response:

$$H(f_x, f_y) = \iint\limits_{-\infty}^{\infty} h(x, y)e^{-j2\pi(f_x x + f_y y)}\mathrm{d}x\mathrm{d}y = P(\lambda d_0 f_x, \lambda d_0 f_y) \qquad (10.8.1\text{-}2)$$

$H(f_x, f_y)$ is known as *the coherent transfer function*.

It is observed that a coherent imaging system is equivalent to an ideal low-pass filter, which passes all frequencies within the pupil function's "1" zone and cuts off all frequencies outside this zone.

10.8.2 Incoherent Imaging System

Incoherent imaging systems are linear in intensity. The visual quality of an image is largely determined by the contrast of the relative intensity of the information-bearing details of the image to the ever-present background. The output image and the input ideal image can be normalized by the total image energy to reflect this property:

$$I'(x_0, y_0, d_0) = \frac{I(x_0, y_0, d_0)}{\iint\limits_{-\infty}^{\infty} I(x_0, y_0, d_0)\mathrm{d}x_0\mathrm{d}y_0}$$

$$\qquad (10.8.2\text{-}1)$$

$$I'_G(x, y) = \frac{I_G(x, y)}{\iint\limits_{-\infty}^{\infty} I_G(x, y)\mathrm{d}x\mathrm{d}y}$$

Let us denote the 2-D FT of $I'(x_0, y_0, d_0)$ and $I'_G(x, y)$ by $J(f_x, f_y)$ and $J_G(f_x, f_y)$, respectively. By convolution theorem, Eq. (10.7.2-2) can be written as

$$J(f_x, f_y) = H_I(f_x, f_y)J_G(f_x, f_y) \qquad (10.8.2\text{-}2)$$

where

$$H_I(f_x, f_y) = \frac{\iint\limits_{-\infty}^{\infty} |h(x, y)|^2 e^{-j2\pi(f_x x + f_y y)}\,\mathrm{d}x\mathrm{d}y}{\iint\limits_{-\infty}^{\infty} |h(x, y)|^2\mathrm{d}x\mathrm{d}y} \qquad (10.8.2\text{-}3)$$

$H_I(f_x, f_y)$ is called the *optical transfer function* (OTF). The *modulation transfer function* (MTF) is defined as $|H_I(f_x, f_y)|$.

It is observed that $H_I(f_x, f_y)$ is the normalized FT of $|h(x, y)|^2$, a nonnegative function. By Property 15 of the FT and Parseval's theorem discussed in Section 2.5, $H_I(f_x, f_y)$ is the normalized autocorrelation of $H(f_x, f_y)$:

$$H_I(f_x, f_y) = \frac{\displaystyle\iint_{-\infty}^{\infty} H(f_x', f_y') H^*(f_x + f_x', f_y + f_y') \mathrm{d}f_x' \mathrm{d}f_y'}{\displaystyle\iint_{-\infty}^{\infty} |H(f_x, f_y)|^2 \mathrm{d}f_x \mathrm{d}f_y} \qquad (10.8.2\text{-}4)$$

The most important properties of the OTF are the following:

A. $H_I(0,0) = 1$
B. $H_I(-f_x, -f_y) = H_I(f_x, f_y)$
C. $|H_I(f_x, f_y)| \le H_I(0,0)$

The last property is a consequence of Schwarz' inequality, which states that for any two complex-valued functions f and g,

$$\left| \iint fg \mathrm{d}A \right|^2 \le \iint |f|^2 \mathrm{d}A \iint |g|^2 \mathrm{d}A \qquad (10.8.2\text{-}5)$$

with equality iff $g = Kf^*$ where K is a complex constant.

Letting f and g be equal to $H(f_x', f_y')$ and $H^*(f_x + f_x', f_y + f_y')$, respectively, and using Eq. (10.8.2-5) yields Property C above.

The coherent transfer function $H(f_x, f_y)$ is given by Eq. (10.8.1-2). Using this result in Eq. (10.8.2-5) gives

$$H_I(f_x, f_y) = \frac{\displaystyle\iint_{-\infty}^{\infty} P(\lambda d_0 f_x', \lambda d_0 f_y') P(\lambda d_0 (f_x + f_x')), \lambda d_0 (f_y + f_y') \mathrm{d}f_x' \mathrm{d}f_y'}{\displaystyle\iint_{-\infty}^{\infty} P(\lambda d_0 f_x, \lambda d_0 f_y) \mathrm{d}f_x \mathrm{d}f_y} \qquad (10.8.2\text{-}6)$$

where the fact $P^2 = P$ is used in the denominator.

Incorporating a change of variables, Eq. (10.8.2-6) can be written as

$$H_I(f_x, f_y) = \frac{\displaystyle\iint_{-\infty}^{\infty} P\left(f_x' + \frac{\lambda d_0 f_x}{2}, f_y' + \frac{\lambda d_0 f_y}{2}\right) P\left(f_x' - \frac{\lambda d_0 f_x}{2}, f_y' - \frac{\lambda d_0 f_y}{2}\right) \mathrm{d}f_x' \mathrm{d}f_y'}{\displaystyle\iint_{-\infty}^{\infty} P(\lambda d_0 f_x', \lambda d_0 f_y')}$$

$$(10.8.2\text{-}7)$$

The two pupil functions in the numerator above are displaced from each other by $(\lambda d_0|f_x|, \lambda d_0|f_y|)$. The integral equals the area of overlap between the two pupil functions. Hence, $H_I(f_x, f_y)$ can be written as

$$H_I(f_x, f_y) = \frac{\text{area of overlap}}{\text{total area}} \qquad (10.8.2-8)$$

where the areas are computed with respect to the scaled pupil function. The OTF is always real and nonnegative.

Note that the incoherent impulse response function $|h(x, y)|^2$ is similar to the power spectrum of a stationary 2-D random field. By the same token, $H_I(f_x, f_y)$ is similar to the autocorrelation function of a 2-D stationary random field [Besag, 1974].

EXAMPLE 10.8 (a) Determine the OTF of a diffraction-limited optical system whose exit pupil is a square of width $2W$, (b) Determine the cutoff frequency f_c of the system.

Solution: The area of the pupil function equals $4W^2$. The area of overlap is illustrated in Figure 10.7.

The area of overlap is computed from Figure 10.7 as

$$A(f_x, f_y) = \begin{cases} (2W - \lambda d_0|f_x|)(2W - \lambda d_0|f_y|) & |f_x| \le 2W/\lambda d_0 \\ & |f_y| \le 2W/\lambda d_0 \\ 0 & \text{otherwise} \end{cases}$$

When $A(f_x, f_y)$ is normalized by $4W^2$, the OTF is given by

$$H_I(f_x, f_y) = \text{tri}\left(\frac{f_x}{2f_c'}\right)\text{tri}\left(\frac{f_y}{2f_c'}\right)$$

where $\text{tri}(\bullet)$ is the triangle function, and f_c' is the cutoff frequency for coherent illumination, equal to $W/\lambda d_0$.

(b) It is obvious that the cutoff frequency f_c is given by

$$f_c = 2f_c'$$

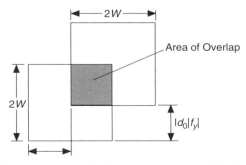

Figure 10.7. The area of overlap for the computation of the OTF of a square aperture.

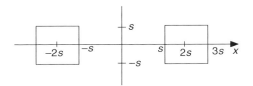

Figure 10.8. The aperture function for Example 10.9.

EXAMPLE 10.9 An exit aperture function consists of two open squares as shown in Figure 10.8.
Determine

    (a) the coherent transfer function

    (b) the coherent cutoff frequencies

    (c) the amplitude impulse response

    (d) the optical transfer function

Solution: (a) The coherent transfer function is the same as the scaled aperture function. Mathematically, the aperture function can be written as

$$P(x,y) = \text{rect}\left(\frac{x-2s}{2s}, \frac{y}{2s}\right) + \text{rect}\left(\frac{x+2s}{2s}, \frac{y}{2s}\right)$$

$H(f_x, f_y)$ is given by

$$H(f_x, f_y) = P(\lambda d_0 f_x, \lambda d_0 f_y)$$
$$= \text{rect}\left(\frac{\lambda d_0 f_x + 2s}{2s}, \frac{\lambda d_0 f_y}{2s}\right) + \text{rect}\left(\frac{\lambda d_0 f_x - 2s}{2s}, \frac{\lambda d_0 f_y}{2s}\right)$$

(b) The cutoff frequencies along the two directions are given by

$$f_{xc} = \frac{3s}{\lambda d_0}, \quad f_{yc} = \frac{s}{\lambda d_0}$$

(c) The amplitude impulse response $h(x, y)$ is the inverse Fourier transform of $H(f_x, f_y)$. Let a be equal to $\lambda d_0 s$. $h(x, y)$ is computed as

$$h(x,y) = \iint\limits_{-\infty}^{\infty} H(f_x, f_y) = e^{j2\pi(f_x x + f_y y)} df_x df_y$$

$$= \left[2\int_0^{3s} \cos(2\pi f_x x) df_x\right] \left[2\int_0^{s} \cos(2\pi f_y y) df_y\right]$$

$$= \frac{1}{\pi^2} \frac{\sin(6\pi s x)}{x} \frac{\sin(2\pi s y)}{y}$$

(d) The total area A under $H(f_x,f_y)$ equals $8s^2$.

The OTF is given by

$$H_I(f_x,f_y) = \frac{\displaystyle\iint\limits_{-\infty}^{\infty} H(f_x',f_y')H^*(f_x'-f_x,f_y'-f_y)\mathrm{d}f_x'\mathrm{d}f_y'}{A}$$

The computation of the above integral is not trivial and can be best done by the computer.

10.9 COMPUTER COMPUTATION OF THE OPTICAL TRANSFER FUNCTION

The easiest way to compute the discretized OTF is by using the FFT. For this purpose, both $h(x,y)$, $H(f_x,f_y)$ and $H_I(f_x,f_y)$, have to be discretized. The discretized coordinates can be written as follows:

$$x = \Delta x n_1 \tag{10.9-1}$$

$$y = \Delta y n_2 \tag{10.9-2}$$

$$f_x = \Delta f_x k_1 \tag{10.9-3}$$

$$f_y = \Delta f_y k_2 \tag{10.9-4}$$

$h(\Delta x n_1, \Delta y n_2)$, $H(\Delta f_x k_1, \Delta f_y k_2)$, and $H_I(\Delta f_x k_1, \Delta f_y k_2)$ will be written as $h(n_1,n_2)$, $H(k_1,k_2)$, and $H_I(k_1,k_2)$, respectively. The size of the matrices involved are assumed to be N_1, by N_2. In order to use the FFT, the following must be satisfied:

$$\Delta x \Delta f_x = \frac{1}{N_1} \tag{10.9-5}$$

$$\Delta y \Delta f_y = \frac{1}{N_2} \tag{10.9-6}$$

$h(n_1,n_2)$ is approximately given by

$$h(n_1,n_2) = \frac{1}{K}h'(n_1,n_2) \tag{10.9-7}$$

where

$$h'(n_1,n_2) = \sum_{k_1=-\frac{N_1}{2}}^{\frac{N_1}{2}-1} \sum_{k_2=-\frac{N_2}{2}}^{\frac{N_2}{2}-1} H(k_1,k_2)e^{j2\pi\left(\frac{n_1 k_1}{N_1}+\frac{n_2 k_2}{N_2}\right)} \tag{10.9-8}$$

and

$$K = \Delta x \Delta y N_1 N_2 \qquad (10.9\text{-}9)$$

$H(k_1, k_2)$ equals $P(-\lambda d_0 \Delta f_x k_1, -\lambda d_0 \Delta f_y k_2)$. N_1 and N_2 should be chosen such that the pupil function is sufficiently represented. For example, the nonzero portion of the pupil function must be completely covered. In order to minimize the effect of periodicity imposed by the FFT, N_1 and N_2 should be at least twice as large the minimum values dictated by the nonzero portion of the pupil function.

Once N_1 and N_2 are properly chosen, $H(k_1, k_2)$ is arranged as discussed in Section 4.4 so that k_1 and k_2 satisfy $0 \leq k_1 < N_1$ and $0 \leq k_2 < N_2$ respectively.

$H_I(k_1, k_2)$ is approximately given by

$$H_I(k_1, k_2) = \frac{\dfrac{\Delta x \Delta y}{K^2} \displaystyle\sum_{n_1=0}^{N_1-1} \sum_{n_2=0}^{N_2-1} |h'(n_1, n_2)|^2 e^{-j2\pi \left(\frac{n_1 k_1}{N_1} + \frac{n_2 k_2}{N_2}\right)}}{\dfrac{1}{K} \displaystyle\sum_{k_1=0}^{N_1-1} \sum_{k_2=0}^{N_2-1} |H(k_1, k_2)|^2} \qquad (10.9\text{-}10)$$

or

$$H_I(k_1, k_2) = \frac{\dfrac{1}{N_1 N_2} \displaystyle\sum_{n_1=0}^{N_1-1} \sum_{n_2=0}^{N_2-1} |h'(n_1, n_2)|^2 e^{-j2\pi \left(\frac{n_1 k_1}{N_1} + \frac{n_2 k_2}{N_2}\right)}}{\displaystyle\sum_{k_1=0}^{N_1-1} \sum_{k_2=0}^{N_2-1} |H(k_1, k_2)|^2} \qquad (10.9\text{-}11)$$

The numerator above is the 2-D DFT of $|h'(n_1, n_2)|^2$. $H_I(k_1, k_2)$ above is actually shifted to positive frequencies because of the underlying periodicity. It should be shifted down again to include negative frequencies.

In summary, $H(k_1, k_2)$ is determined by using the pupil function. $h'(n_1, n_2)$ is obtained by the inverse DFT of $H(k_1, k_2)$ according to Eq. (10.9-8). $H_I(k_1, k_2)$ is given by the DFT of $|h'(n_1, n_2)|^2$ normalized by the area under $|H(k_1, k_2)|^2$.

Note that Δf_x and Δf_y should be chosen small enough so that $h'(n_1, n_2)$ is not aliased when computed by using Eq. (10.9-8). Once Δf_x and Δf_y are chosen, N_1 and N_2 are determined by considering the pupil function as discussed above.

10.9.1 Practical Considerations

In studies of the OTF and MTF, λd_0 is often chosen to be equal to 1, and the negative signs in the pupil function are neglected so that $H(f_x, f_y)$ is simply written as $P(f_x, f_y)$. Normalization by the area of the pupil function may also be neglected.

Another way to generate OTF is by autocorrelating the pupil function with itself.

10.10   ABERRATIONS

A *diffraction-limited system* means the wave of interest is perfect at the exit pupil, and the only imperfection is the finite aperture size. The wave of interest is typically a spherical wave. *Aberrations* are departures of the ideal wavefront within the exit pupil from its ideal form. They are typically phase errors.

In order to include aberrations, the exit pupil function can be modified as

$$P_A(x, y) = P(x, y)e^{jk\phi_A(x,y)} \tag{10.10-1}$$

where $P(x, y)$ is the exit pupil function without aberrations, and $\phi_A(x, y)$ is the phase error due to aberrations.

The theory of coherent and incoherent imaging developed in the previous sections is still applicable with the replacement of $P(x, y)$ by $P_A(x, y)$. For example, the amplitude transfer function becomes

$$\begin{aligned} H(f_x, f_y) &= P_A(-\lambda d_0 f_x, -\lambda d_0 f_y) \\ &= P(-\lambda d_0 f_x, -\lambda d_0 f_y)e^{jk\phi_A(-\lambda d_0 f_x, -\lambda d_0 f_y)} \end{aligned} \tag{10.10-2}$$

The optical transfer function can be similarly written as

$$H_I(f_x, f_y) = \frac{A(f_x, f_y)}{\displaystyle\iint\limits_{-\infty}^{\infty} P(\lambda d_0 f_x, \lambda d_0 f_y)\mathrm{d}f_x\mathrm{d}f_y} \tag{10.10-3}$$

where

$$A(f_x, f_y) = \iint\limits_{-\infty}^{\infty} P_A\left(f_x' + \frac{\lambda d_0 f_x}{2}, f_y' + \frac{\lambda d_0 f_y}{2}\right) P_A^*\left(f_x' - \frac{\lambda d_0 f_x}{2}, f_y' - \frac{\lambda d_0 f_y}{2}\right)\mathrm{d}f_x'\mathrm{d}f_y' \tag{10.10-4}$$

Note that $A(f_x, f_y)$ is to be computed in the area of overlap of the two pupil functions shifted with respect to each other. The ordinary pupil function in this area equals 1. Letting $\iint_{\text{overlap}}$ denote integration in the area of overlap, $A(f_x, f_y)$ can be written as

$$A(f_x, f_y) = \iint\limits_{\text{overlap}} P_1 P_2 \mathrm{d}f_x'\mathrm{d}f_y' \tag{10.10-5}$$

where

$$P_1 = e^{jk\phi_A \left(f_x' + \frac{\lambda d_0 f_x}{2}, f_y' + \frac{\lambda d_0 f_y}{2} \right)} \tag{10.10-6}$$

$$P_2 = e^{-jk\phi_A \left(f_x' - \frac{\lambda d_0 f_x}{2}, f_y' - \frac{\lambda d_0 f_y}{2} \right)} \tag{10.10-7}$$

EXAMPLE 10.10 Show that aberrations do not increase the MTF.
Solution: The MTF is the modulus of the OTF. According to the Schwarz's inequality, it is true that

$$|A(f_x, f_y)|^2 \leq \iint_{\text{overlap}} |P_1|^2 df_x' df_y' \iint_{\text{overlap}} |P_2|^2 df_x' df_y' \tag{10.10-8}$$

Note that

$$|P_1|^2 = |P_2|^2 = 1 \tag{10.10-9}$$

in the area of overlap. Hence, $|A(f_x, f_y)|^2 \leq$ area of overlap, and

$$|H_I(f_x, f_y)|^2 \leq |H_I'(f_x, f_y)|^2 \tag{10.10-10}$$

where $H_I'(f_x, f_y)$ is the optical transfer function without aberrations.

The phase function $\phi_A(x, y)$ is often written in terms of the polar coordinates as $\phi_A(r, \theta)$. What is referred to as *Seidel aberrations* is the representation of $\phi_A(r, \theta)$ as a polynomial in r, for example,

$$\phi_A(r, \theta) = a_{40}r^4 + a_{31}r^3 \cos \theta + a_{20}r^2 + a_{22}r^2 \cos^2 \theta + a_{11}r \cos \theta \tag{10.10-11}$$

Higher order terms can be added to this function. The terms on the right hand side of Eq. (10.9-11) represent the following:

$$
\begin{aligned}
a_{40}r^4 &: \quad \text{spherical aberration} \\
a_{31}r^3 \cos \theta &: \quad \text{coma} \\
a_{20}r^2 &: \quad \text{astigmatism} \\
a_{22}r^2 \cos^2 \theta &: \quad \text{field curvature} \\
a_{11}r \cos \theta &: \quad \text{distortion}
\end{aligned}
$$

10.10.1 Zernike Polynomials

When the exit pupil of the optical system is circular, the aberrations present in an optical system can be represented in terms of Zernike polynomials, which are

Table 10.1.   The Zernike polynomials.

No.	Polynomial
1	1
2	$2\rho\cos(\theta)$
3	$2\rho\sin(\theta)$
	$\sqrt{3}(2\rho^2 - 1)$
4	$\sqrt{3}(2\rho^2 - 1)$
5	$\sqrt{6}\rho^2\cos(2\theta)$
6	$\sqrt{6}\rho^2\sin(2\theta)$
7	$\sqrt{8}(3\rho^2 - 2)\rho\cos(\theta)$
8	$\sqrt{8}(3\rho^2 - 2)\rho\sin(\theta)$
9	$\sqrt{5}(6\rho^4 - 6\rho^2 + 1)$
10	$\sqrt{8}\rho^3\cos(3\theta)$
11	$\sqrt{8}\rho^3\sin(3\theta)$
12	$\sqrt{10}(4\rho^2 - 3)\rho^2\cos(2\theta)$
13	$\sqrt{10}(4\rho^2 - 3)\rho^2\sin(2\theta)$
14	$\sqrt{12}(10\rho^4 - 12\rho^2 + 3)\rho\cos(\theta)$
15	$\sqrt{12}(10\rho^4 - 12\rho^2 + 3)\rho\sin(\theta)$
16	$\sqrt{7}(20\rho^6 - 30\rho^4 + 12\rho^2 - 1)$
17	$\sqrt{10}\rho^4\cos(4\theta)$
18	$\sqrt{10}\rho^4\sin(4\theta)$
19	$\sqrt{12}(5\rho^2 - 4)\rho^3\cos(3\theta)$
20	$\sqrt{12}(5\rho^2 - 4)\rho^3\sin(3\theta)$
21	$\sqrt{14}(15\rho^4 - 20\rho^2 + 6)\rho^2\cos(2\theta)$
22	$\sqrt{14}(15\rho^4 - 20\rho^2 + 6)\rho^2\sin(2\theta)$
23	$4(35\rho^6 - 60\rho^4 + 30\rho^2 - 4)\rho\cos(\theta)$
24	$4(35\rho^6 - 60\rho^4 + 30\rho^2 - 4)\rho\sin(\theta)$
25	$3(70\rho^8 - 140\rho^6 + 90\rho^4 - 20\rho^2 + 1)$
26	$\sqrt{12}\rho^5\cos(5\theta)$
27	$\sqrt{12}\rho^5\sin(5\theta)$
28	$\sqrt{14}(6\rho^2 - 5)\rho^4\cos(4\theta)$
29	$\sqrt{14}(6\rho^2 - 5)\rho^4\sin(4\theta)$
30	$4(21\rho^4 - 30\rho^2 + 10)\rho^3\cos(3\theta)$
31	$4(21\rho^4 - 30\rho^2 + 10)\rho^3\sin(3\theta)$
32	$\sqrt{18}(56\rho^6 - 105\rho^4 + 60\rho^2 - 10)\rho^2\cos(2\theta)$
33	$\sqrt{18}(56\rho^6 - 105\rho^4 + 60\rho^2 - 10)\rho^2\sin(2\theta)$
34	$\sqrt{20}(126\rho^8 - 280\rho^6 + 210\rho^4 - 60\rho^2 + 5)\rho\cos(\theta)$
35	$\sqrt{20}(126\rho^8 - 280\rho^6 + 210\rho^4 - 60\rho^2 + 5)\rho\sin(\theta)$
36	$\sqrt{11}(252\rho^{10} - 630\rho^8 + 560\rho^6 - 210\rho^4 + 30\rho^2 - 1)$
37	$\sqrt{13}(924\rho^{12} - 2772\rho^{10} + 3150\rho^8 - 1680\rho^6 + 420\rho^4 - 42\rho^2 + 1)$

orthogonal and normalized within a circle of unit radius [Kim and Shannon, 1987]. In this process, the phase function $\phi_A(x, y)$ is represented in terms of an expansion in *Zernike polynomials* $z_k(\rho, \theta)$, where ρ is the radial coordinate within the unit circle, and θ is the polar angle.

Table 10.1 shows the Zernike polynomials for $1 \le k \le 37$. Note that each polynomial is of the form

$$z_k(\rho, \theta) = R_n^m(\rho) \cos m\theta \qquad (10.10.1\text{-}1)$$

where n and m are nonnegative integers. $R_n^m(\rho)$ is a polynomial of degree n and contains no power of ρ less than m. In addition, $R_n^m(\rho)$ is even (odd) when m is even (odd), respectively. The representation of $\phi_A(x, y) = \phi_A(\rho, \theta)$ can be written as [Born and Wolf, 1969]

$$\phi_A(\rho, \theta) = A_{00} + \frac{1}{\sqrt{2}} \sum_{n=2}^{\infty} A_{n0} R_n^0(\rho) + \sum_{n=1}^{\infty} \sum_{m=1}^{\infty} A_{nm} R_n^m(\rho) \cos m\theta \qquad (10.10.1\text{-}2)$$

The coefficients A_{nm} are determined for finite values of n and m by least squares. In turn, $\phi_A(\rho, \theta)$ can also be written as

$$\phi_A(\rho, \theta) = \sum_{k=1}^{K} w_k z_k(\rho, \theta) \qquad (10.10.1\text{-}3)$$

where K is an integer such as 37. The coefficients w_k are found by least squares. As each successive Zernike term is normal with respect to every preceding term, each term contributes independently to the mean-square aberration. This means the root-mean square error $\overline{\phi_A}$ due to aberrations can be written as

$$\overline{\phi_A} = \left[\sum_{k=K+1}^{\infty} w_k^2 \right]^{\frac{1}{2}} \qquad (10.10.1\text{-}4)$$

Note that the Zernike representation of aberrations is valid when the exit pupil is circular. Otherwise, the Zernike polynomials are not orthogonal. In some cases, such as aberrations due to atmospheric phase disturbances, the Zernike polynomial representation does not easily give a satisfactory representation with a finite number of terms.

11

Optical Devices Based on Wave Modulation

11.1 INTRODUCTION

Recording and generation of optical waves can be achieved by a number of technologies, using wave modulation algorithms. The oldest, most used technology is the *photographic film*. More recently, *spatial light modulators* have been developed in order to synthesize or control a wave front by optical or electrical control signals in real time.

Another approach is the use of solid state and similar technologies to fabricate *diffractive optical elements*, which control light through diffraction rather than refraction. An exciting development is the combination of refractive and diffractive optical elements in a single device to achieve a number of novel properties such as reduction of aberrations.

In previous chapters, analysis of optical systems was undertaken mostly in terms of linear system theory and Fourier transforms. The same knowledge base will be used in this chapter and succeeding chapters to synthesize optical elements for specific tasks.

This chapter consists of seven sections. Photographic films and plates are the most well-known devices for recording, and their properties especially for coherent recording are described in Sections 11.2 and 11.3. The physical mechanisms for the modulation transfer function of such media are discussed in Section 11.4. Bleaching is an important technique for phase modulation with photographic films and plates, and it is described in Section 11.5.

The implementation of diffractive optical devices discussed in detail in Chapters 15 and 16 is usually done with other technologies, especially the ones used in VLSI and integrated optics. Fundamentals of such devices are covered in Section 11.6. A particular implementation technology is e-beam lithography and reactive ion etching. It is described in Section 11.7.

11.2 PHOTOGRAPHIC FILMS AND PLATES

Photographic film is a very low-cost optical device for detecting optical radiation, storing images, and spatially controlling light [Goodman]. It is made up of an emulsion containing light-sensitive silver halide (usually AgBr) particles. The

Diffraction, Fourier Optics and Imaging, by Okan K. Ersoy
Copyright © 2007 John Wiley & Sons, Inc.

emulsion is sandwiched between a protective layer and a base consisting of acetate or mylar for films and glass for plates. In black-and-white photographic film, there is typically one layer of silver grains. In color film, there are at least three such layers. Dyes added to the silver grains make the crystals sensitive to different colors.

Modulation of light occurs as follows:

1. If a photon is absorbed by a silver halide grain, an electron–hole pair is generated within the grain.
2. A generated electron is in the conduction band and hence moves around in the crystal and may become trapped at a crystal dislocation.
3. A trapped electron attracts a mobile silver ion. This results in a single atom of metallic silver with a lifetime of the order of a few seconds.
4. Several additional silver atoms may be formed at the same site. At least four silver atoms called *silver speck* are needed for the development process.

The development is done in a chemical bath. The developer acts on the silver specks and causes the entire crystal to be reduced to metallic silver in regions where enough light has been absorbed. Those regions that did not turn to metallic silver will eventually do so and hence must be removed by a process called *fixing*. This involves immersing the transparency in a second chemical bath.

The terms used in connection with the usage of photographic films and their definitions are discussed below.

Exposure $E(x,y)$

$$E(x, y) = I_e(x, y)T \tag{11.1-1}$$

where

$I_e(x, y)$ = intensity incident on the film during exposure
T = exposure time
$E(x, y)$ is usually in units of mJ/cm^2.

Intensity Transmittance $\alpha(x,y)$

$$\alpha(x, y) = \text{local average}\left(\frac{I(x, y)}{I_e(x, y)}\right) \tag{11.1-2}$$

where

$I(x, y)$ = intensity transmitted by the transparency after development
$I_i(x, y)$ = incident intensity

Photographic Density

Silver mass per unit area is approximately proportional to the *photographic density* D, which is defined as the logarithm of the reciprocal of the intensity transmittance of a photographic transparency. Thus, D is given by

$$D = \log_{10} \frac{1}{\alpha} \tag{11.1-3}$$

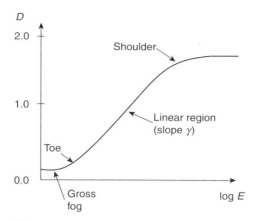

Figure 11.1.   The Hurter–Driffield curve for photographic emulsion.

This can also be written as

$$\alpha = 10^{-D} \tag{11.1-4}$$

Hurter–Driffield (H & D) Curve
The H & D curve is the plot of D *versus* $\log E$. A typical H & D curve for a negative film is shown in Figure 11.1. Note that the density is essentially constant for very low exposures and very high exposures. In between, there is the linear region, which is the most commonly used region in photography.

Gamma
The *gamma* γ of the emulsion is the slope of the H & D curve in the linear region. A *high-contrast film* has a large gamma (typically 2 or 3), and a *low-contrast film* has a small gamma (1 or less). γ is positive for a negative film and negative for a positive film.

11.3   TRANSMITTANCE OF LIGHT BY FILM

In incoherent and coherent optical systems, film is often used to modulate the transmittance of light. How this is achieved depends on whether coherent or incoherent light is used. Both these cases are discussed below.

Incoherent Light
In the linear region of the H & D curve, which is assumed to be used, the density can be written as [Goodman]

$$D = \gamma \log_{10}(E) - D_0 = \log_{10}\left(\frac{1}{\alpha}\right) \tag{11.3-1}$$

or

$$\log_{10}(\alpha) = -\gamma \log_{10}(E) + D_0 \tag{11.3-2}$$

As E equals $I_e T$, we obtain

$$\alpha = K I^{-\gamma} \tag{11.3-3}$$

where $K = 10^{D_0} T^{-\gamma}$ is a positive constant. Equation (11.3-3) shows that transmittance depends nonlinearly on the incident intensity and decreases as increases for positive γ.

EXAMPLE 11.1 Show that two negative films in tandem results in transmittance I increasing with incident intensity.
Solution: The first film has transmittance given by

$$\alpha_1 = K_1 I_1^{-\gamma_1}$$

where I_1 is the intensity of the illuminating beam.

Suppose the first film is placed in contact with the second unexposed film and illuminated with intensity I_2. The intensity incident on the second emulsion equals $\alpha_1 I_2$. The intensity transmittance α_2 can be written as

$$\alpha_2 = K_2(\alpha_1 I_2)^{-\gamma_2} = K I_1^{\gamma_1 \gamma_2}$$

where

$$K = K_1 K_2$$

It is seen that the overall transmittance α_2 increases with I_1.

Coherent Light
With coherent light, phase modulation also becomes important. Phase modulation is caused by variations of the film or plate thickness. The amplitude transmittance of the film can be written as

$$\alpha_c(x, y) = \sqrt{\alpha(x, y)} e^{j\phi(x,y)} \tag{11.3-4}$$

where $\phi(x, y)$ is the phase due to thickness variation.

The phase term $e^{j\phi(x,y)}$ is usually undesirable. It is possible to remove it by using a *liquid gate*. This is schematically shown in Figure 11.2.

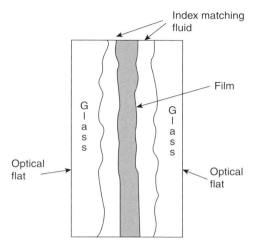

Figure 11.2. Liquid gate to remove thickness variations.

It consists of two pieces of glass, which are optically flat on the outer side. The transparency is placed between the two glass pieces with an index matching fluid (usually oil). With proper care, the phase modulation is removed, and the amplitude transmittance becomes

$$\alpha_{\rm c}(x, y) = \sqrt{\alpha(x, y)} = K\, I_{\rm e}^{-\frac{\gamma}{2}} \tag{11.3-5}$$

As $I_{\rm e} = A_{\rm e}^2$, $A_{\rm e}$ being the amplitude of the coherent light source used during exposure, Eq. (11.3-5) can be written as

$$\alpha_{\rm c}(x, y) = K A_{\rm e}^{-\gamma} \tag{11.3-6}$$

With coherent light, it is sometimes more suitable to graph $\alpha_{\rm c}$ *versus* the exposure E. Such a curve is shown in Figure 11.3.

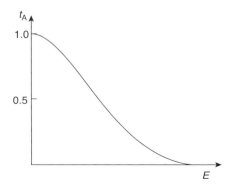

Figure 11.3. Example of the curve of amplitude transmittance *versus* exposure.

Figure 11.4. Full Kelly model of the photographic process.

It is also often the case that the film is used around an operating point whose exposure is E_0. In the linear region, α_c *versus* E can be approximated as

$$\alpha_c = \alpha_{c0} + \beta(E - E_0) = \alpha_{c0} + \beta'(\Delta A_c^2) \qquad (11.3\text{-}7)$$

where β is the slope of the curve, and ΔA_c^2 is the change of squared amplitude during exposure. β and β' are negative (positive) for negative (positive) film or plate, respectively.

11.4   MODULATION TRANSFER FUNCTION

The H & D curve represents a pointwise nonlinearity, transforming input light at each point on the emulsion with respect to its nonlinear curve. In addition, there are other processes that limit the spatial frequency response of a photographic emulsion.

Two major processes are light scattering during exposure and chemical diffusion during development. Both can be modeled as linear systems. However, light scattering is linear in the exposure whereas chemical diffusion is linear in the density. For this reason, the photographic process as a whole is modeled as shown in Figure 11.4. This is called the *Kelley model*. In order to simplify this model, the linear process due to chemical diffusion can be neglected. The simplified model is shown in Figure 11.5.

In order to measure the characteristics of the photographic film, a sinusoidal exposure pattern given by

$$E = E_0 + E_1 \cos(2\pi f x) \qquad (11.4\text{-}1)$$

can be applied. This is relatively easy to do by the holographic technique of mixing two mutually coherent plane waves on the emulsion. The effective exposure on the

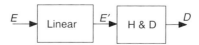

Figure 11.5. Simplified Kelly model of the photographic process.

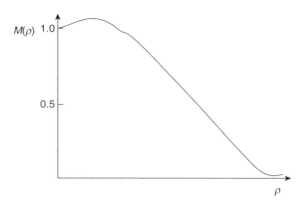

Figure 11.6. An example of MTF curve.

film can be shown to be [Goodman]

$$E' = E_0 + M(f)\cos(2\pi f x) \qquad (11.4\text{-}2)$$

where $M(f)$ is called the *modulation transfer function* (MTF).

The MTF is approximately circularly symmetric. Hence, f can be considered to be radial frequency. A typical MTF curve is shown in Figure 11.6. The cutoff frequency due to the MTF varies from emulsion to emulsion. It is like 50 cycles/mm (line pairs/mm) for a coarse film and 2000 cycles/mm (line pairs/mm) for a fine film.

11.5   BLEACHING

The ordinary development process leads to the creation of metallic silver, which absorbs light. It is often desirable to phase-modulate light so that large diffraction efficiency is obtained. This can be achieved by a process called *bleaching*.

The bleaching process replaces metallic silver in the emulsion by either an emulsion thickness variation or a refractive index variation within the emulsion. A thickness variation is obtained with a *tanning bleach* whereas a refractive index modulation is generated with a *nontanning bleach*.

The tanning bleach results in a surface relief as shown in Figure 11.7. This depends strongly on the spatial frequency content of the density pattern and acts as a

Figure 11.7. An example of surface relief generated by a tanning bleach.

bandpass filter. In other words, no relief is produced at very low or very high spatial frequencies. For a 15 μm thick emulsion, the peak thickness variations of the order of 1–2 μm occur at about 10 cycles/mm [Goodman].

The nontanning bleach results in internal refractive index changes within the emulsion. The resulting transparency represents a spatial modulation of phase through variations in the index of refraction.

11.6   DIFFRACTIVE OPTICS, BINARY OPTICS, AND DIGITAL OPTICS

Conventional optical devices such as lenses, mirrors, and prisms are based on refraction or reflection. By contrast, *diffractive optical elements* (DOEs), for example, in the form of a phase relief, are based on diffraction. The important advantages of DOEs are as follows:

1. A DOE can perform more than one function, for example, have multiple focal points, corresponding to multiple lenses on a single element. It can also be designed for use with multiple wavelengths.
2. DOEs are generally much lighter and occupy less volume than refractive and reflective optical elements.
3. With mass manufacturing, they are expected to be less expensive.

As DOEs are based on diffraction, they are highly dispersive, namely, their properties depend on wavelength. Hence they are more commonly used with monochromatic light. However, the dispersive properties can also be used to advantage. For example, light at one wavelength can be focused at one point, and light at another wavelength can be focused at another point in space. In addition, refractive and diffractive optical elements can be combined in such a way that wavelength dispersion can be removed, and goals such as removal of spherical aberration can be achieved. The topics of diffractive optics and design of DOEs are covered in detail in Chapters 15 and 16.

Diffractive optics, binary optics, and digital optics often refer to the same modulation techniques and technologies of implementation. In all cases, lithography and micromachining techniques are of paramount importance for the implementation of very novel devices for various applications. These are also technologies for VLSI and integrated optics as well as other nano/micro applications. At the other extreme, laser printers can be used to text designs at low resolution [Asakura, Nagashima, 1976].

Even though the word binary is used, quantization, for example, of phase involves a number of levels such as 2, 4, or 8 levels of quantization. Quantization of phase is usually achieved by modulating the thickness of a DOE in quantized steps. 2^N discrete levels of phase are usually used. For example, consider a blazed grating shown in Figure 11.9. Such a grating has 100% diffraction efficiency, meaning that

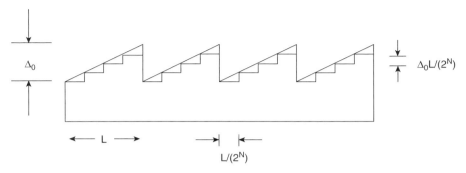

Figure 11.8.   Approximation of a blazed grating by $N = 2^M$ phase levels.

all the wave goes into the first order of diffraction if continuous modulation of thickness to generate the desired phase profile is undertaken. In binary optics, this is approximated by $N = 2^M$ levels, say, $M = 2$ is used as shown in Figure 11.8. E-beam and reactive ion etching techniques of fabricating such an element are discussed in the next section.

In recent years, another technology to fabricate DOE's has been micromachining by femtosecond laser pulses in transparent materials such as silica glass [Watanebe *et al.*, 2002], [Yin *et al.*, 2005]. The intensity of the femtosecond laser pulses in the focal region is high enough to modify the index of refraction while other regions remain unaffected. This phenomenon allows fabrication of a variety of diffractive and photonic devices in glass. A very interesting advantage here is the relative ease of guiding the focused beam within a volume.

11.7   E-BEAM LITHOGRAPHY

Lithography refers to a number of technologies used to print patterns onto a substrate. The most common application is printing the patterns of integrated circuits onto a semiconductor substrate. Advanced lithographic techniques include optical lithography, electron-beam (e-beam) lithography, extreme ultraviolet (EUVL) lithography, and x-ray lithography [Bowden *et al.*, 1983].

There are two basic steps involved in lithography. The first step is to create a desired pattern. The second is to transfer this pattern into the substrate. The quality of a lithographic method is measured primarily by its resolution. Resolution is the minimum feature size that can be obtained. Resolution is determined not only by the lithography method, but also by the ability of the resist to reconstruct the pattern.

E-beam lithography is a high-resolution technology to generate custom-made DOEs. It utilizes an intense and uniform electron source with a small spot size, high stability, and long life. The main attributes of e-beam lithography technology are as follows: (1) It is capable of very high resolution; (2) it works with a variety of materials and patterns; (3) it is slow compared to optical lithography; and (4) it is expensive. Combined with reactive ion etching (RIE), creation of DOEs with very fine resolution and high efficiency is possible.

As an example, consider a binary phase DOE consisting of a 2-D array of small apertures that can be assumed to be sampled points. At each such point, a phase shift of either 0 or 180 degrees is to be generated. At points where a 180 degree phase shift is desired, the substrate is etched to the required depth. The e-beam exposes a pattern onto an e-beam sensitive resist indicating where apertures are to be made, and RIE is used to etch the apertures to the desired depth. This can also be achieved by chemical processing, but RIE is a higher precision technology. This approach can be generalized to any number of phase quantization levels in powers of 2.

A typical e-beam machine is capable of generating a beam with a spot size around 10 nm. Therefore, it is capable of exposing patterns with lines of widths less than 0.1 μm or 100 nm. For binary phase DOEs, each square aperture has a size of approximately 1 μm × 1 μm. For such feature sizes around, a total pattern size of, say, 1 mm × 1 mm can be achieved, while maintaining high resolution. Larger patterns are generated by stitching together a number of such patterns together under interferometric control.

Dry etching systems are capable of generating very fine surface relief. For example, 180 degree phase change needed for binary phase DOEs occurs over a distance of half a wavelength. For visible light, that corresponds to a length of the

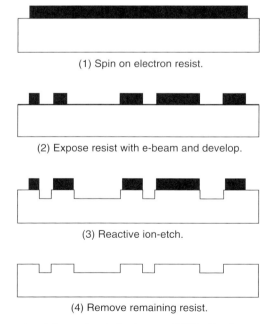

(1) Spin on electron resist.

(2) Expose resist with e-beam and develop.

(3) Reactive ion-etch.

(4) Remove remaining resist.

Figure 11.9. The experimental procedure to implement a DOE with e-beam and reactive ion etching systems.

order of half a micron. RIE makes it possible to induce surface relief of such small dimensions with very high accuracy.

11.7.1 DOE Implementation

The basic procedure for producing a binary phase hologram using e-beam lithography is illustrated in Figure 11.9. First, an e-beam sensitive resist is spin-coated onto a silicon substrate. Next, e-beam lithography is performed to write the desired pattern into the resist. The sample is then developed to remove the exposed resist if the resist is of the positive type such as polymethylmethacrylate (PMMA). Once the exposed resist is removed, reactive ion etching is used to etch the exposed apertures to the desired depth. Finally, the resist is removed by a chemical process. This procedure can be repeated M times to create multilevel DOEs with $N = 2^M$ levels. At each repetition, it is necessary to align the wafer with very high precision so that misplacements are minimized. This is usually done by interferometric control. The electron beam must also be optimally aligned so that the beam diameter is minimized at the surface of the sample. Astigmatism of the beam should also be minimized, as it is the primary limiting factor in performance. Once optimal alignment and focusing is achieved, pattern writing is processed with a CAD program.

12

Wave Propagation in Inhomogeneous Media

12.1 INTRODUCTION

In previous chapters, the index of refraction $n(x, y, z)$ was assumed to be constant, independent of position. In many applications, such as wave propagation in optical fibers, volume diffraction gratings, photorefractive media, and so on, $n(x, y, z)$ is actually not constant. Then, analysis becomes much more difficult, and often numerical methods are used to analyze wave propagation in such media. Some methods used are pseudospectral, such as the *beam propagation method (BPM)* discussed in Sections 12.4 and 12.5, whereas others are usually based on *finite difference* or *finite element methods*.

The BPM method discussed in this chapter is based on the paraxial wave equation for inhomogeneous media, is valid for propagation near the z-axis, and has several other restrictions. Its main advantage is that it is computed fast with the FFT and is sufficiently accurate in a large number of applications. For wide-angle propagation, other BPM algorithms exist, and they are usually based on the finite difference method. This is further discussed in Chapter 19 in the context of dense wavelength division multiplexing/demultiplexing for optical communications and networking.

This chapter consists of five sections. Section 12.2 discusses the Helmholtz equation for inhomogeneous media. The paraxial wave equation for homogeneous media discussed in Section 5.4 is generalized to inhomogeneous media in Section 12.3. The BPM as a prominent numerical method employing the FFT for wave propagation in inhomogeneous media is introduced in Section 12.4. A particular example of how the BPM is used in practice is the directional coupler illustrated in Section 12.5. This is an optical device consisting of two dielectric wave guides placed nearby so that an optical wave launched into one guide can be coupled into the other. Such devices are of common use in optical communications and networking. It is shown that the BPM gives results sufficiently accurate as compared with the rigorous coupled mode theory. As the coupled mode theory cannot be utilized in more complex designs, the BPM is usually the method of choice in the analysis and synthesis of such devices involving wave propagation in inhomogeneous media.

Diffraction, Fourier Optics and Imaging, by Okan K. Ersoy
Copyright © 2007 John Wiley & Sons, Inc.

12.2 HELMHOLTZ EQUATION FOR INHOMOGENEOUS MEDIA

The Helmholtz equation for homogeneous media in which the index of refraction is constant was derived in Section 4.2. In this section, this is extended to inhomogeneous media. Let $u(x, y, z, t)$ represents the wave field at position (x, y, z) and at time t in a medium with the refractive index $n(x, y, z)$ with the wavelength λ and angular frequency ω. According to Maxwell's equations, it must satisfy the scalar wave equation

$$\nabla^2 u(x, y, z, t) + n^2(x, y, z)k^2 u(x, y, z, t) = 0 \qquad (12.2\text{-}1)$$

where $k = \omega/c = 2\pi/\lambda$.

A general solution to the scalar wave equation can be written in the form of

$$u(x, y, z, t) = U(x, y, z) \cos(\omega t + \Theta(x, y, z)) \qquad (12.2\text{-}2)$$

where $U(x, y, z)$ is the amplitude and $\Theta(x, y, z)$ is the phase at position (x, y, z).

In complex notation, Eq. (12.2-2) is written as

$$u(x, y, z, t) = \mathrm{Re}[U(x, y, z) \exp(-j\omega t)] \qquad (12.2\text{-}3)$$

where $U(x, y, z)$ is the complex amplitude equal to $|U(x, y, z)| \exp(-j\Theta(x, y, z))$.

Substituting $u(x, y, z, t)$ from Eq. (12.2-3) into the wave equation (12.2-1) yields the *Helmholtz equation* in inhomogeneous media:

$$(\nabla^2 + k^2(x, y, z))U(x, y, z) = 0 \qquad (12.2\text{-}4)$$

where the position-dependent wave number $k(x, y, z)$ is given by

$$k(x, y, z) = n(x, y, z)k_0 \qquad (12.2\text{-}5)$$

12.3 PARAXIAL WAVE EQUATION FOR INHOMOGENEOUS MEDIA

The paraxial wave equation for homogenous media was discussed in Section 5.4. In this section, it is extended to inhomogeneous media. Suppose that the variation of index of refraction is given by

$$n(x, y, z) = \bar{n} + \Delta n(x, y, z) \qquad (12.3\text{-}1)$$

where $\bar{n}$ is the average index of refraction. The Helmholtz equation (12.2-4) becomes

$$[\nabla^2 + \bar{n}^2 k_0^2 + 2\bar{n}\Delta n k_0^2]U = 0 \qquad (12.3\text{-}2)$$

where $(\Delta n)^2 k^2$ term has been neglected.

If the field is assumed to be propagating mainly along the z-direction, it can be expressed as

$$U(x, y, z) = U'(x, y, z)e^{j\bar{k}z} \tag{12.3-3}$$

where $U'(x, y, z)$ is assumed to be a slowly varying function of z, and $\bar{k}$ equals $\bar{n}k_0$. Substituting Eq. (12.3-3) in the Helmholtz equation yields

$$\nabla^2 U' + 2j\bar{k}\frac{\delta}{\delta z}U' + (k^2 - \bar{k}^2)U' = 0 \tag{12.3-4}$$

As $U'(x, y, z)$ is a slowly varying function of z, Eq. (12.3-3) can be approximated as

$$\frac{\delta}{\delta z}U' = \frac{j}{2\bar{k}}\left[\frac{\delta^2 U'}{\delta x^2} + \frac{\delta^2 U'}{\delta y^2} + (k^2 - \bar{k}^2)U'\right] \tag{12.3-5}$$

Equation (12.3-5) is the *paraxial wave equation* for inhomogeneous media.

Equation (12.3-3) allows rapid phase variation with respect to the z-variable to be factored out. Then, the slowly varying field in the z-direction can be numerically represented along z that can be much coarser than the wavelength for many problems. The elimination of the second derivative term in z in Eq. (12.3-5) also reduces the problem from a second order boundary value problem requiring iteration or eigenvalue analysis, to a first order initial value problem that can be solved by simple "integration" along the z-direction. This also means a time reduction of computation by a factor of at least the number of longitudinal grid points.

The disadvantages of this approach are limiting consideration to fields that propagate primarily along the z-axis (i.e., paraxiality) and also placing restrictions on the index variation, especially the rate of change of index with z. The elimination of the second derivative with respect to z also eliminates the possibility for backward traveling wave solutions. This means devices for which reflection is significant cannot be sufficiently modeled.

12.4 BEAM PROPAGATION METHOD

The *beam propagation method* is a numerical method for approximately simulating optical wave propagation in inhomogeneous media. There are a number of different versions of the BPM method. The one discussed in this section has the constraints that reflected waves can be neglected and all refractive index differences are small [Marcuse].

The BPM can be derived by starting with the angular spectrum method (ASM) discussed in Section 4.3, which is for constant index of refraction, say, $\bar{n}$. The BPM

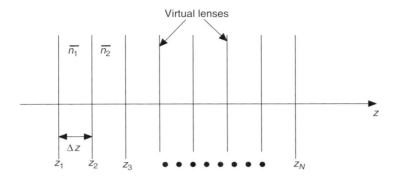

Figure 12.1. Modeling of the inhomogeneous medium for implementing BPM.

can be conceived as a model consisting of homogeneous sections terminated by virtual lenses as shown in Figure 12.1.

The BPM is derived by assuming that the wave is governed in small sections of width Δz by diffraction as in a homogeneous medium of index $\bar{n}$, but with different amounts of phase shift between the sections generated by virtual lenses, modeling the refractive index inhomogeneity. The additional phase shift can be considered to be due to $[n(x, y, z) - \bar{n}]$. By allowing Δz to be small, two steps are generated: (1) Wave propagation in a homogeneous medium of index $\bar{n}$, which is the average index in the current section, computed with the ASM and (2) the virtual lens effect, which is discussed below.

The effectiveness of this method depends on the choice of Δz at each step to be small enough to achieve the desired degree of accuracy within a reasonable amount of time.

12.4.1 Wave Propagation in Homogeneous Medium with Index $\bar{n}$

In each section, the medium is initially assumed to be homogeneous with index of refraction $\bar{n}$. Then, k also equals $\bar{k}$. The wave propagation in such a section, say, with initial input assumed to be at $z - \Delta z$ is computed by the ASM method of Section 4.3.

For $\bar{k}^2 \geq 4\pi^2(f_x^2 + f_y^2)$ so that the evanescent waves are excluded, $U(x, y, z)$ is expressed as

$$U(x, y, z) = \iint\limits_{-\infty}^{\infty} A(f_x, f_y, z - \Delta z) \exp(+j\sqrt{\bar{k}^2 - 4\pi^2(f_x^2 + f_y^2)}z)$$

$$\times \exp(j2\pi(f_x x + f_y y)) \mathrm{d}f_x \mathrm{d}f_y \tag{12.4-1}$$

where $A(f_x, f_y, z - \Delta z)$ is the initial plane wave spectrum given by Eq. (4.3-2).

The whole process can be written as

$$U(x, y, z) = F^{-1}\{F\{U(x, y, 0)\} \exp(j\mu z)\} \tag{12.4-2}$$

where F and F^{-1} indicate forward and inverse Fourier transforms, respectively, and μ is given by

$$\mu = \sqrt{\bar{n}^2 k^2 - 4\pi^2 (f_x^2 + f_y^2)} \tag{12.4-3}$$

To separate the rapid z-variations caused by large μ, it is rewritten in the equivalent form as follows:

$$\mu = \bar{k} - \frac{4\pi^2 (f_x^2 + f_y^2)}{\bar{k} + \sqrt{\bar{k}^2 - 4\pi^2 (f_x^2 + f_y^2)}} \tag{12.4-4}$$

Hence, Eq. (12.4-2) becomes

$$U(x, y, z)$$
$$= F^{-1} \left\{ F\{U(x, y, z - \Delta z)\} \exp\left(-j\Delta z \frac{4\pi^2 (f_x^2 + f_y^2)}{\bar{k} + \sqrt{\bar{k}^2 - 4\pi^2 (f_x^2 + f_y^2)}} \right) \right\} \exp(j\bar{k}\Delta z)$$
$$\tag{12.4-5}$$

12.4.2   The Virtual Lens Effect

This is the process of correcting the phase of the complex amplitude $U(x, y, z)$ at the end of each section with a phase function, which is chosen as

$$G(z) = \exp[jk(x, y, z)\Delta z] \tag{12.4-6}$$

$G(z)$ satisfies

$$(\nabla^2 + k^2(x, y, z))\mathbf{G}(z) = 0 \tag{12.4-7}$$

This choice of $G(z)$ yields a remarkably accurate phase change for waves traveling nearly parallel to the z-axis.

As $n(x, y, z) = \bar{n} + \Delta n(x, y, z)$ and the contribution of the homogeneous medium, $\exp(j\bar{k}\Delta z)$, has already been included in Eq. (12.4-5), the phase correction factor $G(z)$ can be written as

$$G(z) = \exp(j[k(x, y, z) - \bar{k}]\Delta z) \tag{12.4-8}$$

Finally, the wave field at z is expressed as

$$U(x, y, z) = F^{-1} \left\{ F\{U(x, y, z - \Delta z)\} \exp\left(-j\Delta z \frac{4\pi^2 (f_x^2 + f_y^2)}{\bar{k} + \sqrt{\bar{k}^2 - 4\pi^2 (f_x^2 + f_y^2)}} \right) \right\} G(z)$$
$$\tag{12.4-9}$$

In some applications such as in integrated optics, the 2-D BPM is used. This is achieved simply by skipping the y-variable related terms.

It is observed that Eq. (12.4-9) is similar to the ASM equation (12.4-5) except for the phase factor $G(z)$. Hence, it can be implemented fast with the FFT as discussed in Section 4.4.

It can be shown that the approximation involved in deriving Eq. (12.4-9) is equivalent to using the paraxial wave equation (12.3-5) derived in Section 12.3 [Feit and Fleck, 1978].

12.5   WAVE PROPAGATION IN A DIRECTIONAL COUPLER

As an example of simulation of wave propagation using the BPM, a directional coupler will be discussed. A directional coupler is an optical device used in optical communications, consisting of two parallel dielectric wave guides placed nearby each other so that an optical wave launched into one guide can be coupled into the other. The directional coupler structure is illustrated in Figure 12.2.

12.5.1   A Summary of Coupled Mode Theory

To describe waveforms that travel in a directional coupler, the theory of coupled mode equations is used [Saleh and Teich, 1991]. In this section, a brief summary of this theory is given so that the BPM simulation results can be compared with the analytical results obtained with the coupled mode theory.

The peak amplitude variations along z in wave guide i is defined as $a_i(z)$ for $i = 1, 2$. The beam intensities are given by $|a_i|^2$. The coupled mode equations show the power exchange between the two wave guides due to their proximity. They are given by

$$\frac{\partial a_1}{\partial z} = -jC_{21}\exp(j\Delta\beta z)a_2(z)$$

$$\frac{\partial a_2}{\partial z} = -jC_{12}\exp(-j\Delta\beta z)a_1(z)$$

(12.5-1)

where the coefficients β_1 and β_2 are the uncoupled propagation constants depending on the refractive index in each guide, and $\Delta\beta = \beta_1 - \beta_2$; C_{12} and C_{21} are the

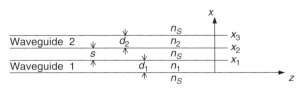

Figure 12.2.   The directional coupler structure.

coupling coefficients defined as follows:

$$C_{12} = \frac{1}{2}(n_1^2 - n^2)\frac{k_0^2}{\beta_2}\int\limits_{guide2} u_2(x)u_1(x)dx \tag{12.5-2}$$

$$C_{21} = \frac{1}{2}(n_2^2 - n^2)\frac{k_0^2}{\beta_1}\int\limits_{guide1} u_1(x)u_2(x)dx \tag{12.5-3}$$

The general solutions for can be derived as follows [Saleh and Teich, 1991]:

$$a_1(z) = \left\{A_1\left[\cos(\gamma z) - j\frac{\Delta\beta}{2\gamma}\sin(\gamma z)\right] - A_2\left[\frac{C_{21}}{j\gamma}\sin(\gamma z)\right]\right\}\exp\left(+j\frac{\Delta\beta}{2}z\right) \tag{12.5-4}$$

$$a_2(z) = \left\{A_1\left[\frac{C_{21}}{j\gamma}\sin(\gamma z)\right] + A_2\left[\cos(\gamma z) + j\frac{\Delta\beta}{2\gamma}\sin(\gamma z)\right]\right\}\exp\left(-j\frac{\Delta\beta}{2}z\right) \tag{12.5-5}$$

where A_i's are the initial peak amplitudes in the wave guides, and

$$\gamma^2 = \left(\frac{\Delta\beta}{2}\right)^2 + C^2 \tag{12.5-6}$$

$$C = \sqrt{C_{12}C_{21}} \tag{12.5-7}$$

12.5.2 Comparison of Coupled Mode Theory and BPM Computations

To show that the BPM is sufficiently accurate for simulation, the wave intensities in two different directional coupler structures were computed with the BPM and compared with the analytical results obtained from the coupled mode theory [Pojanasomboon, Ersoy, 2001]. According to Eqs. (12.5-4) and (12.5-5), the analytical results for the case of zero initial intensity in wave guide 2 ($A_2 = 0$) can be expressed as

$$|a_1(z)|^2 = |A_1|^2\left[\cos^2(\gamma z) - \left(\frac{\Delta\beta}{2\gamma}\right)^2\sin^2(\gamma z)\right] \tag{12.5-7}$$

$$|a_2(z)|^2 = |A_1|^2\left[\left(\frac{|C_{21}|}{\gamma}\right)^2\sin^2(\gamma z)\right] \tag{12.5-8}$$

where

$$\Delta\beta = \beta_1 - \beta_2, \quad \gamma^2 = \left(\frac{\Delta\beta}{2}\right)^2 + C^2 \tag{12.5-9}$$

$$C = \sqrt{C_{12}C_{21}} \tag{12.5-10}$$

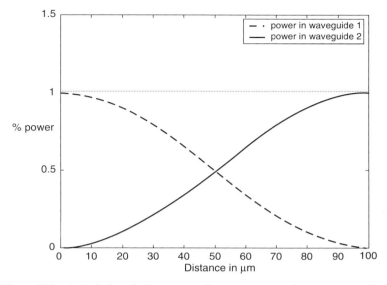

Figure 12.3. Analytical results for power exchange between synchronous wave guides.

12.5.2.1 Case 1: Synchronous Wave guides.
The directional coupler in this case has the same refractive index values in both guides, namely, $n_1 = n_2$. Because β_1 and β_2 depend on the refractive index in the guides, this case yields $\beta_1 = \beta_2$ or $\Delta\beta = 0$. The intensities in the wave guides are given by

$$|a_1(z)|^2 = |A_1|^2 \cos^2(\gamma z) \tag{12.5-6}$$

$$|a_2(z)|^2 = |A_1|^2 \left(\frac{|C_{21}|}{\gamma}\right)^2 \sin^2(\gamma z) \tag{12.5-7}$$

The parameters used in the simulations were $A_1 = 1$, $n_1 = n_2 = 1.1$, $d_1 = d_2 = s = 1$. The power exchange determined from the analytical expressions is illustrated in Figure 12.3. According to Eqs. (12.5-6) and (12.5-7), complete power exchange can be achieved at $z = \pi/2\gamma$.

In the BPM simulation, power at any z in each wave guide is calculated as

$$\int_{\text{guide}} |U(x, z)|^2 dx.$$

The results are shown in Figure 12.4. It is observed that the BPM simulation results with the same parameters yield the same general response as in Figure 12.3.

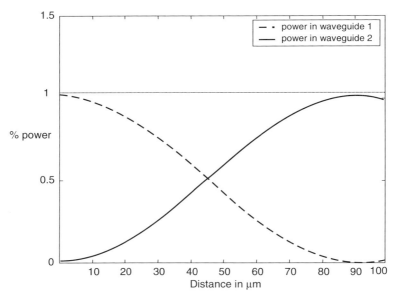

Figure 12.4. The BPM simulation of power exchange in directional coupler with synchronous wave guides.

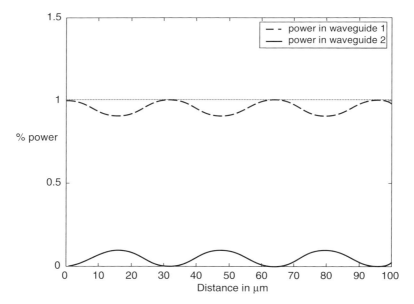

Figure 12.5. Analytical results for power exchange between nonsynchronous wave guides.

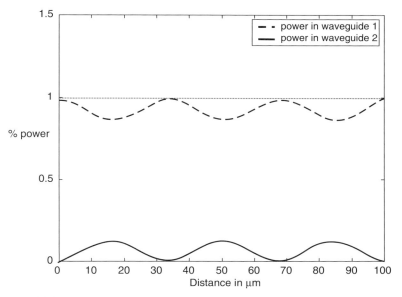

Figure 12.6. BPM simulation of power exchange between nonsynchronous wave guides.

12.5.2.2 Case 2: Nonsynchronous Wave guides. In the nonsynchronous case, the mismatch $n_1 \neq n_2$ makes $\Delta\beta = \beta_1 - \beta_2 \neq 0$. The power of the propagating wave in wave guide 2 is obtained as

$$|a_2(z)|^2 = |A_1|^2 \left[\left(\frac{|C_{21}|}{\gamma} \right)^2 \sin^2(\gamma z) \right] \tag{12.5-8}$$

The term $\gamma = \sqrt{(\Delta\beta/2)^2 + C^2}$ does not allow the quantity in the bracket to be equal to 1. Hence, complete power exchange between the two guides cannot be accomplished at any z as shown in Figure 12.5. The corresponding BPM simulation gives the same results as shown in Figure 12.6.

These results show that the BPM gives highly accurate results as compared with the coupled mode theory in these applications. In more complicated designs such as the nonperiodic grating-assisted directional coupler, the coupled mode theory cannot be used, and the BPM is the method of choice for reliable analysis and subsequent design [Pojanasomboon, Ersoy, 2001].

13

Holography

13.1 INTRODUCTION

Holography involves recording a modulated form of a desired (object) wave. It is also known as *wave front reconstruction.* The resulting device is called a *hologram.* Two major types of holography can be called *analog* and *digital holography.* Analog holography deals with continuous-space waves [Farhat, 1975], [Stroke, 1975]. Digital holography discussed in Chapters 15 and 16 results when the wave fields are sampled, and the information carried in amplitude and/or phase of the wave is coded with special algorithms. Digital holography is more commonly known as *diffractive optics.* Some other terminologies used for diffractive optics are *computer-generated holography, diffractive optical elements* (*DOEs*), and *binary optics.*

Holography was first discovered by Dennis Gabor in 1948, which is before the invention of the laser [Gabor]. Being a communications engineer, he recognized that the intensity resulting from the sum of a desired wave and a reference wave carries the information on both the amplitude and the phase of the object wave. After the invention of the laser as a coherent source, Gabor's ideas became a practical reality.

This chapter consists of six sections. The basic mechanism of holography also called coherent wave front recording and the Leith–Upatnieks hologram, the first type of hologram successfully implemented with a laser setup, are discussed in Section 13.2. A number of different types of holograms are described in Section 13.3.

As holography is a well-defined mathematical process, it can be simulated in the computer, and the results of holographic reconstruction can be displayed graphically. How this can be done is described in Section 13.4. Holographic imaging depends on a number of parameters such as wavelength and size. If these change, so do the properties of the reconstructed images. Analysis of holographic imaging and magnification as a function of these parameters are discussed in Section 13.5. As in optical imaging systems, aberrations limit the quality of holographic images. Different types of aberrations in the case of holographic imaging are discussed in Section 13.6.

Diffraction, Fourier Optics and Imaging, by Okan K. Ersoy
Copyright © 2007 John Wiley & Sons, Inc.

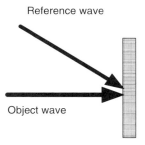

Figure 13.1. Geometry 1 for recording a hologram.

13.2 COHERENT WAVE FRONT RECORDING

Suppose that an object (desired) wave $U(x, y)$ is expressed as

$$U(x, y) = A(x, y)e^{j\phi(x,y)} \tag{13.2-1}$$

Another reference wave $R_r(x, y)$ is expressed as

$$R_r(x, y) = B(x, y)e^{j\psi(x,y)} \tag{13.2-2}$$

The two waves will be incident on a recording medium that is sensitive to intensity as shown in Figure 13.1 or Figure 13.2.

It is important that the waves are propagating at an angle with each other as shown. The intensity resulting from the sum of the two waves is given by

$$I(x, y) = |A(x, y)|^2 + |B(x, y)|^2 + 2A(x, y)B(x, y)\cos(\psi(x, y) - \phi(x, y)) \tag{13.2-3}$$

where the last term equals $AB^* + A^*B$ and includes both $A(x, y)$ and $\phi(x, y)$.

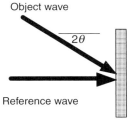

Figure 13.2. Geometry 2 for recording a hologram.

The transmission function of optical recording devices including photographic film is sensitive to intensity. We will assume that the sensitivity is linear in intensity. $B(x, y)$ will be assumed to be constant, equal to B, say, a plane wave incident perpendicular to the hologram as shown in Figure 13.2 and approximately a spherical wave far away from its origin. The transmission function of such a device can be written as

$$t(x, y) = C + \beta \left[|U|^2 + UR_r^* + U^*R_r \right] \tag{13.2-4}$$

where C and β are constants. $t(x, y)$ represents stored information. Now suppose that the generated hologram is illuminated by another reference wave R. The wave emanating from the hologram can be written as

$$Rt = U_1 + U_2 + U_3 + U_4 \tag{13.2-5}$$

where

$$U_1 = CR \tag{13.2-6}$$

$$U_2 = \beta |U|^2 R \tag{13.2-7}$$

$$U_3 = \beta R_r^* R U \tag{13.2-8}$$

$$U_4 = \beta R_r R U^* \tag{13.2-9}$$

Note that U_3 is the same as U except for a multiplicative term. Thus, U_3 appears coming from the original object and is called the *virtual image*. U_4 is the same as U^* except for a multiplicative term. U_4 converges toward an image usually located at the opposite side of the hologram with respect to U_3. This is illustrated in Figure 13.3. U_3 and U_4 are also called *twin images*. Usually U_3 forms an image to the left of the hologram, and U_4 forms an image to the right of the hologram. Then, these images are called virtual image and real image, respectively. However, which image is virtual and which image is real actually depend on the properties of the reference waves used during recording and reconstruction. These issues are further discussed in Section 13.6.

Suppose that R_r and R are the same, and they are constant, as in a plane wave perpendicular to the direction of propagation. Then, U_3 is proportional to U, and U_4 is proportional to U^*. In this case, U and U^* would be overlapping in space, and a

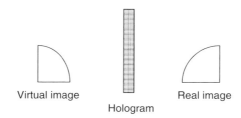

Virtual image   Real image
Hologram

Figure 13.3. Formation of virtual and real images.

viewer would not see the original image due to U only. This was the case when Gabor first discussed his holography method.

13.2.1   Leith–Upatnieks Hologram

Instead of a simple plane perpendicular plane reference wave, R can be chosen as

$$R(x, y) = De^{-j2\pi\alpha y} \qquad (13.2\text{-}10)$$

Now U_4 can be written as

$$U_4(x, y) = \beta De^{-j4\pi\alpha y} U^*(x, y) \qquad (13.2\text{-}11)$$

The term $e^{-j4\pi\alpha y}$ causes U_4 to propagate in a direction that does not overlap with the direction in which U_1, U_2, and U_3 propagate if α is chosen large enough.

Alternatively, the geometry shown in Figure 13.2 can be used. In this case, the reference wave illumination is normally incident on the hologram, but the wave coming from the object is at an average angle 2θ with the z-axis. The reference wave equals D, a constant, whereas the object wave can be written as

$$U(x, y) = A(x, y)e^{-j2\pi\alpha y} \qquad (13.2\text{-}12)$$

where $A(x, y)$ is now complex and contains the information-bearing part of the phase of $U(x, y)$. Further analysis gives the same results as in Eqs. (13.6)–(13.9). We note that U_3 is proportional to $A(x, y)e^{-j2\pi\alpha y}$ whereas U_4 is proportional to $A^*(x, y)e^{j2\pi\alpha y}$. Hence, they propagate in different directions.

EXAMPLE 13.1 Determine the minimum angle $2\theta_{\min}$ so that all the wave components are separated from each other at a distance sufficiently far away from the hologram.

Solution: At a sufficient distance from the hologram, for example, in the Fraunhofer region, the wave components will be similar to their Fourier spectra. Denoting the Fourier transform by FT, we have

$$\mathrm{FT}[U_1] = C_1\delta(f_x, f_y)$$
$$\mathrm{FT}[U_2] = \beta_1 A_F * A_F$$
$$\mathrm{FT}[U_3] = \beta_2 A_F(f_x, f_y - \alpha)$$
$$\mathrm{FT}[U_4] = \beta_3 A_F(-f_x, -f_y - \alpha)$$

where $C_1, \beta_1, \beta_2, \beta_3$ are constants, $*$ indicates autocorrelation, and A_F is the FT of $A(x, y)$.

Suppose that the bandwidth of $A(x, y)$ equals $f_{\max}$ in cycles/mm. $\mathrm{FT}[U_1]$ is simply a delta function at the origin. $\mathrm{FT}[U_2]$ has a bandwidth of $2f_{\max}$ because it is proportional to the autocorrelation of A_F with itself. U_3 is centered at α, and U_4 is

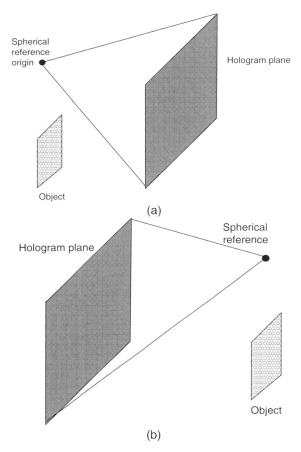

Figure 13.4. Two geometries for lensless Fourier transform holography.

centered at $-\alpha$ along the f_y direction. These relationships are shown in Figure 13.4. In order to have no overlap between the terms, we must satisfy

$$\alpha \geq 3f_{max}$$

or

$$2\theta_{min} = \sin^{-1}(3\lambda f_{max})$$

13.3 TYPES OF HOLOGRAMS

There are many types of holograms. Transmission holograms transmit light such that the information is viewed through the transmitted light. With reflection holograms, the information is viewed as a result of reflection from the hologram. Most types of holograms discussed below are transmission holograms.

13.3.1 Fresnel and Fraunhofer Holograms

Fresnel and Fraunhofer approximations were studied in Chapter 5. A hologram is a *Fresnel hologram* if the object to be reconstructed is in the Fresnel region with respect to the hologram. A hologram is a *Fraunhofer hologram* if the object to be reconstructed is in the Fraunhofer region with respect to the hologram.

13.3.2 Image and Fourier Holograms

In *an image hologram*, the middle of the object to be reconstructed is brought to focus on the plane of the imaging device. In a *Fourier hologram*, the plane of the imaging device is the same as the plane at which the Fourier transform of the object transmittance is generated. A Fourier hologram looks like a modulated diffraction grating, and such holograms typically have less need for a high-resolution recording device.

A variant of the Fourier hologram is the *lensless Fourier transform hologram*. There are two possible geometries for recording such a hologram as shown in Figure 13.4. The main property of such a hologram is that the focus of the spherical reference wave is on the same plane as the object image, either virtual (Figure 13.4(a)) or real (Figure 13.4(b)). Suppose that both the object wave and the reference wave are spherical waves with their sources on the same plane. On the hologram plane, the pattern of intensity has the same type of fringe patterns as in a real Fourier hologram. Hence the name lensless Fourier transform hologram.

As compared with Fourier and lensless Fourier transform holograms, Fresnel holograms have higher density fringe patterns and therefore require higher resolution recording devices. For this reason, Fourier holograms are usually preferred when recording with electronic recording devices.

13.3.3 Volume Holograms

Volume holograms are recorded in a thick medium. A major advantage gained is that the hologram is wavelength selective and thereby also works with white light after recording. Such a hologram also has very high diffraction efficiency.

In order to explain how a volume hologram is generated, the simple case of the interference of two plane waves in a thick medium will be discussed. One is the object wave with wave vector $\mathbf{k}_0$, and the other one is the reference wave with wave vector $\mathbf{k}_r$. They can be written as

$$U_0(\mathbf{r}) = A_0 e^{-j\mathbf{k}_0 \bullet \mathbf{r}} \qquad (13.3\text{-}1)$$

$$U_r(\mathbf{r}) = A_r e^{-j\mathbf{k}_r \bullet \mathbf{r}} \qquad (13.3\text{-}2)$$

The intensity due to the interference of the two waves in the medium can be written as

$$I(\mathbf{r}) = |U_0(\mathbf{r}) + U_r(\mathbf{r})|^2$$
$$= I_0 + I_r + 2A_0A_r \cos(\mathbf{k}_s \bullet \mathbf{r}) \qquad (13.3\text{-}3)$$

where

$$I_0 = |A_0|^2 \qquad (13.3\text{-}4)$$

$$I_r = |A_r|^2 \qquad (13.3\text{-}5)$$

$$\mathbf{k}_s = \mathbf{k}_0 - \mathbf{k}_r \qquad (13.3\text{-}6)$$

Equation 13.3 represents a sinusoidal pattern with a period equal to $p = 2\pi/|\mathbf{k}_s|$.

Consider the case of $\mathbf{k}_r$ pointing along the z-direction, and $\mathbf{k}_0$ being at an angle θ with the z-axis, as shown in Figure 13.5. Then, the period can be written as

$$p = \frac{\lambda}{2 \sin(\theta/2)} \qquad (13.3\text{-}7)$$

The resulting pattern recorded in a thick emulsion represents a volume grating. If the same reference wave is used during reconstruction, it can be shown that the object wave gets reconstructed perfectly provided that the reference wave is incident at the same angle to the hologram as before, and its wavelength is also the same as before [Saleh and Teich, 1991]. If the wavelength of the reference wave changes, the object wave does not get reconstructed.

If white light is used during reconstruction, the component of the white light that is at the same wavelength as the original reference wave provides the reconstruction

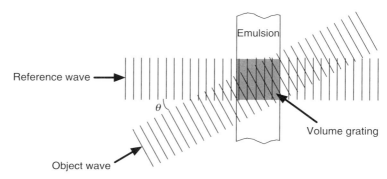

Figure 13.5.   Interference pattern of two plane waves generating a volume grating in a thick emulsion.

of the object wave. This is a very useful feature because white light can be used for reconstruction.

13.3.4   Embossed Holograms

Embossing is the process used in replicating compact Disks and DVDs with precision of the order of an optical wavelength. The same process is used to replicate a holographic recording in the form of a surface relief hologram. In this way, surface relief holograms are recorded in photoresists or photothermoplastics. Cheap mass reproduction leads to their wide usage in a number of markets, for example, as security features on credit cards or quality merchandise.

The first step in this process is to make a hologram with a photoresist or photothermoplastic. A metal master hologram is next made from the photoresist hologram by electroforming, leading to a thin layer of nickel on top of the hologram. The layer of nickel is next separated from the photoresist master and forms the metal submaster, which can be easily duplicated a number of times. The resulting metal submasters are used in the embossing process to replicate the hologram many times. For this purpose, the metal submaster is heated to a high temperature and stamped into a material such as polyester. The shape is retained when the film is cooled and removed from the press. The embossed pattern is usually metallized with aluminum to create the final reflection hologram.

13.4   COMPUTER SIMULATION OF HOLOGRAPHIC RECONSTRUCTION

The holographic equations can be computed, and the images can be reconstructed from a hologram existing in the computer, for example, by using the NFFA discussed in Chapter 7. Considering the most important terms in Eq. (13.2-3), a hologram can be modeled as

$$h(x, y) = u(x, y) + u^*(x, y) + c \qquad (13.4\text{-}1)$$

where $u(x, y)$ is the object wave, and c is a constant. Equation (13.4-1) includes the twin images and a constant term resulting in a central peak.

Figure 13.6 shows the image of an object whose hologram is to be generated by using the inverse NFFA propagation to yield $u(x, y)$ with the following physical parameters:

$z = 80$ mm, $\lambda = 0.6328\,\mu$, hologram size: $400\,\mu \times 400\,\mu$, output size: 64.8 mm $\times\ 64.8$ mm

The resulting hologram is shown in Figure 13.7. The reconstruction from the hologram is obtained by using the forward NFFA propagation, and the result is shown in Figure 13.8.

Figure 13.6. The object to be reconstructed.

13.5 ANALYSIS OF HOLOGRAPHIC IMAGING AND MAGNIFICATION

The important terms for imaging are U_3 and U_4 given by Eqs. (13.2-8) and (13.2-9), respectively. The following will be assumed:

- The object and recording reference waves are spherical waves originating from the points $[x_r, y_r, z_r]$ and $[x_o, y_o, z_o]$, respectively.
- The spherical reconstruction wave may be different from the recording wave and its origin is at $[x_c, y_c, z_c]$.

Figure 13.7. The hologram of the object.

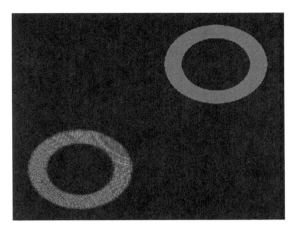

Figure 13.8. Holographic reconstruction of the twin images.

- The recording and reconstruction wavelengths are λ_1 and λ_2, respectively. λ_1 may be different from λ_2.
- The final hologram may be of a different size from the initial hologram generated by recording. A point on the initial hologram at $z = 0$ will be denoted by $[x, y]$ whereas the corresponding point on the final hologram will be denoted by $[x', y']$.

The phase due to the object wave at a hologram point $[x, y]$ can be written relative to the origin as

$$\phi_o(x, y) = \frac{2\pi}{\lambda_1} \left\{ \left[(x - x_0)^2 + (y - y_0)^2 + (z - z_0)^2 \right]^{1/2} - [x_o^2 + y_o^2 + z_o^2]^{1/2} \right\}$$

$$(13.5\text{-}1)$$

Keeping the first two terms of the Taylor series expansion, the last equation can be written for nonconstant terms as

$$\phi_o(x, y) = \frac{2\pi}{\lambda_1} \left[\frac{1}{2z_0} \left(x^2 + y^2 - 2xx_0 - 2yy_0 + z_o^2 \right) + \beta(x, y) \right] \qquad (13.5\text{-}2)$$

Where the third-order term $\beta(x, y)$ is given by

$$\beta(x, y) = -\frac{1}{8z^3} \begin{pmatrix} x^4 + y^4 + 2x^2y^2 - 4x^3x_0 \\ -4y^3y_0 - 4x^2yy_0 - 4xy^2x_0 \\ +6x^2x_o^2 + 6y^2y_o^2 + 2x^2y_o^2 \\ +2y^2x_o^2 + 8xyx_0y_0 - 4xx_o^3 \\ -4yy_o^3 - 4xx_0y_o^2 - 4xx_o^2y_0 \end{pmatrix} \qquad (13.5\text{-}3)$$

$\beta(x, y)$ is related to the aberrations discussed in Section 13.5. Similar equations can be written at the hologram point (x, y) relative to the origin for $\phi_c(x, y)$, the phase due to the reconstruction wave at wavelength λ_2, and $\phi_r(x, y)$, the phase due to the recording reference wave at wavelength λ_1.

With respect to Eqs. (13.2-8) and (13.2-9), the important phase terms for U_3 and U_4 can be written as

$$\phi_V = \phi_c + \phi_o - \phi_r \tag{13.5-4}$$

$$\phi_R = \phi_c - \phi_o + \phi_r \tag{13.5-5}$$

At this point, terms of order higher than 1 in $1/z$ or $1/z_c$ or $1/z_0$ are neglected. Writing ϕ_I for ϕ_3 or ϕ_4, we get

$$\phi_I(x, y) = \frac{2\pi}{\lambda_2} \frac{1}{2z_c} (x'^2 + y'^2 \pm 2x'x_r \pm 2y'y_r) + \frac{2\pi}{\lambda_1} \frac{1}{2z_0} (x^2 + y^2 - 2xx_o - 2yy_o)$$
$$- \frac{2\pi}{\lambda_1} \frac{1}{2z_r} (x^2 + y^2 \pm 2xx_r \pm 2yy_r) \tag{13.5-6}$$

where $+$ sign is for ϕ_3 and $-$ sign is for ϕ_4.

The hologram magnification M_h is defined by

$$M_h = \frac{x'}{x} = \frac{y'}{y} \tag{13.5-7}$$

and the wavelength ratio is given by

$$\mu = \frac{\lambda_2}{\lambda_1} \tag{13.5-8}$$

Then, Eq. (13.5-6) can be written as

$$\phi_I(x, y) = \frac{\pi}{\lambda_2} \left[\begin{array}{l} (x'^2 + y'^2)\left(\dfrac{1}{z_c} + \dfrac{\mu}{M_h^2 z_0} - \dfrac{\mu}{M_h^2 z_r}\right) \\[2mm] -2x\left(\dfrac{x_c}{z_c} + \dfrac{\mu x_o}{M_h z_0} \pm \dfrac{\mu x_r}{M_h z_r}\right) \\[2mm] -2y\left(\dfrac{y_c}{z_c} + \dfrac{\mu y_o}{M_h z_0} \pm \dfrac{\mu y_r}{M_h z_r}\right) \end{array} \right] \tag{13.5-9}$$

Equation (13.5-9) can be interpreted as the phase corresponding to another spherical wave originating from the point (x_I, y_I, z_I). The relevant phase for this wave within the Fresnel approximation is given by

$$\phi'_I = \frac{\pi}{\lambda_2 z_I} (x'^2 + y'^2 - 2x'x_I - 2y'y_I) \tag{13.5-10}$$

Setting $\phi_I' = \phi_I$ yields

$$z_I = \frac{M_h^2 z_c z_o z_r}{M_h^2 z_o z_r + \mu z_c z_r - \mu z_c z_o} \tag{13.5-11}$$

$$x_I = \frac{M_h^2 x_c z_o z_r + \mu M_h x_o z_c z_r \pm \mu M_h x_r z_c z_o}{M_h^2 x_c z_o z_r + \mu z_c z_r - \mu z_c z_o} \tag{13.5-12}$$

$$y_I = \frac{M_h^2 y_c z_o z_r + \mu M_h y_o z_c z_r \pm \mu M_h y_r z_c z_o}{M_h^2 y_c z_o z_r + \mu z_c z_r - \mu z_c z_o} \tag{13.5-13}$$

Which image is virtual and which image is real is determined by the signs of z_r and z_c, respectively. The transverse magnification is given by

$$M_t = \frac{\partial x_I}{\partial x_o} = \frac{M_h}{\left[1 + \dfrac{M_h^2 z_o}{\mu z_c} - \dfrac{z_o}{z_r}\right]} \tag{13.5-14}$$

$$M_a = \frac{\partial z_I}{\partial z_o} = -\frac{M_h^2}{\mu}\frac{d}{dz_o}\left[\frac{z_o}{1 - z_o\left(\dfrac{M_h^2}{\mu z_c} + \dfrac{1}{z_r}\right)}\right]$$

$$= -\frac{1}{\mu}\frac{M_h^2}{\left[1 + \dfrac{M_h^2 z_o}{\mu z_c} - \dfrac{z_o}{z_r}\right]^2} \tag{13.5-15}$$

$$= -\frac{1}{\mu}M_t^2$$

EXAMPLE 13.2 Determine μ so that $M_a = M_t$ in magnitude.
Solution: We set

$$|M_a| = \frac{1}{\mu}M_t^2 = M_t$$

Hence,

$$\mu = M_t \tag{13.5-16}$$

This means the changes in hologram size and the reference wave origin can be compensated by choosing a new wavelength satisfying Eq. (13.5-16). By the same token, if the wavelength is changed, the hologram size and/or the reference wave origin can be changed to make the two types of magnification equal to each other as much as possible.

13.6   ABERRATIONS

In discussing aberrations, it is more convenient to replace the rectangular coordinates x and y by the polar coordinates r and θ. Wave front aberrations are defined by the phase difference $\phi_A(r, \theta)$ between the ideal spherical wave front and the actual wave front with source at (x_I, y_I, z_I).

We will consider the aberrations of the image U_3 due to the third-order terms as in Eq. (13.5-3). The third-order terms due to ϕ_c, ϕ_o, ϕ_r are combined to give the third-order term in the actual wave front. The aberration wave function becomes

$$\phi_A(r, \theta) = \frac{2\pi}{\lambda_2} \left[\begin{array}{c} -\dfrac{1}{8} r^4 S + \dfrac{1}{2} r^3 (C_x \cos\theta + C_y \sin\theta) \\[2mm] -\dfrac{1}{2} r^2 (A_x \cos^2\theta + A_y \sin^2\theta + 2A_x A_y \cos\theta \sin\theta) \\[2mm] -\dfrac{1}{4} r^2 F + \dfrac{1}{2} r(D_x \cos\theta + D_y \sin\theta) \end{array} \right] \tag{13.6-1}$$

where the parameters belong to the following aberrations:

S: spherical aberration

C_x, C_y: coma

A_x, A_y: astigmatism

F: field curvature

D_x, D_y: distortion

They are given by the following equations:

$$S = \frac{1}{z_c^3} - \frac{\mu}{M_h^4 z_o^3} + \frac{\mu}{M_h^4 z_r^3} - \frac{1}{z_I^3} \tag{13.6-2}$$

$$C_x = \frac{x_c}{z_c^3} - \frac{\mu x_o}{M_h^3 z_o^3} + \frac{\mu x_r}{M_h^3 z_r^3} - \frac{x_I}{z_I^3} \tag{13.6-3}$$

$$C_y = \frac{y_c}{z_c^3} - \frac{\mu y_o}{M_h^3 z_o^3} + \frac{\mu y_r}{M_h^3 z_r^3} - \frac{y_I}{z_I^3} \tag{13.6-4}$$

$$A_x = \frac{x_c^2}{z_c^3} - \frac{\mu x_o^2}{M_h^2 z_o^3} + \frac{\mu x_r^2}{M_h^2 z_r^3} - \frac{x_I^2}{z_I^3} \tag{13.6-5}$$

$$A_y = \frac{y_c^2}{z_c^3} - \frac{\mu y_o^2}{M_h^2 z_o^3} + \frac{\mu y_r^2}{M_h^2 z_r^3} - \frac{y_I^2}{z_I^3} \tag{13.6-6}$$

$$F = \frac{x_c^2 + y_c^2}{z_c^3} - \frac{\mu(x_o^2 + y_o^2)}{M_h^2 z_o^3} + \frac{\mu(x_r^2 + y_r^2)}{M_h^2 z_r^3} - \frac{x_I^2 + y_I^2}{z_I^3} \tag{13.6-7}$$

$$D_x = \frac{x_c^3 + x_c y_c^2}{z_c^3} - \frac{\mu(x_o^3 + x_o y_o^2)}{M_h z_o^3} + \frac{\mu(x_r^3 + x_r y_r^2)}{M_h z_r^3} - \frac{x_I^3 + x_I y_I^2}{z_I^3} \tag{13.6-8}$$

$$D_y = \frac{y_c^3 + y_c x_c^2}{z_c^3} - \frac{\mu(y_o^3 + y_o x_o^2)}{M_h z_o^3} + \frac{\mu(y_r^3 + y_r x_r^2)}{M_h z_r^3} - \frac{y_I^3 + y_I x_I^2}{z_I^3} \tag{13.6-9}$$

EXAMPLE 13.3 Determine S, C_x, and A_x when the recording and reconstruction waves are plane waves.

Solution: In this case, z_r and z_c are infinite. Therefore S, c_x, and Ax become

$$S = \frac{\mu}{M_h^4 z_o^3}\left[\frac{\mu^2}{M_h^2} - 1\right]$$

$$C_x = \frac{\mu}{M_h^3 z_o^2}\left[\frac{x_c}{z_c}\frac{\mu}{M_h} - \frac{x_o}{z_o}\left(\frac{\mu^2}{M_h^2} - 1\right) + \frac{x_r}{z_r}\frac{\mu^2}{M_h^2}\right]$$

$$A_x = \frac{\mu}{M_h^2 z_o}\left[\frac{x_o^2}{z_o^2}\left(\frac{\mu^2}{M_h^2} - 1\right) - \frac{2\mu}{M_h}\frac{x_o}{z_o}\left(\frac{x_c}{z_c} + \frac{\mu x_r}{M_h z_r}\right) + \left(\frac{x_c}{z_c} + \frac{\mu x_r}{M_h z_r}\right)^2\right]$$

EXAMPLE 13.4 Determine C_x and A_x when $z_r = z_o$

Solution: After a little algebra, we find

$$C_x = \frac{\mu}{M_h}\left[-\frac{x_o}{z_o}\left(\frac{1}{M_h^2 z_o^2} - \frac{1}{z_c^2}\right) + \frac{x_r}{z_r}\left(\frac{1}{M_h^2 z_o^2} - \frac{1}{z_c^2}\right)\right]$$

$$A_x = \frac{\mu}{M_h}\left[-\frac{x_o^2}{z_o^2}\left(\frac{1}{M_h z_o} + \frac{\mu}{M_h z_c}\right) + \frac{2}{z_c}\frac{x_o}{z_o}\left(\frac{x_c}{z_c} + \frac{\mu x_r}{M_h z_r}\right) + \frac{x_r^2}{z_r^2}\left(\frac{1}{M_h z_o} - \frac{\mu}{M_h z_c}\right) - \frac{2}{z_c}\frac{x_c}{z_c}\frac{x_r}{z_r}\right]$$

14

Apodization, Superresolution, and Recovery of Missing Information

14.1   INTRODUCTION

In optical imaging, diffraction is the major phenomenon limiting the resolving power of the system. For example, consider imaging of a distant star through a telescope. Because of the distance of the star, the image should form on the focal plane, ideally as a point. Instead, the image intensity is proportional to the square of the Fourier transform of the exit pupil of the telescope. As a result, the image consists of a central maximum in radiance surrounded by other optima (secondary maxima or minima) which may be mistaken for other sources. For the same reason, a second and weaker star nearby may be missed altogether.

The techniques of *apodization* and *superresolution* have the goal of minimizing the effects of finite aperture size. Superresolution is also closely related to the topic of recovery of missing information.

The algorithms discussed in this chapter are with discretized signals so that they can be directly implemented in the computer. The algorithms can also be easily converted to analog representation if so desired.

This chapter consists of eighteen sections. Section 14.2 is on apodization or windowing techniques. Various windowing sequences are highlighted in this section. Two-point resolution, which is inherent in optical imaging devices, and the general guidelines for the recovery of signals are covered in Section 14.3.

In succeeding sections, the basic theory needed to develop recovery algorithms is first described. Section 14.4 describes the basic theory for contractions and shows under what conditions iterations with a mapping transformation of vectors in a linear vector space leads to a fixed point. This is followed in Section 14.5 by an iterative method of contractions called the method of constrained iterative signal restoration for signal recovery. Section 14.5 follows up with the same theme in the case of deconvolution with a method called iterative constrained deconvolution.

Diffraction, Fourier Optics and Imaging,   by Okan K. Ersoy

A subset of contractions is projections which are introduced in Section 14.7. There are a number of different types of projections. The method of projections on to convex sets (POCS) is the one which is relatively easier to handle, and it is covered in Section 14.8. This method is further discussed in succeeding chapters as a tool which is used within other algorithms for the design of diffractive optical elements and other diffractive optical devices. In Section 14.9, the Gerchberg–Papoulis algorithm is discussed as a special case of the POCS algorithm in which the signal and Fourier domains are used together with certain convex sets. Section 14.10 gives other examples of POCS algorithms.

Two applications that have always been of major interest are restoration from phase and restoration from magnitude. The second application is usually much more difficult to deal with than the first application. An iterative optimization method for restoration from phase is discussed in Section 14.11. Recovery from a discretized phase function with the DFT is the topic of Section 14.12.

A more difficult type of projection is generalized projections in which constraints do not need to be convex. In such cases, iterative optimization may or may not lead to a solution. This topic is covered in Section 14.13. Restoration from magnitude, also called phase retrieval, can usually be formulated as a generalized projection problem. A particular iterative optimization approach for this purpose is discussed in Section 14.14.

The remaining sections discuss other types of algorithms for signal and image recovery. Section 14.15 highlights the method of least squares and the generalized inverse for image recovery. The singular value decomposition for the computation of the generalized inverse is described in Section 14.16. The steepest descent algorithm and the conjugate gradient method for the same purpose are covered in Sections 14.17 and 14.18, respectively.

14.2 APODIZATION

Apodization is the same topic as *windowing* in signal processing. Without loss of generality, it is discussed in one dimension below.

The major effect of the exit pupil function is the truncation of the input wave field. In signal theory, this corresponds to the truncation of the input signal. The *Gibbs phenomenon* also called *"ringing"* is produced in a spectrum when an input signal is truncated. This means oscillations near points of discontinuity. This is also true when the spectrum is truncated, this time ringing occurring in the reconstruction of the signal.

In order to be able to simulate the behavior of the truncated wave field, it can be sampled, and the resulting sequence can be digitally processed. Let $z[n]$ be the truncated sequence due to the exit pupil, and $Z(f)$ be the DTFT of $z[n]$.

Also let the sampling interval of $u[n]$ be T_s. Truncation is equivalent to multiplying $u[n]$ by a rectangular sequence $\omega[n]$ in the form

$$u'[n] = u[n]\omega[n] \tag{14.2-1}$$

where

$$\omega[n] = \begin{cases} 1 & |n| \leq N \\ 0 & \text{otherwise} \end{cases} \qquad (14.2\text{-}2)$$

The DTFT of $\omega[n]$ (see Appendix B for a discussion of DTFT) is given by

$$W(f) = \sum_{n=-N}^{N} e^{-j2\pi f n T_s}$$
$$= \frac{\sin(\pi f T_s (2N+1))}{\sin(\pi f T_s)} \qquad (14.2\text{-}3)$$

$W(f)/N$ is plotted in Figure 14.1 for T_s equal to 1 and N equal to 6, 8, 15, and 25, respectively. Rather wide main lobe and sidelobes with considerable amplitudes are observed. As N increases, the width of the main lobe and the area under the sidelobes decrease, but the height of the main lobe remains equal to N.

By convolution theorem, Eq. (14.2-1) is equivalent in the DTFT domain to

$$U'(f) = U(f) * W(f) \qquad (14.2\text{-}4)$$

An optimal $W(f)$ is needed, with minimal sidelobes and a very narrow main lobe width to make $Z'(k)$ as close as possible to $U(f)$. Obviously, the optimal $W(f)$ is $\delta(f)$, which means no windowing. A compromise is searched for, in which the window is nonzero for $|n| \leq N$ and its $W(f)$ is as similar as possible to $\delta(f)$. In this search, the law of no free lunch exhibits itself. Sidelobes are removed at the expense of widening of the main lobe, and vice versa. Widening of the main lobe is equivalent to low-pass filtering, smoothing out the fast variations of $U(f)$.

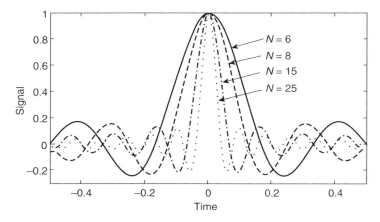

Figure 14.1. The DTFT of rectangular windows of length $N = 6$, 8, 15, and 25.

There are many windowing functions discussed in the literature. The first among these historically are *Fejer's arithmetic mean method* and *Lanczos' σ factors* [Lanczos]. In the analog domain, the windowing functions are also analog. Discrete-time windows are obtained from them by sampling. Some popular discrete-time windows are reviewed below.

14.2.1 Discrete-Time Windows

The windows discussed below are noncausal. The causal windows are obtained by replacing $\omega[n]$ by $\omega[n - N]$.

14.2.1.1 Bartlet Triangular Window

$$\omega[n] = \begin{cases} 1 - \dfrac{n}{N} & 0 \le n \le N \\ 1 + \dfrac{n}{N} & -N \le n \le 0 \\ 0 & \text{otherwise} \end{cases} \tag{14.2-5}$$

14.2.1.2 Generalized Cosine Windows

$$\omega[n] = \begin{cases} a + b\cos\left(\dfrac{\pi n}{N}\right) + c\cos\left(\dfrac{2\pi n}{N}\right) & -N \le n \le N \\ 0 & \text{otherwise} \end{cases} \tag{14.2-6}$$

where the constants a, b, c are as follows for three particular windows:

Window	a	b	c
Hanning	0.5	0.5	0
Hamming	0.54	0.46	0
Blackman	0.42	0.5	0.08

14.2.1.3 The Gaussian Window

$$\omega[n] = \exp\left(-\frac{1}{2}\left[\frac{\alpha n}{N}\right]^2\right) \tag{14.2-7}$$

where α is a parameter to be chosen by the user.

14.2.1.4 Dolph–Chebyshev Window.
This window is obtained by minimizing the main lobe width for a given sidelobe level. Its DFT is given by

$$W[k] = W[-k] = (-1)^n \frac{\cos\left[N\cos^{-1}\left[\beta\cos\left(\dfrac{\pi k}{N}\right)\right]\right]}{\cosh(N\cosh^{-1}(\beta))} \quad 0 \le n \le N-1 \tag{14.2-8}$$

where

$$\beta = \cosh\left[\frac{1}{N}\cosh^{-1}(10^{\alpha})\right] \tag{14.2-9}$$

The corresponding time window $\omega[n]$ is obtained by computing the inverse DFT of $W(n)$ and scaling for unity peak amplitude. The parameter α represents the log of the ratio of the main-lobe level to the sidelobe level. For example, α equal to 3 means sidelobes are 3 decades down from the main lobe, or sidelobes are 60 dB below the main lobe.

Among the windows discussed so far, the simplest window is the rectangular window. The Bartlett window reduces the overshoot at the discontinuity, but causes excessive smoothing. The Hanning, Hamming, and Blackman windows achieve a smooth truncation of $u[n]$ and gives little sidelobes in the other domain, with the Blackman window having the best performance. The Hamming and Gaussian windows do not reach zero at $|n| = N$.

Often the best window is the Kaiser window discussed next.

14.2.1.5 Kaiser Window

$$\omega[n] = \begin{cases} \dfrac{I_0\left(\alpha\left[1 - \left(\dfrac{n}{N}\right)^2\right]^{1/2}\right)}{I_0(\alpha)} & 0 \le |n| \le N \\ 0 & \text{otherwise} \end{cases} \tag{14.2-10}$$

where $I_0(x)$ is the zero-order modified Bessel function of the first kind given by

$$I_0(x) = \sum_{m=0}^{\infty}\left[\frac{\left(\frac{x}{2}\right)^m}{m!}\right]^2 \tag{14.2-11}$$

and α is a shape parameter which is chosen to make a compromise between the width of the main lobe and sidelobe behavior.

The Kaiser window is the same as the rectangular window when α equals 0. $\omega[n]$ for $N = 20$ and $\alpha = 2$, 4, 8, and 20 are shown in Figure 14.2. Increase of α at constant N reduces the sidelobes, but also increases the main lobe width. Increase of N at constant α reduces the main lobe width without appreciable change of sidelobes.

EXAMPLE 14.1 Rerun example 5.8 with apodization using the Hanning window along the x-direction. Compare the results along the x-direction and the y-direction.
Solution: The Fraunhofer result obtained when apodization with the Hanning window along the x-direction is used with the same square aperture as shown in Figure 14.3. It is observed that there are less intense secondary maxima along the

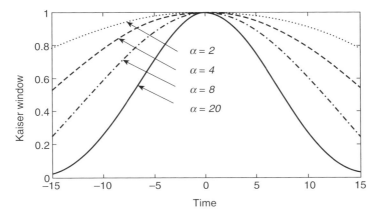

Figure 14.2. Kaiser windows for $N = 20$, and $\alpha = 2$, 4, 8, and 20.

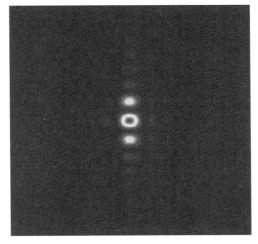

Figure 14.3. The intensity diffraction pattern from a square aperture in the Fraunhofer region when apodization with the Hanning window is used along the x-direction.

x-direction than along the y-direction. This is a consequence of the apodization along the x-direction.

14.3 TWO-POINT RESOLUTION AND RECOVERY OF SIGNALS

Consider the Fraunhofer diffraction pattern from a circular aperture of radius R. As discussed in Example 5.8, the intensity distribution due to a plane wave incident on the circular aperture has a central lobe whose radius is given by

$$\delta = \frac{0.61\,\lambda d_0}{R} \qquad\qquad (14.3\text{-}1)$$

where d_0 is the distance from the circular aperture. δ is called the *Rayleigh distance* and is a measure of the limit by which two point sources can be resolved. Note that δ is due to the exit pupil of the optical system, which is a *finite aperture*.

Finite aperture means the same thing as *missing information* since the wave field is truncated outside the aperture. Below the wave field is referred to as the signal since the topic discussed is the same as *signal recovery* in signal processing. In addition, the window function is generalized to be a *distortion* or *transformation operator D,* as commonly used in the literature on signal recovery. The coverage will still be in 1-*D* without loss of generality.

For a space-limited signal, the exact recovery of the signal is theoretically possible since the Fourier transform of such a signal is an analytic function [Stark, 1987]. If an analytic function is exactly known in an arbitrarily small region, the entire function can be found by analytic continuation. Consequently, if only a small part of the spectrum of such a signal is known exactly, the total spectrum and the signal can be recovered.

Unfortunately, especially due to measurement noise, the exact knowledge of part of the spectrum is often impossible. Instead, imperfect measurements and *a priori* information about the signal, such as positivity, finite extent, etc. are the available information.

The measured signal v can be written in terms of the desired signal u as

$$v = Du \qquad\qquad (14.3\text{-}2)$$

where D is a distortion or transformation operator. With known D, the problem is to recover z (in system identification, the problem is to estimate D when v and u are known).

The straightforward approach to solve for u is to find D^{-1} such that

$$u = D^{-1}v \quad v = Du \qquad\qquad (14.3\text{-}3)$$

Unfortunately, D is often a difficult transformation. For example, when more than one u corresponds to the same v, D^{-1} does not exist. Such is the case, for example, when D corresponds to a low-pass filter. Even if D^{-1} can be approximated, it is often ill-conditioned such that slight errors cause large errors in the estimation of u. The problem of recovery of u under such conditions is often referred to as the *inverse problem*. In the following sections, *signal recovery by contractions* and *projections* will be studied as potential methods for solving inverse problems. These methods are usually implemented in terms of *a priori* information in the signal and spectral domains.

14.4   CONTRACTIONS

In what is discussed below, signals are assumed to be vectors in a complete normed linear vector space. See Appendix B for a discussion of linear vector spaces. Vector norms are discussed in Section B.3. The Euclidian norms to be used below are as follows:

Continuous Signal

$$|u| = \left[\int_{-\infty}^{\infty} |z(t)|^2 dt \right]^{1/2} \tag{14.4-1}$$

Discrete Signal

$$|u| = \left[\sum_{n=-\infty}^{\infty} |u[n]|^2 \right]^{1/2} \tag{14.4-2}$$

In the following, the signal u will be assumed to be discrete and to belong to a finite-dimensional vector space S, which can be considered to be a subset of the real space R^n or the complex space C^n, n being the dimension of the space.

Let S_0 be a subspace of S. A mapping by an operator A of the vectors in S_0 is defined to be *nonexpansive* if

$$|Au - Av| \leq |u - v| \tag{14.4-3}$$

for every u and v which belong to S_0. The mapping is *strictly nonexpansive* if inequality holds whenever $u \neq v$.

Linear operators on u are usually represented as matrices. The norm of a matrix can be defined in a number of ways. For example, the spectral norm of the linear operator A is given by

$$|A| = \left[R(A^H A) \right]^{1/2} \tag{14.4-4}$$

where A^H denotes the complex conjugate transpose of A, and $R(B)$ is the spectral radius of B, which is the largest eigenvalue of B in absolute value. A matrix norm is said to be consistent with a vector norm if

$$|Au| \leq |A||x| \tag{14.4-5}$$

For example, the spectral norm given by Eq. (14.4-4) is consistent. For a nonexpansive mapping, A must satisfy

$$|A| \leq 1 \tag{14.4-6}$$

due to the inequality given by Eq. (14.4-3).

Another topic of importance is the fixed point(s) of the linear operator A. Consider

$$Au = u \qquad (14.4\text{-}7)$$

Any vector u^* which satisfies this equation is called a *fixed point* of A.

If the mapping A is strictly nonexpansive, and there are two fixed points u^* and v^*, then

$$|u^* - v^*| = |Au^* - Av^*| < |u^* - v^*| \qquad (14.4\text{-}8)$$

is a contradiction. It is concluded that there cannot be more than one fixed point in a strictly nonexpansive contraction.

14.4.1   Contraction Mapping Theorem

When the mapping by A is strictly nonexpansive so that

$$|Au - Av| \le |u - v| \qquad (14.4\text{-}9)$$

where $0 < \alpha < 1$, then, A has a unique fixed point in S_0.

Proof:
Let u_0 be an arbitrary point in S_0. The sequence $u_k = Au_{k-1}, k = 1, 2, \ldots$ is formed. The sequence $[u_k]$ belongs to S_0. We have

$$|u_{k+1} - u_k| = |Au_k - Au_{k-1}| \le \alpha |u_k - u_{k-1}| \qquad (14.4\text{-}10)$$

and, by repeated use of Schwarz inequality,

$$|u_{k+p} - u_k| = |u_{k+p} - u_{k+p-1} + u_{k+p-1} - u_{k+p-2} + u_{k+p+2} \cdots - u_k|$$

$$= \sum_{i=1}^{p} |u_{k+i} - u_{k+i-1}| \le (\alpha^{p-1} + \alpha^{p-2} + \cdots + 1)|u_{k+1} - u_k| \quad (14.4\text{-}11)$$

Since

$$|z_{k+1} - z_k| \le \alpha^k |z_1 - z_0| \qquad (14.4\text{-}12)$$

and

$$\alpha^{p-1} + \alpha^{p-2} + \cdots + 1 = \frac{1 - \alpha^p}{1 - \alpha} < \frac{1}{1 - \alpha} \,, \qquad (14.4\text{-}13)$$

the relation (14.4-11) can be written as

$$\left| u_{k+p} - u_k \right| \leq \frac{\alpha^k}{1 - \alpha} \left| u_1 - u_0 \right| \tag{14.4-14}$$

A sequence $[a_n]$ is called a *Cauchy sequence* if, given any $\in > 0$, there is a positive integer $N(\in)$ such that if $n, m > N(\in)$, it follows that

$$\left| a_n - a_m \right| < \in \tag{14.4-15}$$

A basic theorem of analysis is that a sequence converges to a fixed point if it is a Cauchy sequence.

The relation (14.4-14) shows that $[u_k]$ is a Cauchy sequence. It follows that

$$\lim_{k \to \infty} A u_k = A u^* = u^* \tag{14.4-16}$$

where u^* is the fixed point. In summary, if the mapping by A is strictly nonexpansive, the sequence created by $u_{k+1} = A u_k$ converges to the unique fixed point u^*.

14.5 AN ITERATIVE METHOD OF CONTRACTIONS FOR SIGNAL RECOVERY

Assume that there is a measured signal v, which is distorted in some fashion. From v, a more accurate signal is desired to be obtained through iterations. The iterated signal at iteration k is u_k, k being an integer for counting iterations. As k increases, u_k approaches the recovered signal.

The method to be discussed for this purpose is also known as the *method of constrained iterative signal restoration* [Schafer]. It is based on the following iterative equation:

$$u_{k+1} = F u_k \tag{14.5-1}$$

where u_ks are the successive approximations to the recovered signal, and F is an operator. F is obtained both from a distortion operator and *a priori* knowledge in the form of certain constraints. In general, F is not unique. A simple technique of estimating F is given below.

Prior knowledge can be expressed by a constraint operator:

$$C u \to u \tag{14.5-2}$$

For example, with a multidimensional signal u, $C[u]$ may be given as

$$C u[n] = \begin{cases} u[n] & 0 \leq u[n] \leq a \\ 0 & u[n] < 0 \\ a & u[n] > a \end{cases} \tag{14.5-3}$$

As another example, if $u(t)$ is an analog signal bandlimited by a system to frequencies below f_c, $Cu(t)$ can be written as

$$Cu(t) = \int_{-\infty}^{\infty} u(t) \frac{\sin 2\pi f_c(t - \tau)}{\pi(t - \tau)} \, d\tau \qquad (14.5\text{-}4)$$

Every estimate u_k of u should be constrained by C. Cu_k is interpreted as an approximation to u_{k+1}. The result Cu_k is further processed with a distortion operator D, if any, to yield DCu_k. Interpreting DCu_k as an estimate of v, the output error Δv is defined as

$$\Delta v = v - DCu_k \qquad (14.5\text{-}5)$$

The recovered signal error between the $(k + 1)$th and kth iterations can be defined as

$$\Delta u = u_{k+1} - Cu_k \qquad (14.5\text{-}6)$$

How does Δu and Δv relate? The simplest assumption is that they are proportional, say, $\Delta u = \lambda \Delta v$. Then,

$$Fu_k = u_{k+1} = Cu_k + \Delta u = Cu_k + \lambda(v - DCu_k) \qquad (14.5\text{-}7)$$

This can also be written as

$$Fu_k = \lambda v + Gu_k \qquad (14.5\text{-}8)$$

where

$$G = (I - \lambda D)C \qquad (14.5\text{-}9)$$

and I is the identity operator. The initial approximation to u can be chosen as $u_0 = \lambda v$.

If F is chosen contractive, u_k converges to a fixed-point u^*.

Equation (14.5-8) shows that

$$|Fu_k - Fu_n| = |Gu_k - Gu_n| \qquad (14.5\text{-}10)$$

Thus, if F is contractive, so is G. With $u_0 = \lambda v$, Eq. (14.5-8) becomes

$$Fu_k = u_{k+1} = u_0 + Gu_k \qquad (14.5\text{-}11)$$

The procedure at the kth iteration according to Eq. (14.5-11) is as follows:

1. Apply the constraint operator C on u_k, obtaining u'_k.
2. Apply the distortion operator D on u'_k, obtaining u''_k.
3. Compute u_{k+1} as $u_0 + \lambda(u_k - u''_k)$.

An example of this method is given below for deconvolution.

14.6   ITERATIVE CONSTRAINED DECONVOLUTION

Given v and h, *deconvolution* involves determining u when v is the convolution of u and h:

$$v = h * u \qquad (14.6\text{-}1)$$

When the iterations start with $u_0 = \lambda v$, Eq. (14.5-8) can be written as

$$u_{k+1} = \lambda v + (\delta - \lambda h) * C u_k \qquad (14.6\text{-}2)$$

where the constraint operator is simply the identity operator for the time being.
The FT of Eq. (14.6-2) yields

$$U_{k+1} = \lambda V + U_k - \lambda H U_k \qquad (14.6\text{-}3)$$

where the capital letters indicate the FTs of the corresponding lowercase-lettered variables in Eq. (14.6-2)
Eq. (14.6-3) is a first-order difference equation in the index k. It can be written as

$$U_{k+1} - (1 - \lambda H)U_k = \lambda V \qquad (14.6\text{-}4)$$

whose solution is

$$U_k = \frac{V}{H}[1 - \lambda H]^{k+1} u[k] \qquad (14.6\text{-}5)$$

where $u[k]$ is the 1-D unit step sequence. As $k \to \infty$, the solution becomes

$$U_\infty = \frac{V}{H} \qquad (14.6\text{-}6)$$

if

$$|1 - \lambda H| < 1 \qquad (14.6\text{-}7)$$

Equation (14.6-6) is the same result obtained with the inverse filter. The advantage of the interactive procedure is that stopping the iterations after a finite number of steps may yield a better result than the actual inverse filter.

Equation (14.6-7) shows that the procedure will break down if H equals 0 at some frequencies. Equation (14.6-7) can also be written as

$$\text{Re}[H] > 0 \qquad (14.6\text{-}8)$$

In order to avoid testing the validity of Eq. (14.6-8) at all frequencies, a better procedure involves convolving both sides of Eq. (14.6-1) with inverted h^* ($h^*[-m]$ in 1-D and $h^*[-m, -n]$ in 2-D with discrete signals). Then, we get

$$v' = x * h' \qquad (14.6\text{-}9)$$

where h' is a new impulse response sequence whose transfer function is given by

$$H' = |H|^2 \qquad (14.6\text{-}10)$$

and v' is the convolution of v with the inverted h^*.

In solving Eq. (14.6-9) by the iterative method, Eq. (14.6-8), namely, $\text{Re}[H'] > 0$ is automatically satisfied provided that $H \neq 0$ at any frequency.

Since the iterative solution converges toward the inverse filter solution, it may also end up in an undesirable behavior. For example, a unique result may not be obtained if $H \rightarrow 0$. The use of a constraint operator often circumvents this problem. For example, u may be assumed to be positive over a finite support. In 2-D, the constraint operator for this purpose is given by

$$Cu_k[m, n] = \begin{cases} u_k[m, n] & M_1 \leq m < M_2, \ N_1 \leq n \leq N_2 \\ u_k[m, n] \geq 0 \\ 0 & \text{otherwise} \end{cases} \qquad (14.6\text{-}11)$$

Letting $u'_k = Cu_k$, Eq. (14.6-2) in the DTFT domain becomes

$$U'_{k+1} = \lambda V + U'_k - \lambda H U'_k \qquad (14.6\text{-}12)$$

The iterations defined by Eq. (14.6-12) can be done by using the DFT. However, $u'_k = Cu_k$ is to be implemented in the signal domain. The resulting block diagram is shown in Figure 14.4.

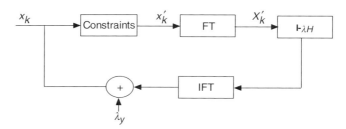

Figure 14.4.   The block diagram for iterative constrained deconvolution utilizing the Fourier transform.

In many applications the impulse response is shift variant. The iterative method discussed above can still be used in the time or space domain, but the algebraic equations in the transform domain are no longer valid.

EXAMPLE 14.2 A 2-D Gaussian blurring impulse response sequence is given by

$$h[m,n] = \exp\left[-\frac{m^2 + n^2}{100}\right]$$

The image sequence is assumed to be

$$u[m,n] = [\delta(m-24)\delta(n-32) + \delta(m-34)\delta(n-32)]$$

(a) Determine the blurred image $v = u * h$. Visualize h and v.
(b) Compute the deconvolved image with $\lambda = 2$, 65 iterations, and a positivity constraint.

Solution: (a) The convolution of u and h is given by [Schafer, Mersereau, Richards]

$$v[m,n] = \exp\left[-\frac{(m-24)^2 + (n-32)^2}{100}\right] + \exp\left[-\frac{(m-34)^2 + (n-32)^2}{100}\right]$$

Plots of h and v are shown in Figure 14.5(a) and (b), respectively.
(b) Figure 14.5(c) shows the deconvolution results after 65 iterations.

14.7   METHOD OF PROJECTIONS

A subset of contractions is projections. An operator P is called a *projection operator* or *projector* if the mapping Pu satisfies

$$|Pu - u| = \inf_{v \in S_0} |v - u| \qquad (14.7\text{-}1)$$

where $u \in S$, and S_0 is a subset of S. In the context of signal recovery, S_0 is the subset to which the desired signal belongs to. $\mathbf{g} = Pu$ is called the projection of u onto S_0. This is interpreted as the projection operator P generating a vector $\mathbf{g} \in S_0$ which is closest to $u \in S$. G is unique if S_0 is a *convex set*. A very brief review of a convex set is given below.

Let u_1 and u_2 belong to S_0. S_0 is convex, if for all λ, $0 \leq \lambda \leq 1$, $\lambda u_1 + (1 - \lambda)u_2$ also belongs to S_0. A convex set and a nonconvex set are visualized in Figure 14.6.

It can be shown that if S_0 is convex, then P is contractive [Youla]. Hence, it is desirable that S_0 is indeed convex. If so, the resulting method is called *projections onto convex sets*. We will consider this case first.

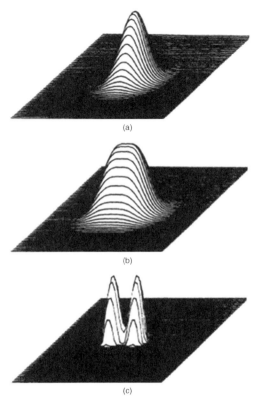

Figure 14.5.   Deconvolution of blurred 2-D signal: (a) Gaussian blurring function, (b) blurred signal, (c) recovered signal by deconvolution [Courtesy of Schafer, Mersereau, Richards].

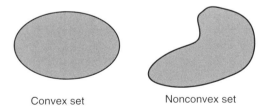

Convex set   Nonconvex set

Figure 14.6.   Examples of a convex set and a nonconvex set.

EXAMPLE 14.3 Consider the set S_3 of all real nonnegative functions $u(x, y)$ (such as optical field amplitude) that satisfy the energy constraint

$$|u|^2 = \iint |u(x, y)|^2 \mathrm{d}x\mathrm{d}y \leq E$$

Show that S_3 is convex.

Solution: Consider $u_1, u_2 \in S_3$. For $0 \le \lambda \le 1$. The Cauchy–Schwarz inequality to will be used to write

$$|\lambda u_1 + (1 - \lambda)u_2| \le \lambda|u_1| + (1 - \lambda)|u_2| \le \lambda E^{1/2}$$

Hence, the set is convex.

EXAMPLE 14.4 Consider the set S_4 of all real functions in S whose amplitudes must lie in the closed interval $[a,b]$, $a < b$, and $a \ge 0, b > 0$. Show that S_4 is convex.
Solution: For $0 \le \lambda \le 1$, we write

$$u_3 = \lambda u_1 + (1 - \lambda)u_2 \ge \lambda a + (1 - \lambda)a = a$$

for the lower bound, and

$$u_3 = \lambda u_1 + (1 - \lambda)u_2 \le \lambda b + (1 - \lambda)b = b$$

for the upper bound. Hence, $u_3 \in S_0$, and S_4 is convex.

14.8   METHOD OF PROJECTIONS ONTO CONVEX SETS

The method of POCS has been successfully used in signal and image recovery. The viewpoint is that every known property of an unknown signal u is considered as a constraint that restricts the signal to be a member of a closed convex set S_i. For M properties, there are M such sets. The signal belongs to S_0, the intersection of these sets:

$$S_0 = \bigcap_{i=1}^{M} S_i \tag{14.8-1}$$

The signal restoration problem is defined as projecting v, the corrupted version of u, onto S_0.

Let P_i denote the projector for the set S_i. Let us also define the operator T_i by

$$T_i = 1 + \lambda_i(P_i - 1) \tag{14.8-2}$$

If u_i^* is the fixed-point of P_i in S_i, we get

$$T_i u_i^* = u_i^* + \lambda_i(P_i u_i^* - u_i^*) = u_i^* \tag{14.8-3}$$

Thus, u_i^* is also the fixed point of T_i. It can also be shown that T_i is contractive if $0 < \lambda_i < 2$. λ_is are called the relaxation parameters. Note that $\lambda_i = 1$ makes T_i equal to P_i. Thus, T_i is a more general operator than P_i. Proper choice of λ_i usually speeds up convergence.

Let T be defined by

$$T = T_m T_{m-1} \dots T_1 \qquad (14.8\text{-}4)$$

The following can be shown to be true [Stark]:

1. The iterations defined by

$$u_0 \quad \text{arbitrary}$$

$$u_{k+1} = T u_k$$

   converge weakly to a fixed point in S_0. Weak convergence means $\lim_{k \to \infty} (u_k, w) = (u^*, w), \forall w \in S_0$ where (a, b) is the inner product of the vectors $\mathbf{a}$ and $\mathbf{b}$. Under certain conditions, the convergence is strong, meaning $\lim_{k \to \infty} |u_n - u^*| = 0$.
2. λ_i can be changed at each iteration. Let λ_{ik} denote the value of λ_i at iteration k. For every choice of λ_{ik}, provided that $0 < \lambda_{ik} < 2$, the iterations defined by

$$u_0 \quad \text{arbitrary}$$

$$u_{k+1} = T u_k$$

converge weakly to a fixed point u^*.

Quite often, the POCS algorithm is used when $M = 2$. This is especially true when one projection is in the signal domain and the other one in the transform domain. This case is shown in Figure 14.7.

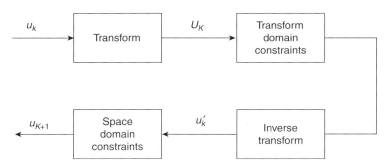

Figure 14.7. The POCS algorithm in the transform and signal domains.

There are two possible ways to estimate the optimal values of λ_{1k} and λ_{2k} for fast convergence. In the first case, λ_{1k} is first optimized, and T_1 is applied. This is followed by the optimization of λ_{2k} and the application of T_2. In general, it can be shown that for a linear projector $\lambda_{ik} = 1$, and $\lambda_{ik} \geq 0$ otherwise. In the second case, λ_{1k} and λ_{2k} can be optimized simultaneously so that $T = T_2 T_1$ results in minimum error [Stark].

14.9   GERCHBERG–PAPOULIS (GP) ALGORITHM

The original GP algorithm is a special case of the POCS algorithm, and uses projectors P_1 and P_2 in the signal and the Fourier domains, respectively [Gerchberg]. Hence

$$u_{k+1} = P_2 P_1 u_k \tag{14.9-1}$$

where P_1 and P_2 are defined as follows:

Let A be a region in R^2, and S_1 be the set of all functions in S that vanish almost everywhere (except, possibly, over a point set of measure zero) outside A. P_1 is defined by

$$P_1 u = \begin{cases} u & (x, y) \in A \\ 0 & \text{otherwise} \end{cases} \tag{14.9-2}$$

Let S_2 be the set of all functions whose Fourier transforms assume a prescribed function $V(f_x, f_y)$ over a closed region B in the Fourier plane. P_2 is defined by

$$P_2 u \leftrightarrow \begin{cases} U(f_x, f_y) & (f_x, f_y) \in B \\ V(f_x, f_y) & (f_x, f_y) \notin B \end{cases} \tag{14.9-3}$$

It can be shown that both S_1 and S_2 are convex sets.

14.10   OTHER POCS ALGORITHMS

In the GP algorithm, the two projection operators P_1 and P_2 are with respect to the convex sets S_1 and S_2 discussed above, respectively. Now, we introduce three more projection operators utilizing convex sets.

Consider the set S_3 introduced in Example 14.3. Assume u is complex. It can be written as

$$u = u_R + j u_I \tag{14.10-1}$$

where u_R, u_I denote the real and complex parts of u, respectively. x_R^+ is also defined as

$$u_R^+ = \begin{cases} u_R & u_R > 0 \\ 0 & \text{otherwise} \end{cases} \tag{14.10-2}$$

The projection operator P_3 is defined by

$$P_3 u = \begin{cases} 0 & u_R < 0 \\ u_R^+ & u_R > 0, E_R^+ \leq E \\ \sqrt{E/E_R^+}\, u_R^+ & u_R > 0, E_R^+ > E \end{cases} \tag{14.10-3}$$

where E_R^+ is given by

$$E_R^+ = \iint [u_R^+(x, y)]^2 dx dy \tag{14.10-4}$$

and E is the estimated or known energy of the desired image. E_R^+ is bounded.

Consider the set S_4 defined in Example 14.4. The projection operator P_4 is defined by

$$P_4 u = \begin{cases} a & u(x, y) < a \\ u(x, y) & a \leq u(x, y) \leq b \\ b & u(x, y) \geq b \end{cases} \tag{14.10-5}$$

Finally, consider the subset S_5 of all real nonnegative functions in S. S_5 can be shown to be convex. The projection operator P_5 is defined by

$$P_5 u = \begin{cases} u_R & u_R \geq 0 \\ 0 & \text{otherwise} \end{cases} \tag{14.10-6}$$

where $u = u_R + j u_I$ as before.

Some examples of the applications of the theory presented above is given in the following sections.

14.11   RESTORATION FROM PHASE

For the sake of simplicity, the 1-D case will be discussed below. The signal $u(x)$ will be assumed to have a compact support $[-a, a]$. The known phase of the FT of $u(x)$ is $\phi(f)$. The amplitude of the signal is to be estimated. It can be shown that the necessary condition for $u(x)$ to be completely specified by $\phi(f)$ within a scale factor is that there is no point of symmetry x_0 such that

$$u(x_0 + x) = \pm u(x_0 - x) \tag{14.11-1}$$

The two projections of interest can be written as follows:

$$P_1 u = \begin{cases} u(x) & |u(x)| \leq a \\ 0 & \text{otherwise} \end{cases} \tag{14.11-2}$$

$$P_2 u \leftrightarrow \begin{cases} A(f)\cos[\phi(f) - \psi(f)]e^{j\phi(f)} & |\phi(f) - \psi(f)| < \dfrac{\pi}{2} \\ 0 & \text{otherwise} \end{cases} \tag{14.11-3}$$

where $A(f)$ is the FT amplitude, and $\psi(f)$ is the FT phase. It is observed that $A(f)\cos(\phi(f) - \psi(f))$ is the projection of $U(f)$ in the direction given by $\phi(f)$. In a number of papers, the relative directions given by $\phi(f)$ and $\psi(f)$ are not considered, and P_2 is expressed as

$$P_2 u \leftrightarrow A(f)e^{j\phi(f)} \tag{14.11-4}$$

It is possible that $\phi(f)$ is not known for all f. Suppose that $\phi(f)$ is known for f in a set S. Then P_2 can be expressed as

$$P_2 u \leftrightarrow \begin{cases} A(f)\cos[\phi(f) - \psi(f)]e^{j\phi(f)} & |\phi(f) - \psi(f)| < \dfrac{\pi}{2} \text{ and } f \in S \\ 0 & \text{otherwise}, f \in S \\ U(f) & f \notin S \end{cases} \tag{14.11-5}$$

where $U(f)$ is the FT of $u(t)$. It is observed that P_1 is a linear operator. Then, the optimal λ_{1k} is 1. A lower bound for λ_{2k} can be derived as [Stark]

$$\lambda_{2k} \geq \lambda_L = 1 + \frac{|P_2 u_k - P_1 P_2 u_k|^2}{|P_1 P_2 u_k - u_k|^2} \tag{14.11-6}$$

λ_L can be used as an estimate of λ_{2k}. However, λ_L may be greater than 2. In later iterations, it is necessary to restrict λ_{2k} to be less than 2 to guarantee convergence. For example, $\min(\lambda_L, 1.99)$ can be chosen.

Since λ_{1k} equals 1, one cycle of iterations can be written as

$$\begin{aligned} u_{k+1} &= P_1[1 + \lambda_{2k}(P_2 - 1)]u_k \\ &= (1 - \lambda_{2k})u_k + \lambda_{2k}P_1 P_2 u_k \end{aligned} \tag{14.11-7}$$

where $P_1 u_k$ is replaced by u_k since $u_k = P_1 T_2 u_{k-1}$ and therefore $u_k \in D_1$.

14.12   RECONSTRUCTION FROM A DISCRETIZED PHASE FUNCTION BY USING THE DFT

The discrete signal of length N will be denoted by $u[n]$. The phase of the DFT of $u[n]$ as well as the support of $u[n]$ are assumed to be known. The support of $u[n]$ will be defined as follows:

$$u[n] = 0 \quad \text{outside } 0 \leq n < M \quad \text{and } u[0] \neq 0$$

It can be shown that $N \geq 2M$ for the reconstruction algorithm to be satisfactory [Hayes *et al.*, 1980]. The DFT of $u[n]$ is given by

$$U[k] = A[k]e^{j\varphi[k]} \tag{14.12-1}$$

$\varphi[k]$ is assumed to be known.

The procedure of the algorithm is as follows:

1. Initialize $A[k]$ with an initial guess, $A_1[k]$. Then, the initial $U_1[k]$ is given by

$$U_1[k] = A_1[k]e^{j\varphi[k]} \tag{14.12-2}$$

2. Compute the inverse DFT of $U_1[k]$, yielding the first estimate $u_1[n]$. Let i equal 1.

3. Define a new sequence $v_i[n]$ as

$$v_i[n] = \begin{cases} u_i[n] & 0 \leq n < M \\ 0 & M \leq n < N \end{cases} \tag{14.12-3}$$

4. Compute the DFT $V_i[k]$ of $v_i[n]$. $V_i[k]$ can be written as

$$V_i[k] = A_i[k]e^{j\psi_i[k]} \tag{14.12-4}$$

The rest of the procedure can be written in two ways with respect to two different algorithms as follows, respectively.

Algorithm 1 [Levi and Stark]

5. Project $V_i[k]$ as follows:

$$U_{i+1}[k] = \begin{cases} A_i[k]\cos(\varphi[k] - \psi_i[k])e^{j\phi[k]} & \text{if } |[k] - \psi_i[k]| < \dfrac{\pi}{2} \\ 0 & \text{otherwise} \end{cases} \tag{14.12-5}$$

This operation corresponds to P_2. This equation can be modified so that T_2 is used by

$$U_{i+1}[k] = [1 + \lambda_{2i}(P_2 - 1)][V_i[k]] \tag{14.12-6}$$

λ_{2i} is computed by using the relation (14.11-6).

6. Generate a new estimate of $u_{i+1}[n]$ by computing the inverse DFT of $U_{i+1}[k]$.
7. Repeat steps 2–6 until convergence.

Algorithm 2 [Hayes *et al.*]
The only different step is the following:
   8. Project $V_i[k]$ as follows:

$$U_{i+1}[k] = A_i[k]e^{j\phi[k]} \tag{14.11-7}$$

All the other steps are the same as in Algorithm 1.

EXAMPLE 14.5 Let the error E_i at the $(i+1)$th iteration be defined as

$$E_{i+1} = \sum_{n=0}^{N-1} \left| u[n] - u_{i+1}[n] \right|^2$$

where $u[n]$ is the desired signal to be reconstructed. Prove that E_i is nonincreasing as i increases.
Solution: By Parseval's theorem, the following is true:

$$E_i = \frac{1}{N} \sum_{k=0}^{N-1} \left| U[k] - U_i[k] \right|^2$$

Since $Z(k)$ and $Z_i(k)$ have the same phase, the last equation can be written as

$$E_{i+1} = \frac{1}{N} \sum_{k=0}^{N-1} \left| U[k] - U_{i+1}[k] \right|^2$$

We also define

$$E_i' = \frac{1}{N} \sum_{k=0}^{N-1} \left| U[k] - V_i[k] \right|^2$$

The triangle equation for vector differences can be used to note the following:

$$E_{i+1} \le E_i'$$

with equality if $U_{i+1}[k] = V_i[k]$. The inequality above is due to the fact that $U_{i+1}[k]$ has the same phase as $U[k]$ whereas $V_i[k]$ has some other phase.

Again using the Parseval's theorem on this inequality results in

$$E_i \leq \sum_{n=0}^{N-1} |u[n] - v_i[n]|^2$$

Since $v_i[n]$ is before the projection to create $u_{i+1}[n]$, the above equation shows that the projection reduces the error.

14.13   GENERALIZED PROJECTIONS

There are many signal recovery problems involving nonconvex constraints. For example, digitizing a signal and constraining a signal or its Fourier transform to have a specified magnitude are nonconvex projections.

When S_i is a nonconvex set, there may be more than one point satisfying the definition of a projector P_i. This discrepancy can be removed by specifying additional constraint(s). Another problem with nonconvex sets is that a proof for the existence of a projection has not been found.

In the discussion below, the number of projections M will be limited to 2. In this case, the following error measure is useful:

$$E(u_k) = |P_1 u_k - u_k| + |P_2 u_k - u_k| \qquad (14.13\text{-}1)$$

The following theorem determines the convergence properties of generalized projections with $m = 2$ [Levi–Stark]:

Theorem: The recursion given by

$$u_{k+1} = T_1 T_2 u_k, \quad u_0 \text{ arbitrary} \qquad (14.13\text{-}2)$$

has the property

$$E(u_{k+1}) \leq E(u_k) \qquad (14.13\text{-}3)$$

for every λ_1 and λ_2 which satisfy

$$0 \leq \lambda_i \leq \frac{A_i^2 + A_i}{A_i^2 + A_i - \dfrac{1}{2}(A_i + B_i)} \qquad (14.13\text{-}4)$$

$$A_1 = \frac{|P_1 T_2 u_k - T_2 u_k|}{|P_2 T_2 u_k - T_2 u_k|} \qquad (14.13\text{-}5)$$

$$A_2 = \frac{|P_2 u_k - u_k|}{|P_1 u_k - u_k|} \qquad (14.13\text{-}6)$$

$$B_1 = \frac{(P_2 T_2 u_k - T_2 u_k, P_1 T_2 u_k - T_2 u_k)}{|P_2 T_2 u_k - T_2 u_k|^2} \tag{14.13-7}$$

$$B_2 = \frac{(P_1 u_k - u_k, P_2 u_k - u_k)}{|P_1 u_k - u_k|^2} \tag{14.13-8}$$

It can be shown that the upper limit in Eq. (14.13-4) includes 1. Hence, the theorem applies with P_1, P_2 instead of T_1, T_2.

The projections P_1 and P_2 can involve multiple constraints. Quite often, one set of constraints is in the signal domain, and the other set is in the transform domain.

EXAMPLE 14.6 A projector P maps a function $u(x)$ such that the resulting function $g(x)$ satisfies

$$g(x) = \begin{cases} s(x) & x \in D \subset D_0 \\ 0 & x \notin D_0 \end{cases}$$

where $s(x)$ *for all* x. Specify P.
Solution:
   P Satisfies

$$|Pu - u| = \inf_v |v - u|$$

Since $g(x) = Pu(x)$, we have

$$|Pu - u| = |g - u| = \int_{x \notin D_0} |g(x) - u(x)|^2 dx + \int_{x \in D_0 \cap D^c} |g(x) - u(x)|^2 dx$$

$$+ \int_{x \in D, s(x) > 0} |g(x) - u(x)|^2 dx + \int_{x \in D, s(x) < 0} |g(x) - u(x)|^2 dx$$

$|g - u|$ is minimum if $g(x)$ is chosen as

$$P_1 u = \begin{cases} 0 & x \notin D_0 \\ u(x) & x \in D_0 \cap D^c \\ s(x) & x \in D, x(t) > 0 \\ -s(x) & x \in D, x(t) \leq 0 \end{cases}$$

14.14   RESTORATION FROM MAGNITUDE

This problem is also called phase retrieval. It occurs in a number of fields such as astronomy, optics, and x-ray crystallography, where only the intensity equal to the square of the amplitude is available at each measurement point, and the phase is to be determined.

Phase retrieval is a much more difficult problem than restoration from phase. In the 2-D case, all the functions $u(x, y)$, $-u(x, y)$, $u(x - x_0, y - y_0)$, and $u(-x, -y)$ have the same Fourier transform magnitude. Apart from these four cases, it can be shown that a time- or space-limited sequence can be uniquely determined from its Fourier magnitude if the z-transform of the sequence is irreducible (cannot be written in terms of first-degree polynomials in z or z^{-1}) [Hayes *et al.*]. This result is not interesting in the 1-D case since all polynomials can be factored in terms of first-degree polynomials according to the fundamental theorem of algebra. However, in the 2-D case, most polynomials are irreducible. Consequently, restoration of images from Fourier magnitude information is feasible.

Let the known Fourier transform magnitude be $|U(f)|$. The two projections can be written as

$$P_1 u = \begin{cases} u & |u| < a \\ 0 & \text{otherwise} \end{cases} \tag{14.14-5}$$

$$P_2 u \leftrightarrow |U(f)| e^{j\phi(f)} \tag{14.14-6}$$

where $\phi(f)$ is the currently available Fourier phase. We observe that P_1 is convex whereas P_2 is not. P_1 may also involve other constraints.

The general restoration algorithm can be written as

$$u_{k+1} + T_1 T_2 u_k, \quad u_0 \text{ arbitrary} \tag{14.14-7}$$

where $T_i = 1 + \lambda_i(P_i - 1)$, and T_1 can be chosen equal to P_1 since P_1 is linear. When both λ_i are chosen equal to 1, the algorithm is known as Gerchberg–Saxton algorithm as discussed in Section 14.9.

When $u_k = P_1 T_2 u_{k-1}$, u_k belongs to the space D_1 whose projector is P_1. Hence, $P_1 u_k = u_k$, and $E(u_k)$ of Eq. (14.13-1) reduces to

$$E(u_k) = |P_2 u_{k+1} - u_{k+1}| \tag{14.14-8}$$

Equation (14.13-2) also reduces to

$$u_{k+1} = (1 - \lambda_{2k}) u_k + \lambda_{2k} P_1 P_2 u_k \tag{14.14-9}$$

Computing the Fourier transform of this equation and denoting FTs with the operator $F(\cdot)$ yields

$$U_{k+1} = (1 - \lambda_{2k}) U_k + \lambda_{2k} F(P_1 P_2 u_k) \tag{14.14-10}$$

Due to Parseval's relation, Eq. (14.14-8) can be written as

$$E(u_{k+1}) = \int\limits_{-\infty}^{\infty} [U(f) - U_{k+1}]^2 \mathrm{d}f \qquad (14.14\text{-}11)$$

λ_{2k} can be estimated by a simple search in the range $0 < \lambda_{2k} < a$ (where a is typically a number such as 3) so that $E(u_{k+1})$ is minimized.

14.14.1 Traps and Tunnels

When at least one of the projections is nonconvex, the convergence may occur to a point which is not a fixed point of every individual T_i. Such a point u_m is called a *trap*. x_t fails to satisfy one or more of the *a priori* constraints but satisfies

$$u_m = T_1 T_2 u_m \qquad (14.14\text{-}12)$$

Traps cannot occur with convex sets. On the other hand, *tunnels* may occur with box convex and nonconvex sets. A tunnel occurs when u_k changes very little from iteration to iteration.

A trap can be detected by observing no change in $E(u_k) > 0$ from iteration to iteration. If P_1 is linear and $P_1 T_2 u_k = u_k$, it can be shown that the correct solution u^* lies in a hyperplane orthogonal to $P_2 u_k - u_k$ [Stark].

14.15 IMAGE RECOVERY BY LEAST SQUARES AND THE GENERALIZED INVERSE

The techniques discussed to this point involved iterative optimization. Another powerful approach for image recovery involves the method of least squares. For this approach, the measured image is modeled as

$$v = \mathbf{H}u + n \qquad (14.15\text{-}1)$$

where $\mathbf{H}$ is a matrix of size $N_1 \times N_2$, u is the desired image, and n is the noise image. In Eq. (14.15-1), the images are row or column-ordered.

The unconstrained least squares estimate of u is obtained by minimizing

$$E = |v - Hu|^2 = (v - Hu)^{t*}(v - Hx) \qquad (14.15\text{-}2)$$

where $|\cdot|^2$ denotes the L^2 norm.

Computing the partial derivative of E with respect to u and setting it equal to zero results in

$$\mathbf{H}^t v = \mathbf{H}^t \mathbf{H} u \qquad (14.15\text{-}3)$$

If $\mathbf{H}^t[\mathbf{H}$ of size $N_2 \times N_2$ is nonsingular, the least-square solution is given by

$$\hat{u} = \mathbf{H}^+ v \qquad (14.15\text{-}4)$$

where $\mathbf{H}^+$ is called the *generalized inverse* (pseudo inverse) of $\mathbf{H}$, and is given by

$$\mathbf{H}^+ = (\mathbf{H}^t\mathbf{H})^{-1}\mathbf{H}^t \qquad (14.15\text{-}5)$$

$\mathbf{H}^+$ is of size $N_2 \times N_1$. $\mathbf{H}^+$ satisfies

$$\mathbf{H}^+\mathbf{H} = I \qquad (14.15\text{-}6)$$

However, it is not true that $\mathbf{H}\mathbf{H}^+$ equals I. A necessary condition for $\mathbf{H}^t\mathbf{H}$ to be nonsingular is that $N_1 \geq N_2$ and the rank r of $\mathbf{H}$ is N_2.

If $N_1 < N_2$, and the rank of $\mathbf{H}$ is N_1, H^+ is defined such that

$$\mathbf{H}\mathbf{H}^+ = I \qquad (14.15\text{-}7)$$

$\mathbf{H}^+$ satisfying Eq. (14.15-7) is not unique. Uniqueness is achieved by constraining the solution given by Eq. (14.15-4) to have minimum norm. In other words, among all possible solutions, the one with minimum $|\hat{x}|^2$ is chosen. Then, the pseudo inverse is given by

$$\mathbf{H}^+ = \mathbf{H}^t(\mathbf{H}\mathbf{H}^t)^{-1} \qquad (14.15\text{-}8)$$

In conclusion, $\mathbf{H}^+$ always exists uniquely as discussed above, and the least squares solution is $\mathbf{H}^+ v$.

$\mathbf{H}^+$ has the following additional properties:

1. $\mathbf{H}\mathbf{H}^+ = (\mathbf{H}\mathbf{H}^+)^t$ In other words, $\mathbf{H}\mathbf{H}^+$ is symmetric.
2. $\mathbf{H}^+\mathbf{H} = (\mathbf{H}^+\mathbf{H})^t$ In other words, $\mathbf{H}^+\mathbf{H}$ is symmetric.
3. $\mathbf{H}\mathbf{H}^+\mathbf{H} = \mathbf{H}$
4. $\mathbf{H}^+\mathbf{H}\mathbf{H}^t = \mathbf{H}^t$

14.16   COMPUTATION OF $\mathbf{H}^+$ BY SINGULAR VALUE DECOMPOSITION (SVD)

The SVD representation of the matrix $\mathbf{H}$ of size $N_1 \times N_2$ and rank r can be written as

$$H = \sum_{m=1}^{r} \lambda_m^{\frac{1}{2}} \psi_m \phi_m^t \qquad (14.16\text{-}1)$$

where ψ_m and ϕ_m are the eigenvectors of $\mathbf{H}\mathbf{H}^t$ and $\mathbf{H}^t\mathbf{H}$, respectively, with the singular values λ_m, $1 \leq m \leq r$.

The SVD representation of the pseudoinverse $\mathbf{H}^+$ is given by

$$\mathbf{H}^+ = \sum_{m=1}^{r} \lambda_m^{\frac{1}{2}} \phi_m \psi_m^t \qquad (14.16\text{-}2)$$

Using Eq. (14.16-2) in Eq. (14.15-4), the pseudoinverse solution can be written as

$$\hat{u} = \sum_{m=1}^{r} \lambda_m^{-\frac{1}{2}} \phi_m \psi_m^t v \qquad (14.16\text{-}3)$$

which is the same as

$$\hat{u} = \sum_{m=1}^{r} \omega_m \phi_m \qquad (14.16\text{-}4)$$

where

$$\omega_m = \frac{\psi_m^t v}{\lambda_m^{1/2}} \qquad (14.16\text{-}5)$$

Equation (14.16-4) shows that $\hat{u}$ is in the vector space spanned by the eigenvectors of $\mathbf{H}'\mathbf{H}$.

The use of Eqs. (14.16-4) and (14.16-5) is relatively simple for small problems. However, for large scale problems, it becomes very difficult to do so. For $N_1 = N_2 = 256$, $\mathbf{H}$ is of size $65{,}536 \times 65{,}536$. Then, it becomes prohibitive.

If the degradation point spread function (PSF) is separable so that $v \doteq \mathbf{H}_1 u \mathbf{H}_2$, the generalized inverse is also separable, and

$$\hat{u} = \mathbf{H}_1^+ v \mathbf{H}_2^+ \qquad (14.16\text{-}6)$$

EXAMPLE 14.7 Determine the least squares solution of

$$\begin{bmatrix} 1 & 2 \\ 2 & 1 \\ 1 & 3 \end{bmatrix} \begin{bmatrix} x_1 \\ x_2 \end{bmatrix} = \begin{bmatrix} 1 \\ 2 \\ 3 \end{bmatrix}$$

by determining the pseudoinverse.

Solution: $\mathbf{H}$ and $\mathbf{A} = \mathbf{H}'\mathbf{H}$ are given by

$$\mathbf{H} = \begin{bmatrix} 1 & 2 \\ 2 & 1 \\ 1 & 3 \end{bmatrix} \quad \mathbf{A} = \begin{bmatrix} 6 & 7 \\ 7 & 14 \end{bmatrix}$$

The eigenvalues of $\mathbf{A}$ are $\lambda_1 = 10 + \sqrt{65}$ and $\lambda_2 = 10 - \sqrt{65}$. Since $\mathbf{A}$ is non-singular, $\mathbf{H}^+$ is given by

$$\mathbf{H}^+ = \mathbf{A}^{-1}\mathbf{H}^t = \frac{1}{35}\begin{bmatrix} 14 & -7 \\ -7 & 6 \end{bmatrix}\begin{bmatrix} 1 & 2 & 1 \\ 2 & 1 & 3 \end{bmatrix} = \frac{1}{35}\begin{bmatrix} 0 & 21 & -7 \\ 5 & -8 & 11 \end{bmatrix}$$

The least-squares solution is given by

$$\hat{u} = \mathbf{H}^+ y = \begin{bmatrix} 0.6 \\ 0.628 \end{bmatrix}$$

EXAMPLE 14.8 Suppose that the solution is constrained in some sense in the signal domain and the constraint is applied as a matrix acting on $\hat{u}$. For example, bounding the image to be nonzero in a certain region would be such an operator. Determine the least squares solution in this case.
Solution: The error measure to be minimized can be written as

$$E = |v - \mathbf{HT}u|^2$$

where $\mathbf{T}$ is the constraint operator in matrix form. In the case of boundedness, $\mathbf{T}$ is a diagonal matrix whose diagonal elements are 1s and 0s. We have

$$E = (v - \mathbf{HT}u)^t(v - \mathbf{HT}u)$$
$$= v^t v - 2u^t \mathbf{T}^t \mathbf{H}^t v + u^t \mathbf{T}^t \mathbf{H}^t \mathbf{HT}u$$

The gradient of E is given by

$$g = -2T^t \mathbf{H}^t v + 2\mathbf{T}^t \mathbf{H}^t \mathbf{HT}u$$

Setting g equal to 0 yields

$$\hat{u} = [\mathbf{T}^t \mathbf{H}^t \mathbf{HT}]^{-1} \mathbf{T}^t \mathbf{H}^t v$$

provided that the matrix inverse exists.

14.17   THE STEEPEST DESCENT ALGORITHM

When the goal is to obtain $\hat{u}$ rather than $\mathbf{H}^+$, the iterative gradient method can be used. This method is based on the fact that the steepest descent of $E = |v - \mathbf{H}u|^2$ is in the direction of the negative gradient of E with respect to u.

Suppose that u_k is the vector u at the kth iteration. The gradient of E with respect to u_k is given by

$$g_k = -2\mathbf{H}'(v - \mathbf{H}u_k) \qquad (14.17\text{-}1)$$

The iterative gradient algorithm is given by

$$u_{k+1} = u_k - \alpha_k g_k \quad \text{with} \quad u_0 = 0 \qquad (14.17\text{-}2)$$

where α_k is a scalar. Using Eq. (14.17-2) in Eq. (14.17-1), g_k can be written as

$$g_k = g_{k-1} - \alpha_{k-1}\mathbf{H}'\mathbf{H}g_{k-1} \qquad (14.17\text{-}3)$$

It can be shown that u_k converges to $\hat{u}$ if

$$0 \le \alpha_k \le \frac{2}{\lambda_{\max}} \qquad (14.17\text{-}4)$$

where $\lambda_{\max}$ is the maximum eigenvalue of $A = \mathbf{H}'\mathbf{H}$.

If α_k is chosen constant, the method is called the *one-step gradient algorithm*. The optimum value of α for fastest convergence is given by

$$\alpha_{\text{opt}} = \frac{2}{\lambda_{\max} + \lambda_{\min}} \qquad (14.17\text{-}5)$$

where $\lambda_{\min}$ is the minimum eigenvalue of $\mathbf{A}$.

If $\mathbf{A}$ is highly ill-conditioned, the *condition number* of $\mathbf{A}$ equal to $\lambda_{\max}/\lambda_{\min}$ is large, and convergence may be very slow.

If α_k is optimized at each iteration, the method is called the *method of steepest descent*. The optimal value of α_k is given by

$$\alpha_k = \frac{g_k^t g_k}{g_k^t A g_k} \qquad (14.17\text{-}6)$$

When $\lambda_{\max}/\lambda_{\min}$ is large, convergence may still be slow when α_k is used.

EXAMPLE 14.9 Determine the approximate least squares solution of Example 14.8 by (a) the one-step gradient algorithm in 12 iterations, (b) the steepest descent algorithm in three iterations.

Solution: (a) In the one-step gradient algorithm, σ_{opt} is found as

$$\alpha_{\text{opt}} = \frac{2}{\lambda_1 + \lambda_2} = 0.1$$

In the first iteration, we have

$$g_0 = -\begin{bmatrix} -8 \\ -13 \end{bmatrix}, \quad u_1 = \begin{bmatrix} 0.8 \\ 1.3 \end{bmatrix}$$

and the error $E_1 = |v - \mathbf{H}u_1|^2$ equals 9.46.

After 12 such iterations, we get

$$u_{11} = \begin{bmatrix} 0.685 \\ 1.253 \end{bmatrix}, \quad u_{12} = \begin{bmatrix} 0.555 \\ 0.581 \end{bmatrix}$$

with the error $E_{12} = |v - \mathbf{H}u_{12}|^2 = 1.101$.

(b) With the steepest descent algorithm, α_k is optimized at each iteration according to Eq. (14.17-6). The result after three iterations is

$$u_3 = \begin{bmatrix} 0.5592 \\ 0.629 \end{bmatrix}$$

where the error $E_3 = |v - Hu_3|^2 = 1.0285$.

14.18   THE CONJUGATE GRADIENT METHOD

In the conjugate gradient method, the scalar α_k as well as the correction vector is optimized at each iteration. This allows faster convergence. For this purpose, Eq. (14.17-2) is replaced by

$$u_{k+1} = u_k + \alpha_k r_k \qquad (14.18\text{-}1)$$

where the correction vector r_k replaces the gradient vector g_k.

The correction vector r_k has the following properties:

$$g_{k+1}^t r_k = 0 \qquad (14.18\text{-}2)$$

$$[\mathbf{H}r_k]^t [\mathbf{H}r_m] = r_k^t \mathbf{A} r_m = 0 \; k \neq m, \quad 0 \leq k, \, m < N \qquad (14.18\text{-}3)$$

where N is the rank of $\mathbf{A}$.

Due to these properties, the algorithm converges in N iterations. The iterative equations in addition to Eq. (14.18-1) are as follows:

$$r_k = -g_k + \beta_k r_{k-1} \quad r_0 = -g_0 \qquad (14.18\text{-}4)$$

$$\beta_k = \frac{g_k^t \mathbf{A} r_{k-1}}{r_{k-1}^t \mathbf{A} r_{k-1}} \qquad (14.18\text{-}5)$$

$$\alpha_k = \frac{g_k^t r_k}{r_k^t \mathbf{A} r_k} \tag{14.18-6}$$

$$g_k = -\mathbf{H}^t v + \mathbf{A} u_k = g_{k-1} + \alpha_{k-1} \mathbf{A} r_{k-1} \tag{14.18-7}$$

The final solution $\hat{u}$ can also be written as

$$\hat{u} = \sum_{m=0}^{N-1} \alpha_m r_m \tag{14.18-9}$$

When N is a large number, the algorithm can be stopped earlier than N iterations. If the rank of $\mathbf{H}$ is $M < N$, it is satisfactory to run the algorithm with M iterations.

EXAMPLE 14.10 Determine the conjugate gradient method for the least squares problem of Example 14.8.
Solution: The least-squares solution satisfies

$$\mathbf{T}^t \mathbf{H}^t \mathbf{H} \mathbf{T} u = \mathbf{T}^t \mathbf{H}^t v$$

This is in the same form as before with the replacement of $\mathbf{H}$ by $\mathbf{H}' = \mathbf{HT}$. The only iterative equation which needs to be modified is the equation for the gradient, which becomes

$$g_k = g_{k-1} - \alpha_{k-1} \mathbf{T}^t \mathbf{H}^t \mathbf{H} r_{k-1}$$

The initial image u_0 is chosen as satisfying the constraint represented by T:

$$\mathbf{T} u_0 = u_0$$

g_0 rm and r_0 are given by

$$g_0 = r_0 = \mathbf{T}^t \mathbf{H}^t (v - \mathbf{H} u_0)$$

15

Diffractive Optics I

15.1 INTRODUCTION

Diffractive optics involves creation of holograms by computation in a digital computer, followed by actual implementation by a recording system such as a laser film writer, scanning electron beam writer, high resolution printer, usually followed by photoreduction, and the like. Diffractive optics is very versatile since any type of wave can be considered for computation within the computer. Digital holograms created with such technology are more commonly called *diffractive optical element* (DOE). Another name for DOE is *computer-generated hologram* (CGH). Because of these equivalent terminologies, the words hologram, DOE, and CGH will be used equivalently in this chapter.

In communication engineering, it is well known that a complex signal must be coded in some fashion before it can be transmitted over a single channel. This can be done by modulating the amplitude of a carrier by the modulus of the complex signal and/or by modulating the phase of the carrier by the phase of the complex signal. The real part of this modulated carrier then can be transmitted, and through proper demodulation procedures, the complex signal can be recovered at the receiver. Both holography and diffractive optics are based on such modulation and demodulation principles.

DOEs have found applications in many areas such as wave shaping, laser machining, 3-D displays [Yatagai, 1975], optical pattern recognition, optical interconnects, security devices [Dittman, *et al.*, 2001], optical communications/networking, and spectroscopy. Some important advantages of DOEs are as follows:

- A DOE can perform more than one function, for example, have multiple focal points, corresponding to multiple lenses on a single element. It can also be designed for use with multiple wavelengths.
- DOEs are generally much lighter and occupy less volume than refractive and reflective optical elements.
- With mass manufacturing, they can be manufactured less expensively for a given task.

Diffraction, Fourier Optics and Imaging,   by Okan K. Ersoy
Copyright © 2007 John Wiley & Sons, Inc.

Since DOEs are based on diffraction, they are highly dispersive, namely, their properties depend on wavelength. Hence they are more commonly used with monochromatic light. However, the dispersive properties can also be used to advantage. For example, light at one wavelength can be focused at one point, and light at another wavelength can be focused at another point in space. In devices such as *optical phased arrays* used in optical communications and networking, dispersion is used to separate different wavelengths at different focal points. In addition, refractive and diffractive optical elements can be combined in such a way that wavelength dispersion can be removed, and goals such as removal of spherical aberration can be achieved.

There are several steps in the creation of a DOE. These are sampling of waves, computation of wave propagation, usually with the FFT, coding of complex wave information on the hologram, for example, as a real and positive image on the hologram plane, and recording of the resulting DOE.

In 3-D wave coding, phase is usually much more important to be represented correctly than amplitude. Hence, phase is given much more attention. In applications in which the output image intensity is used, output amplitude variations can be significantly reduced by multiplying the output image with a random phase factor (diffuser). This does not change the intensity of the image [Burckhardt, 1970].

This chapter consists of ten sections. The first few methods discussed utilizes the Fourier transformation property of a lens and generate a DOE design for a Fourier transform hologram. The first such method is the Lohmann method discussed in Section 15.2. In this method, amplitude modulation is achieved by controlling the sizes of the apertures, and phase modulation is achieved by controlling the positions of the apertures generated. Section 15.3 discusses the approximations involved in implementing DOEs based on the Lohmann method. These approximations are compensated for in a method called the Lohmann-ODIFIIT method in Section 16.8. Section 15.4 simplifies the Lohmann method by choosing constant aperture size, implying constant amplitude. In order to compensate for the error generated, iterative optimization discussed in Chapter 14 is utilized. Section 15.5 describes computer experiments with the Lohmann methods. Another Fourier method based on hard-clipping of the hologram transmittance function resulting in a binary hologram is discussed in Section 15.6.

The methods discussed so far assume that the reconstructed image is on an output plane. The method discussed in Section 15.7 is capable of generating output image points at arbitrary locations in 3-D. It is a simple method to use. It was the method used in demonstrating for the first time how to make DOEs with a scanning electron microscope system. Section 15.8 extends the method of Section 15.7 and makes it capable of generating many object points in 3-D space.

Essentially all DOE methods involve nonlinear coding of amplitude and phase. As a result, they generate harmonic images. When the FFT is used, the implicit assumption is that the object is periodic. This also causes the images to repeat in a periodic fashion on the reconstruction plane. The one-image-only holography method discussed in Section 15.9 generates a single image by using semi-irregular sampling and a spherical reference wave whose origin is located near the hologram

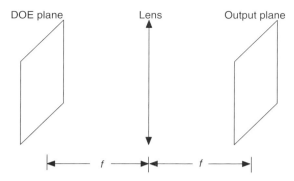

Figure 15.1. The Fourier transform system used in DOE design.

plane. This method is later utilized in Chapter 19 for the design of phasars for dense wavelength division multiplexing in order to achieve many more wavelength channels in optical communications and networking than what is possible with the present phasar methods.

The final section goes back in time and introduces a classical DOE that functions as a flat lens. It is called a *binary Fresnel zone plate*.

15.2 LOHMANN METHOD

The Lohmann method as well as a number of other methods for DOEs are developed for the Fourier transform arrangement shown in Figure 15.1.

In the Lohmann method (also called the *detour phase method*), the hologram plane is divided into smaller rectangles each containing an aperture [Lohmann, 1970]. An example of a Lohmann cell is shown in Figure 15.2. The size of the aperture is used to control amplitude, and its position is changed to adjust phase. This results in a binary transmission pattern.

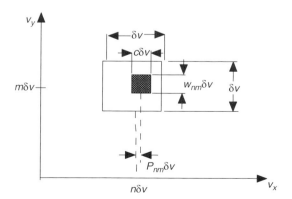

Figure 15.2. The (n,m)th cell of a Lohmann hologram.

Let $h(x, y)$ be the output amplitude from the hologram $H(v_x, v_y)$. It should be proportional to the desired image $f(x, y)$. The binary transmission function of the hologram can be written as

$$H(v_x, v_y) = \sum_n \sum_m \text{rect}\left[\frac{v_x - (n + P_{nm})\delta v}{c\delta v}\right] \text{rect}\left[\frac{v_y - m\delta v}{W_{nm}\delta v}\right] \qquad (15.2\text{-}1)$$

When a tilted plane wave, $\exp(2\pi i x_0 v_x)$ is incident on the binary hologram, the complex amplitude after the hologram is $H(v_x, v_y)\exp(2\pi i x_0 v_x)$. The complex amplitude at the image plane is the Fourier transform of $H(v_x, v_y)\exp(2\pi i x_0 v_x)$:

$$\iint H(v_x, v_y)e^{2\pi i[(x+x_0)v_x + yv_y]}\mathrm{d}v_x\mathrm{d}v_y$$
$$= c(\delta v)^2\text{sinc}[c\delta v(x + x_0)]\sum_n \sum_m W_{nm}\text{sinc}(yW_{nm}\delta v) \qquad (15.2\text{-}2)$$
$$\times \exp\{2\pi i[\delta v((x + x_0)(n + P_{nm}) + ym)]\}$$

The parameters W_{nm}, P_{nm}, and the two constants x_0, c are chosen such that the complex amplitude in the image plane matches the desired image $f(x, y)$.

Equation (15.2-2) can be compared to the desired image by writing $f(x, y)$ in the form

$$f(x, y) = \iint F(v_x, v_y)e^{2\pi i(xv_x + yv_y)}\mathrm{d}v_x\mathrm{d}v_y = \sum_n \sum_m F(n\delta v, m\delta v)e^{2\pi i[\delta v(xn + ym)]}$$

$$(15.2\text{-}3)$$

The two sinc functions and the factor $\exp[2\pi i(xP_{nm}\delta v)]$ in Eq. (15.2-2) can be assumed to be close to unity. The validity of this assumption will be discussed in Section 15.3. Equating the Fourier coefficients in Eqs. (15.2-2) and (15.2-3) yields

$$c(\delta v)^2 W_{nm}\exp\{2\pi i[x_0\delta v(n + P_{nm})]\} \propto F(n\delta v, m\delta v);$$
$$F(n\delta v, m\delta v) \propto c(\delta v)^2 A_{nm}\exp[2\pi i(\phi_{nm}/2\pi)]; \quad (15.2\text{-}4)$$
$$W_{nm} \approx A_{nm}; \ P_{nm} + n \approx \phi_{nm}/2\pi x_0\delta v.$$

By choosing $x_0\delta v$ equal to an integer M, we obtain

$$P_{nm} \approx \phi_{nm}/2\pi M. \qquad (15.2\text{-}5)$$

Equations (15.2-4) and (15.2-5) show that the height W and the position P of the aperture in each cell can be chosen proportional to the amplitude A and the phase ϕ of the complex amplitude F at the cell, respectively.

15.3   APPROXIMATIONS IN THE LOHMANN METHOD

In the previous section, the three approximations (a) $\text{sinc}[c\delta v(x + x_0)] \approx \text{const}$, (b) $\text{sinc}(yW_{nm}\delta v) \approx 1$, and (c) $\exp[2\pi i(xP_{nm}\delta_x)] \approx 1$ were made for simplicity. The effects of these approximations on the reconstructed image depend on several factors, and with proper design, can be minimized [Lohmann, 1970].

The sinc function $\text{sinc}[c\delta v(x + x_0)]$ creates a drop-off in intensity in the x-direction proportional to the distance from the center of the image plane. Approximation (a) considers this sinc factor to be nearly constant inside the image region. If the size of the image region is $\Delta x \times \Delta y$, then at the edges $x = \pm(\Delta x/2)$ where the effects are most severe, we have $\text{sinc}[cM \pm c/2]$. This implies that a small aperture size c results in less drop-off in intensity. However, this also reduces the brightness of the image. For $cM = 1/2$, the brightness ratio between the center and the edge of the image region is 9:1. By reducing this product to $cM = 1/3$, the ratio drops to 2:1. Thus, brightness can be sacrificed for a reduction in intensity drop-off at the edges of the reconstructed image.

The sinc function $\text{sinc}(yW_{nm}\delta v)$ indicates a drop-off in intensity similar to (a), but in the y-direction. This approximation is a little less dangerous because $\Delta x \delta v = 1$, which means $|yW\delta v| < 1/2$. It is well known that amplitude errors in apertures of coherent imaging systems have little effect on the image because they do not deviate rays like phase errors do [Lohmann, 1970]. To reduce the effects of this approximation, every W could be reduced by a constant factor. However, some brightness would then be sacrificed.

A possible solution to the sinc roll-off in the x direction is to divide the desired image by $\text{sinc}[c\delta v(x + x_0)]$. The desired image $f(x, y)$ becomes

$$\frac{f(x, y)}{\text{sinc}[c\delta v(x + x_0)]}.$$

The same thing cannot be done for the y direction because $\text{sinc}(yW_{nm}\delta v)$ depends on the aperture parameter W_{nm} which is yet to be determined. Luckily, this sin c factor is less influential than the x dependent sinc, and design of the hologram can be altered to reduce its effects.

The phase shift $\exp[2\pi i(xP_{nm}\delta_x)]$ causes a phase error that varies with x location in the image plane. The range of this phase error depends on x and P. Since $|x| \leq \Delta x/2 = 1/2\delta v$ and $|P| \leq 1/2M$ the phase error ranges from zero to $\pi/2M$. At its maximum, the phase error corresponds to an optical path length of $\lambda/4M$. For $M = 1$, this is within the $\lambda/4$ Rayleigh criterion for wave aberrations [Lohmann, 1970].

The detrimental effects of these approximations are less when the size of the image region is restricted. The reconstruction errors increase with distance from the center of the image plane. If the reconstruction image region is smaller, the errors are also less.

15.4   CONSTANT AMPLITUDE LOHMANN METHOD

The previous section showed that the sinc oscillations due to aperture size is a difficult approximation to deal with. Therefore, it would be desirable, also for the purpose of simpler implementation, to make the height of each aperture constant

[Kuhl, Ersoy]. This would allow for the desired image to be divided in the y-direction by the y-dependent sinc drop-off just as was done in the x-direction for the sinc factor associated with the constant width of the aperture.

Logically, if every aperture has the same size, then only the positioning of the apertures is affecting the output. This means that all the information is contained in the phase. The method discussed below is used to "shift" information in the hologram plane from the amplitude to the phase.

If only the magnitude of the desired image is of concern, the phase at each sampling point in the observation plane is a free parameter. Then, the range of height W_{nm} values can be reduced by iterative methods discussed in Chapter 14. Suppose the sampled desired image has amplitudes a_{nm} with an unspecified corresponding phase θ_{nm}. The discrete Fourier transform of the image is $\{W_{nm} \exp(i2\pi P_{nm})\}_{nm}$, where $\{\ldots\}_{nm}$ indicates the sequence for all points n and m. The first step in reducing the range of W_{nm} values is to assign values of θ_{nm} to the initial desired image which are independent and identically distributed phase samples with a uniform distribution over $(-\pi, \pi)$ [Gallagher and Sweeney, 1979]. The resulting DFT of the image samples is denoted by $\mathrm{DFT}[\{a_{nm} \exp(i\theta_{nm})\}_{nm}] = \{A_{nm} \exp(i\psi_{nm})\}_{nm}$. Then, the spectral amplitudes A_{nm} are set equal to any positive constant A. The inverse DFT of the spectrum with adjusted amplitudes is $\mathrm{DFT}[\{A \exp(i\psi_{nm})\}] = \{\tilde{a}_{nm} \exp(i\tilde{\theta}_{nm})\}$. The original image amplitudes a_{nm} are now combined with the new phase values $\tilde{\theta}_{nm}$ to form the new desired image samples. This process is repeated for a prescribed number of iterations. The image phase obtained from the last iteration becomes the new image phase. The final image phase values are used with the original image amplitudes to generate $W_{nm} \exp(i2\pi P_{nm})$ used for designing the hologram.

By constraining the amplitude in the hologram domain and performing iterations, information in the hologram plane is transferred from the amplitude to the phase. Therefore, this reduces the negative effects caused by making all the apertures the same height. If all the apertures have the same height, then approximation (b) in Section 15.3 can be handled the same way as approximation (a). Further generalization of this approach is discussed in Section 16.6.

15.5 QUANTIZED LOHMANN METHOD

The Lohmann method discussed in Section 15.2 allows for an infinite number of aperture sizes and positions, which is not practical for many methods of implementation. To overcome this obstacle, a discrete method can be used to quantize the size and position of the apertures in each cell. In the modified method, the size of each aperture still controls amplitude, and phase is controlled by shifting the aperture position. However, the possible values of amplitude and phase are now quantized.

In the quantized Lohmann method, each hologram cell is divided into an $N \times N$ array. This will be referred to as N-level quantization. This restricts the possible center positions and heights of each aperture. Thus, the phase and amplitude at each cell is quantized. Specifically, there are N possible positions for the center of the aperture (phase), and $N/2 + 1$ potential height values (amplitudes) including zero amplitude, since the cell is symmetric in the y direction. A large N produces a pattern close to that of the exact hologram.

For example, we can consider a Lohmann cell divided into 4×4 smaller squares. For a value of $c = 1/2$ in the Lohmann algorithm, which means that the width of the aperture is fixed at half the width of the entire cell, this cell permits three values of normalized amplitude $(0, 1/2,$ and $1)$ and three values of phase $(-\pi, -\pi/2, 0,$ and $\pi/2)$ for a total of eight possible combinations. Quantization means there will be error when coding amplitude and phase. Therefore, it makes sense to incorporate an optimization algorithm such as the POCS and to design subholo-grams iteratively until a convergence condition is met. This topic is further discussed in Section 16.6.

Quantizing aperture positions is useful in practical implementations. For example, spatial light modulators (SLMs) can be used in real time to control amplitude and phase modulation at each hologram point. Unfortunately, the SLM cannot accurately generate an exact Lohmann cell, and quantization is necessary to make realization practical. Similarly, technologies used to make integrated circuits can be used for realizing DOEs as discussed in Sections 11.6 and 11.7. Since precise, continuous surface relief is very difficult, quantization of phase and amplitude is also necessary in all such technologies.

15.6   COMPUTER SIMULATIONS WITH THE LOHMANN METHOD

The binary transmission pattern obtained from the Lohmann method can be displayed in one of two ways:

Method 1: To display the exact Lohmann hologram (i.e., aperture size and positions are exactly as specified), the pattern is drawn into a CAD layout, for example using AutoCAD.

Method 2: This is the same as the quantized Lohmann method.

All holograms were designed using parameter values of $c = 1/2$ and $M = 1$ [Kuhl, Ersoy]. Setting $c = 1/2$ means that each aperture has a width equal to half that of the cell. It can also be shown that this value of c maximizes the brightness of the image [Gallagher and Sweeney, 1979]. After choosing $c = 1/2$, M has to be chosen equal to 1 [Lohmann, 1970]. Also, each approximation discussed in Section 15.3 was assumed to be valid.

Figure 15.3 shows a binary image E of size 64×64, which is placed completely on one side of the image plane.

Because the field in the hologram plane is real, the Fourier transform of the image has Hermitian symmetry:

$$U[n, m] = U^*[N - n, M - m] \qquad (15.6\text{-}1)$$

Figure 15.4 shows the amplitude of the DFT of the image.

Figure 15.3. The image E used in the generation of the Lohmann hologram.

Figure 15.5 shows the Lohmann hologram generated with Method 2 discussed above with 16 levels of quantization.

Figure 15.6 shows the simulated reconstruction. The twin images are clearly visible.

The 512×512 gray-scale image shown in Figure 15.8 was also experimented with.

Figure 15.4. The amplitude of the DFT of the image of Figure 15.3.

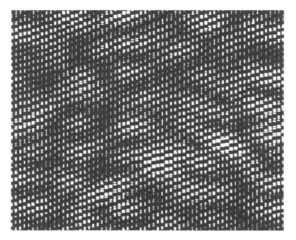

Figure 15.5.   The Lohmann hologram of the image of Figure 15.3.

The computer reconstruction from its Lohmann hologram is shown in Figure 15.8.

The constant amplitude Lohmann method discussed in Section 15.4 was also experimentally investigated with both images, using the iterative optimization approach. The constant amplitude Lohmann hologram for the image E is shown in Figure 15.9. The computer reconstruction obtained from it is shown in Figure 15.10. Similarly, the computer reconstruction obtained from the constant amplitude Lohmann hologram of the gray-level image of Figure 15.8 is shown in Figure 15.11.

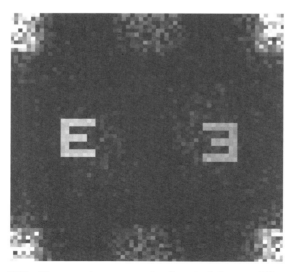

Figure 15.6.   The computer reconstruction from the hologram of Figure 15.5.

Figure 15.7. The gray-level image of size 512×512.

Comparing Figures 15.6 and 15.10, the desired image E is brighter in the constant amplitude case, but the sharpness of the image E appears to have decreased slightly. The constant amplitude hologram also produced significantly less noise in the corners of the image plane.

Comparing Figures 15.8 and 15.11, the desired image was again brighter and exhibited less noise in the corners of the image plane when compared to the original method.

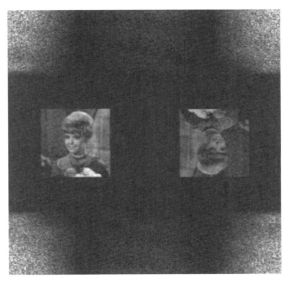

Figure 15.8. The computer reconstruction obtained from the Lohmann hologram of the gray-level image.

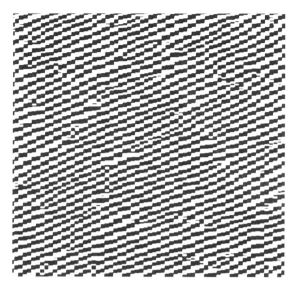

Figure 15.9.   The constant amplitude Lohmann hologram of the image of Figure 15.3.

15.7   A FOURIER METHOD BASED ON HARD-CLIPPING

It is possible to create a binary Fourier transform DOE by hard-clipping its phase function. In this process, the amplitude information is neglected. For the sake of simplicity, we will explain its analysis in 1-D. Consider a complex function

$$A(f)e^{j\phi(f)}$$

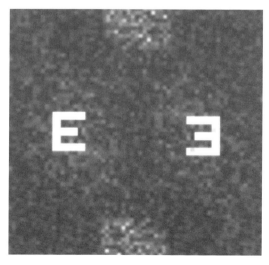

Figure 15.10.   The computer reconstruction obtained from the hologram of Figure 15.9.

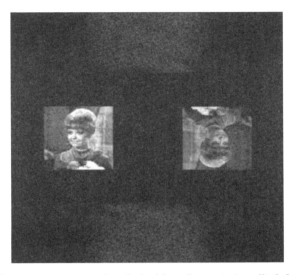

Figure 15.11. The computer reconstruction obtained from the constant amplitude Lohmann hologram of the gray-level image of Figure 15.8.

modulated with a plane reference wave

$$e^{j\alpha f}$$

to generate

$$A(f)e^{j(\alpha f + \phi(f))}.$$

The real part of this signal is passed through the hard-clipping transmission function shown in Figure 15.12.

This function can be modified by adding 0.5 to it so that it changes between 0 and 1. This would have the effect of increasing the DC term of the DFT. The result of hard-clipping the signal with respect to the bipolar transmission function is given by

$$h(f) = \frac{1}{2}\left[1 + \frac{A(x)\cos(\alpha f + \phi(f))}{|A(x)\cos(\alpha f + \phi(f))|}\right] \qquad (15.7\text{-}1)$$

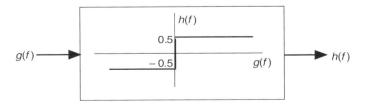

Figure 15.12. The hard-clipping transmission function.

Figure 15.13.   The image used with the hard-clipping method.

$h(f)$ can be expressed in terms of its Fourier series representation as [Kozma and Kelly, 1965]

$$h(f) = \frac{1}{2} + \sum_{m(\text{odd})=1}^{\infty} \frac{2(-1)^{\frac{m-1}{2}}}{m\pi} \cos(m\alpha x + m\phi(f)) \qquad (15.7\text{-}2)$$

Equation (15.7-2) represents infinitely many images at increasing angles of diffraction. The most important ones are the zero-order image due to 1/2 term and the twin images at angles $\pm\alpha$ due to $m = 1$ term.

Figure 15.13 shows the image used in experiments with the hard-clipping method. The inverse DFT of this image was coded with the hard-clipping filter. The resulting DOE is shown in Figure 15.14. The forward DFT of the DOE with the

Figure 15.14.   The hologram generated with the hard-clipping method.

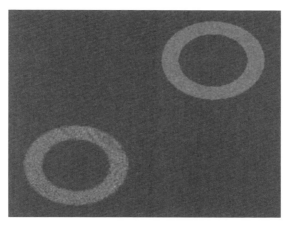

Figure 15.15. Reconstruction from the DOE without random phase diffuser.

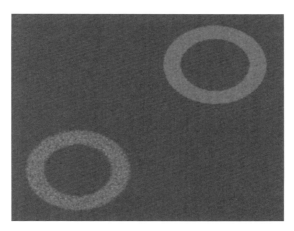

Figure 15.16. Reconstruction from the DOE with random phase diffuser.

zero-order image filtered out is shown in Figure 15.15. The twin images are clearly observed. In order to minimize amplitude variation effects, the image in Figure 15.13 was multiplied with a diffuser, and the process was repeated. The resulting reconstruction is shown in Figure 15.16. It is observed that the result is more satisfactory than before, indicating the significance of using a diffuser with a phase coding method.

15.8 A SIMPLE ALGORITHM FOR CONSTRUCTION OF 3-D POINT IMAGES

Sometimes it is preferable to use a simple but robust algorithm to test new results, new equipment, etc. The algorithm presented below is very useful for such purposes, and was the algorithm which demonstrated for the first time the feasibility of implementing DOEs with scanning electron microscopy [Ersoy, 76]. Even though initially very simple, the algorithm was later developed in more complex ways to generate 3-D images in arbitrary locations in space [Bubb, Ersoy]. The methodology is quite different from other methods since it does not explicitly depend on the Fourier transform. It also leads to one-image only holography discussed in Section 15.9, which is further used in optical phased arrays in Chapter 18.

If (x_o, y_o, z_o) represents the position of an object point to be reconstructed, and (x_i, y_i, z_i) represents the position of a phase-shifting aperture on the hologram, the Huygens–Fresnel principle for a collection of N apertures on a plane $(z = 0)$ leads to

$$U(x_o, y_o, z_o) = \sum_i^N \iint U(x_i, y_i, z_i) \frac{1}{j\lambda} \frac{\exp(jkr_{oi})}{r_{oi}} \cos \delta \cdot dx_i dy_i \qquad (15.8\text{-}1)$$

where δ is the angle between the z-axis and the vector from the center of the aperture to the object point, whose length is r_{oi}, λ is the wavelength, and k is the wave number. $U(x_i, y_i, z_i)$ will be assumed to be a plane wave equal to unity below. For small hologram dimensions, $\cos(\delta)$ can be assumed to be constant. If the phase variations on the hologram plane are also small compared to the phase variations of $\exp(jkr_{oi})$, the above equation can be approximated by

$$U(x_o, y_o, z_o) = \frac{\cos \delta}{j\lambda R} d_x d_y \sum_i^N \iint \frac{\exp(jkr_{oi})}{r_{oi}} dx_i dy_i \qquad (15.8\text{-}2)$$

where θ_i is the phase shift of the reference wave at the ith aperture.

Assume that each aperture is rectangular on the x-y plane with dimensions d_x and d_y, and has a central point $(x_{si}, y_{si}, 0)$ whose radial distance r_{oi} from the observation point satisfies

$$kr_{oi} = 2\pi n + \theta_i \quad n = \text{integer} \qquad (15.8\text{-}3)$$

Using the Fraunhofer approximation, the source field at the apertures can be considered as point sources approximated by narrow sinc functions. Thus, the integral in Eq. (15.8-2) becomes a double summation:

$$U(x_o, y_o, z_o) = \frac{\cos \delta}{j\lambda R} d_x d_y \sum \sum \exp(j\theta_i) \operatorname{sinc}\left(\frac{X_i d_x}{\lambda r_{oi}}\right) \operatorname{sinc}\left(\frac{Y_i d_y}{\lambda r_{oi}}\right) \qquad (15.8\text{-}4)$$

where

$$X_i = x_o - x_{si}$$
$$Y_i = y_o - y_{si} \qquad (15.8\text{-}5)$$
$$R = \text{average value of } r_{oi}$$

if all θ_i are set equal to θ, and x_i, $y_i \ll r_{oi}$, the sinc functions can be replaced by 1, and

$$U(x_o, y_o, z_o) = \frac{\cos \delta}{j\lambda R} d_x d_y N e^{j\theta} \qquad (15.8\text{-}6)$$

Therefore, the amplitude of the field will be proportional to $d_x d_y N$, and its phase will be θ for a plane wave incident on the hologram at right angle, namely, an on-axis plane wave. If the incoming wave has phase variations on the hologram, its phase being Φ_i at each hologram point, then Eq. (15.8-2) should be written as

$$kr_{oi} + \Phi_i = 2\pi n + \theta_i \quad n = \text{integer} \qquad (15.8\text{-}7)$$

The aperture locations (x_i, y_i) can be chosen such that the resulting θ_i will be a constant. An object point is then obtained since all wave fronts generated by the hologram apertures will add up in phase at the specified object point location. Thus, the amplitude of the field will be proportional to $d_x d_y N$, and its phase will be 0 for a plane incoming wave incident on the hologram at right angle. If there are groups of such apertures that satisfy Eq. (15.8-6) at different points in space, a sampled wavefront is essentially created with a certain amplitude and phase at each object point. We note that the modulation of Eq. (15.8-6) is very simple. We vary d_x and/or d_y and/or N for amplitude and θ for phase. The fact that N is normally a large number means that it can be varied almost continuously so that amplitude modulation can be achieved very accurately.

If amplitude modulation accuracy does not need to be very high, the method is valid in the near field as well because r_{oi} is exactly computed for each object point. If the apertures are circular, the sinc functions are replaced by a first-order Bessel function, but Eq. (15.8-6) essentially remains the same.

In practice, each point-aperture on the hologram plane is first chosen randomly and then moved slightly in the x- and/or y-direction so that its center coordinates satisfy Eq. (15.8-7) with constant θ. Overlapping of the apertures is considered negligible so that there is no need for using memory.

15.8.1   Experiments

The method described above was used to test implementation of a DOE with a scanning electron microscope [Ersoy, 1976]. The working area for continuous exposure was 2×2 mm. The number of point apertures with the smallest possible diameter of about 1 μ was 4096×4096. In the experiments performed, the hologram material used was either KPR negative photoresist or PMMA positive photoresist.

Figures 15.17 and 15.18 show the reconstructions from two holograms that were produced. They were calculated at the He–Ne laser wavelength of 0. 6328 μ. In Figure 15.17, eleven points were chosen on a line 3 cm long, satisfying $z = 60$ cm, $x = 4$ cm, $0 \le y \le 3$ cm on the object plane. The number of hologram

Figure 15.17. Reconstruction of eleven points on a line 3 cm long in space from a DOE generated with a scanning electron microscope.

apertures used were 120,000, each being 8×8 adjacent points in size. The image points were chosen of equal intensity. The picture was taken at approximately 60 cm from the hologram, namely the focal plane. The main beam was blocked not to overexpose the film. It is seen that there are both the real and the conjugate images.

In order to show the effect of three-dimensionality, the object shown in Figure 15.18 was utilized. Each of the four letters was chosen on a different plane in space. The distances of the four planes from the hologram plane were 60, 70, 80, and 90 cm, respectively. If all the letters were on a single plane, the distance in the x-direction between them would be 1 cm. In the picture it is seen that this distance is decreasing from the first to the last letter because of the depth effect. The picture was taken at approximately 90 cm from the hologram, namely, the focal plane of the letter E. That is why E is most bright, and L is least bright. The number of hologram apertures used were 100,000, each being 4×4 adjacent points in size. The image points were chosen of equal intensity.

Figure 15.18. Reconstruction of the word LOVE in 3-D space. Each letter is on a different plane in space.

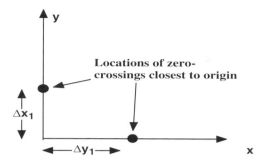

Figure 15.19. Zero-crossings closest to the origin.

15.9 THE FAST WEIGHTED ZERO-CROSSING ALGORITHM

The algorithm discussed in Section 5.7 corresponds to choosing a number of zero crossings of phase for each spherical wave at random positions on the hologram. A disadvantage of this approach is that the hologram quickly saturates as the number of object points increases. Additionally, it is also computationally intensive. The fast weighted zero-crossing algorithm (FWZC) is devised to combat these problems [Bubb, Ersoy].

For each object point to be generated with coordinates (x_o, y_o, z_o), the following procedure is used in the FWZC algorithm:

1. Calculate the zero-crossings $\Delta x_1, \Delta y_1$ of phase closest to the origin on the x-axis and the y-axis, respectively, as shown in Figure 15.19.

   Let $(\Delta x_1, 0.0)$ be the location of the zero-crossing on the x-axis. Then,

$$r_{oi} = \sqrt{(x_o - \Delta x_1)^2 + y_o^2 + z_o^2} = n\lambda \qquad (15.9\text{-}1)$$

$$r'_{oi} = \sqrt{x_o^2 + y_o^2 + z_o^2} = n\lambda + B \qquad (15.9\text{-}2)$$

   Solving these equations for Δx_1 yields

$$\Delta x_1 = x_o \pm \sqrt{x_o^2 + (B^2 - 2Br'_{oi})} \qquad (15.9\text{-}3)$$

   An entirely similar expression can be derived for Δy_1.

2. Using Heron's expression [Ralston], calculate r_x and r_y, the radial distances from these zero-crossings to the object point. For example, r_x is

derived as

$$r_x = z_1 \left[3 + x_1^2 + y_1^2 - \frac{1}{(x_1^2 + y_1^2 + 1)} \right] \qquad (15.9\text{-}4)$$

$$z_1 = \frac{z_o}{2} \qquad (15.9\text{-}5)$$

$$z_2 = \frac{1}{\sqrt{2z_o}} \qquad (15.9\text{-}6)$$

$$x_1 = (x_o - \Delta x_1)z_2 \qquad (15.9\text{-}7)$$

$$y_1 = y_o z_o \qquad (15.9\text{-}8)$$

3. Determine all the zero-crossings on the x-axis. These are calculated by starting at an arbitrary zero-crossing on the x-axis, say, $x = a$, and calculating the movement required to step to the next zero-crossing, say, $x = b$. This is done by approximating the distance function $r_x = \sqrt{(x - x_o)^2 + y_o^2 + z_o^2}$ with the first two terms of its Taylor series. The result is given by

$$\Delta x = b - a = \frac{r(a)\lambda}{x_o - a} + \frac{1}{2(x_o - a)} \left[\left(\frac{r(a)\lambda}{x_o - a} \right)^2 - \lambda^2 \right] \qquad (15.9\text{-}9)$$

When doing the same procedure in the negative x-direction such that $r(b) - r(a) = \lambda$, a similar analysis shows that the first term on the right-hand side of Eq. (15.9-9). changes sign.

4. An identical procedure can be used to find all of the zero-crossings for both the positive and the negative y-axis.

5. It is straightforward to show that for any $x = x_{11}$ and $y = y_{11}$, if $(x_{11}, 0)$ and $(0, y_{11})$ are zero-crossings, then (x_{11}, y_{11}) is also zero-crossing if the Fresnel approximation is valid.

   Utilizing all of the "fast" zero-crossings on the x- and y-axis, we form a grid of zero-crossings on the hologram plane. It is also possible to generate the remaining zero-crossings by interpolation between the fast zero-crossings.

6. The grid of zero-crossings generated as shown in Figure 15.20 indicate the centers of the locations of the apertures to be generated in the recording medium to form the desired hologram.

Once the grid of zero-crossings for each object point is generated, their locations are noted by assigning a "hit" (for example, one) to each location. After zero-crossings for all the object points have been calculated, each aperture location will have accumulated a number of hits, ranging from zero to the number of object points. This is visualized in Figure 15.21.

For object scenes of more than trivial complexity, the great majority of aperture locations will have accumulated a hit. Next, we set a threshold number of hits, and only encode apertures for which the number of hits exceeds the assigned threshold.

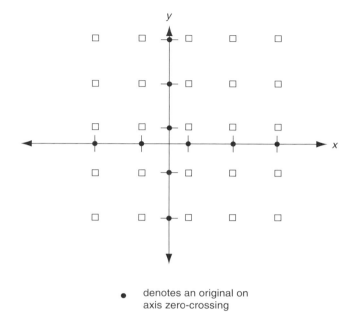

•   denotes an original on
     axis zero-crossing

☐   denotes the new zero-crossings
     at the intersection points

Figure 15.20.   The grid of zero-crossings generated.

Increment counter value	Number of apertures used at least this many times
⋮	⋮
30	46,748
29	64,816
28	71,903
27	86,274
26	95,816
25	106,892
24	145,981
⋮	⋮

Figure 15.21.   Accumulation of zero-crossings at hologram points.

In this way, we choose the most important zero-crossings, for example, the ones that contribute the most to the object to be reconstructed. One scheme that works well in practice is to set the threshold such that the apertures generated cover roughly 50% of the hologram plane.

One problem does exist with this thresholding scheme. Consider the spacing between zero-crossings on the x-axis given by Eq. (15.9-9). As x_o increases, the spacings between zero-crossings decrease. This causes object points distant from the center of the object to have a greater number of zero-crossings associated with them. To attempt to correct for this, the number of hits assigned for each zero-crossing can be varied so that the total number of hits per object point is constant. One possible way to implement this approach is to use

$$\text{no hits per zero-crossing} = \frac{A}{\text{no zero-crossings for the current object point}}$$

$$(15.9\text{-}10)$$

where A is a suitably large constant.

15.9.1 Off-Axis Plane Reference Wave

An off-axis plane-wave tilted with respect to the x-axis can be written as

$$U(x, y) = e^{-j2\pi\alpha x} \qquad (15.9\text{-}11)$$

where

$$\alpha = \frac{\sin\theta}{\lambda} \qquad (15.9\text{-}12)$$

θ being the angle with respect to the optical axis. Using this expression, Eq. (15.9-1) becomes

$$r_{oi} = n\lambda + x\sin\theta \qquad (15.9\text{-}13)$$

The rest of the procedure is similar to the procedure with the on-axis plane wave method.

15.9.2 Experiments

We generated a number of holograms with the methods discussed above [Bubb, Ersoy]. A particular hologram generated is shown in Figure 15.22 as an example. The object reconstruction from it using a He–Ne laser shining on the hologram transparency of reduced size is shown in Figure 15.23.

The reconstruction in Figure 15.23 consists of seven concentric circles, ranging in radius from 0.08 to 0.15 m. The circles are on different z-planes in space: the smallest at 0.75 m, and the largest at 0.6 m. The separation in the z-direction can clearly be seen, in that the effects of perspective and foreshortening are evident. In all the experiments carried out, no practical limit on the number of object points was discovered.

Figure 15.22.   A hologram generated with the FWZC method.

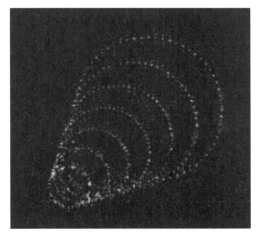

Figure 15.23.   The reconstruction of concentric circles from the hologram of Figure 15.22.

15.10 ONE-IMAGE-ONLY HOLOGRAPHY

For a plane perpendicular reference wave incident on the hologram, it is easy to observe that the existence of the twin images with various encoding techniques is due to the symmetry of the physical spaces on either side of the plane hologram. If we think of the object wave as a sum of spherical waves coming from individual object points, points that are mirror images of each other with respect to the hologram plane, correspond to the object waves on the hologram with the same amplitude and the opposite phase. Thus, when we choose a hologram aperture i that corresponds to the phase ϕ_i of the virtual image, the same aperture corresponds to the phase $-\phi_i$ of the real image.

The conclusion is that the symmetry of the two physical spaces with respect to the hologram needs to be eliminated to possibly get rid of one of the images. This symmetry can be changed either by choosing a hologram surface that is not planar, or a reference wave that is not a plane-perpendicular wave. However, choosing another simple geometry such as an off-axis plane wave distorts only the symmetry and results in images that are at different positions than before.

An attractive choice is a spherical reference wave because it can easily be achieved with a lens [Ersoy, 1979]. If it becomes possible to reconstruct only the real image, the focal point of the lens can be chosen to be past the hologram so that the main beam and the zero-order wave can be filtered out by a stop placed at the focal point. In the following sections, we are going to evaluate this scheme, as shown in Figure 15.24.

In the encoding technique discussed in Section 15.7, the position of each hologram aperture was chosen according to

$$\varphi(x_i, y_i) + kr_{oi} = 2n\pi + \phi_0 \qquad (15.10\text{-}1)$$

In the present case, $\varphi(x_i, y_i)$ is the phase shift caused by the wave propagation from the origin of the reference wave front at (x_c, y_c, z_c) to the hologram aperture at (x_i, y_i); kr_{oi} is the phase shift caused by the wave propagation from the aperture at (x_i, y_i) on hologram to an object point located at (x_o, y_o, z_o). The radial distance r_{oi} is given by

$$r_{oi} = \pm\sqrt{(x_o - x_i)^2 + (y_o - y_i)^2 + z_o^2} \qquad (15.10\text{-}2)$$

$+ (-)$ sign is to be used if the object is desired to be real (virtual).

For a spherical reference wave with its focal point at the position (x_c, y_c, z_c), the phase of the reference wave $\varphi(x_i, y_i)$ can be written as

$$\varphi(x_i, y_i) = kr_{ci} \qquad (15.10\text{-}3)$$

where

$$r_{ci} = \pm\sqrt{(x_c - x_i)^2 + (y_c - y_i)^2 + z_c^2} \qquad (15.10\text{-}4)$$

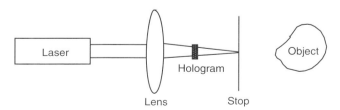

Figure 15.24.   The setup for one-image- only holography.

where $+ (-)$ sign is to be used if the focal point of the lens is in the front (back) of the hologram.

Equation (15.10-1) can be written as

$$r_{ci} + r_{oi} = n\lambda + \frac{\phi_0\lambda}{2\pi} \tag{15.10-5}$$

Suppose that the position of the aperture is to be moved a distance Δ in a defined direction, say the x-direction, such that Eq. (15.10-5) is satisfied. Then, the new lengths of the radius vectors r'_{oi} and r'_{ci} are given by

$$r'_{oi} = \sqrt{r_{oi}^2 - 2\Delta(x_0 - x_i) + \Delta^2} \tag{15.10-6}$$

$$r'_{ci} = \sqrt{r_{ci}^2 - 2\Delta(x_c - x_i) + \Delta^2} \tag{15.10-7}$$

$r'_{ci} + r'_{oi}$ satisfy

$$r'_{ci} + r'_{oi} = r_{ci} + r_{oi} - B \tag{15.10-8}$$

Solving Eq. (15.10-8) for Δ yields

$$\Delta = C_1[1 - \sqrt{1 - C_2/C_1}] \tag{15.10-9}$$

where

$$C_1 = \frac{F_1}{F_2} \tag{15.10-10}$$

$$C_2 = \frac{F_3}{F_1} \tag{15.10-11}$$

$$F_1 = F_4(X + X_c) + Xr_{ci}^2 + X_c r_{oi}^2 \tag{15.10-12}$$

$$F_2 = (r_{ci} + r_{oi})^2 - (X - X_c)^2 - 2F_5 \tag{15.10-13}$$

$$F_3 = 2F_5 r_{oi} r_{ci} - F_5^2 \tag{15.10-14}$$

$$F_4 = r_{oi} r_{ci} - F_5 \tag{15.10-15}$$

$$F_5 = -\frac{B^2}{2} + B(r_{ci} + r_{oi}) \tag{15.10-16}$$

We note that amplitude modulation of the object points can still be achieved by varying the number of hologram apertures for each object point proportional to the desired amplitude at the object point. Even though the amplitude of the spherical wave varies slowly at the hologram, the average effect from randomly distributed apertures can be considered to be constant.

15.10.1 Analysis of Image Formation

In order to analyze the method, the lengths of the radii will be expanded in the paraxial approximation [Meier, 1966]. Instead of a spherical reference wave, suppose that a plane reference wave given by

$$U_r = A e^{jk(\alpha x + \beta y + \gamma z)} \qquad (15.10\text{-}17)$$

is used. Then, Eq. (15.10-5) can be written in the paraxial approximation as

$$x_i \left(\alpha - \frac{x_0}{z_0} \right) + y_i \left(\beta - \frac{y_0}{z_0} \right) + \frac{x_i^2 + y_i^2}{2z_0} = n\lambda + \frac{\phi_0 \lambda}{2\pi} \qquad (15.10\text{-}18)$$

where the sign of z_0 should be chosen the same as the sign used in Eq. (15.10-2).

If a spherical reference wave as determined by Eq. (15.9-3) is used instead, Eq. (15.10-5) can be written as

$$-x_i \left(\frac{x_c}{z_c} + \frac{x_0}{z_0} \right) - y_i \left(\frac{y_c}{z_c} + \frac{y_0}{z_0} \right) + \left(\frac{x_i^2 + y_i^2}{2} \right) \left(\frac{1}{z_0} + \frac{1}{z_c} \right) = n\lambda + \frac{\phi_0 \lambda}{2\pi} - F_{0c}$$

$$(15.10\text{-}19)$$

where

$$F_{0c} = z_0 + \frac{x_0^2 + y_0^2}{2z_0} + z_c + \frac{x_c^2 + y_c^2}{2z_c} \qquad (15.10\text{-}20)$$

Above the signs of z_0 and z_c should be chosen the same as the signs used in Eqs. (15.10-2) and (15.10-3).

An image forms whenever Eq. (15.10-18) or (15.10-19) is valid for arbitrary n, disregarding the constant terms $\phi_0 \lambda / 2\pi$ and F_{0c}. It is observed that if Eq. (15.10-18) is valid for the object point at (x_0, y_0, z_0), the reference wave with direction cosines (α, β, γ), and wavelength λ, it is also valid for an object point at (x_0', y_0', z_0'), the reference wave with direction cosines $(\alpha', \beta', \gamma')$, and wavelength λ' such that

$$z_0' = \frac{\lambda}{\lambda'} \frac{z_0}{m} \qquad (15.10\text{-}21)$$

$$x_0' = \frac{\lambda}{\lambda'} \left[\frac{\alpha' - m\alpha}{m} z_0 + x_0 \right] \qquad (15.10\text{-}22)$$

$$y_0' = \frac{\lambda}{\lambda'} \left[\frac{\beta' - m\beta}{m} z_0 + y_0 \right] \qquad (15.10\text{-}23)$$

where m is an integer. m equal to 1 and –1 corresponds to real and virtual images while m equal to 0 corresponds to the zeroth order wave, a plane wave traveling in the same direction as the original reference wave. Other values of m correspond to higher order images.

By the same token, if Eq. (15.10-19) is valid for the object point at (x_0, y_0, z_0) and the reference wave with focal point at (x_c, y_c, z_c), whose wavelength is λ, it is also valid for the object point at (x_0', y_0', z_0') and the reference wave with focal point at (x_c', y_c', z_c'), whose wavelength is λ' such that

$$\frac{1}{z_0'} = \frac{\lambda'}{\lambda} m \left(\frac{1}{z_0} + \frac{1}{z_c} \right) \tag{15.10-24}$$

$$x_0' = z_0' \left[\frac{\lambda}{\lambda'} m \left(\frac{x_c}{z_c} + \frac{x_0}{z_0} \right) - \frac{x_c'}{z_c'} \right] \tag{15.10-25}$$

$$y_0' = z_0' \left[\frac{\lambda}{\lambda'} m \left(\frac{y_c}{z_c} + \frac{y_0}{z_0} \right) - \frac{y_c'}{z_c'} \right] \tag{15.10-26}$$

where m is an integer. We note that $m = 0$ corresponds to a wave that is the same as the reference wave and focuses at the focal point. If $z_0 \gg z_c$, and assuming the initial and the final reference waves are the same, the image positions other than $m = 1$ are approximately given by

$$z_0' \simeq \frac{z_c}{m - 1} \tag{15.10-27}$$

$$x_0' \simeq x_c + m \frac{z_0'}{z_0} x_0 \tag{15.10-28}$$

$$y_0' \simeq y_c + m \frac{z_0'}{z_0} y_0 \tag{15.10-29}$$

In other words, if the focal point of the lens is sufficiently close to the hologram, all the images other than the one for $m = 1$ have to be so close to the hologram that they become completely defocused in the far distance where the desired image is.

However, it is possible to focus one order at a time. If the mth order is desired to be at the position (x_0', y_0', z_0'), the corresponding values of (x_c', y_c', z_c') are simply found such that Eqs. (15.10-24), (15.10-25), and (15.10-26) are satisfied. Then, all the other images become completely defocused and out of view in the far distance where the desired image is, because they are located very close to the hologram.

The above analysis is not always reasonable because the paraxial approximation becomes difficult to justify as the focal point of the spherical reference wave approaches the hologram. Thus, the images other than the one desired may not be well defined. It is also possible that the encoding technique used loses its validity for very close distances to the hologram, especially due to the registration errors of the hologram apertures.

In order to support the above arguments, consider what happens when $z_c' \to \infty$ in Eqs. (15.10-24)–(15.10-26); in other words, a plane wave is used for reconstruction. Again assuming $z_0 \gg z_c$, and λ equal to λ', we get

$$z_0' \simeq \frac{z_c}{m} \tag{15.10-30}$$

$$x_0' \simeq x_c + \frac{z_c}{z_0} x_0 \tag{15.10-31}$$

$$y_0' \simeq y_c + \frac{z_c}{z_0} y_0 \tag{15.10-32}$$

The above equations indicate that the images must focus close to the hologram. Any absence of images in the space of interest indicates that the interference between the harmonic images and the desired image is minimized.

It is also interesting to observe what happens to the positions of the higher order images as z_c gets smaller. From Eq. (15.10-24), for equal wavelengths, we find

$$z'_0 \simeq \frac{z_c z'_c}{m z'_c - z_c} \tag{15.10-33}$$

As m increases, z'_0 approaches z_c/m. This means that it becomes more difficult to observe higher order images as the hologram is designed closer to the focal point of the lens.

A very important consequence of Eq. (15.10-33) is that there is a variable focusing distance; z'_0 can be varied at will by a slight adjustment of the position of the lens.

15.10.2   Experiments

In order to prove the above predictions experimentally, holograms were generated using the scanning electron microscope system, as discussed in Section 15.7.1. In all cases, the sizes of the holograms were 2×2 mm, and the sizes of the apertures were of the order of 1μ.

Figure 15.25 shows the He–Ne laser reconstruction from a regular hologram of one object point with the plane-perpendicular reference wave. At the center, the overexposed main beam is observed; to the right of the main beam, there are the slightly overexposed real image of the object point and higher order images; to the left of the main beam, there are the virtual image and virtual higher order images.

Three holograms of an object point were generated using a spherical reference wave such that the focal point of the lens is 1, 3, and 5 cm to the left of the hologram plane, and its x- and y-coordinates are those of the center point of the hologram. Figure 15.26 shows the result with the 5-cm hologram. There is only one visible object point; the dark square is the enlarged picture of the hologram; the main beam covers the whole figure. Figure 15.27 shows the same object point with the main beam and zero-order image filtered out by putting a stop at the focal point of the lens. The results with the 1- and 3-cm holograms gave the same type of results supporting the arguments given above [Ersoy, 1979].

Figure 15.28 shows the reconstruction of a more complicated object. The hologram is designed to be 3 cm from the focal point of the lens, whose x- and

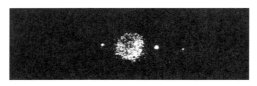

Figure 15.25.   The He–Ne laser reconstruction from a regular hologram of one object point showing different orders.

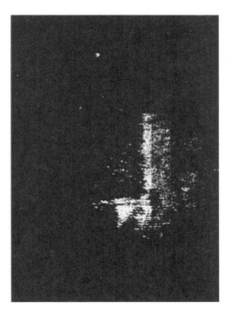

Figure 15.26.   The single visible image from the 5-cm hologram without filtering.

y-coordinates are the same as those of the center point of the hologram. During the experiments, it was observed that it is a simple matter to focus the image at different distances, and thereby control the size of the image as well, by a slight adjustment of the position of the lens, in confirmation of Eq. (15.10-33).

Figure 15.27.   The single visible image from the 5-cm hologram with filtering.

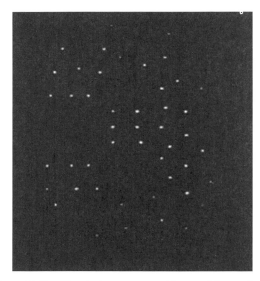

Figure 15.28.   The single reconstructed image of the number 3.

15.11   FRESNEL ZONE PLATES

After covering more complicated designs for a DOE, we go back historically and discuss a classical DOE called Fresnel zone plate (FZP). Such a DOE serves as a flat lens, can be mass-manufactured and is commonly used in many technologies such as overhead projectors. In this section, we will discuss a binary FZP that consists of opaque and transparent circular zones. An example is shown in Figure 15.29.

A side view of a FZP is shown in Figure 15.30.

Figure 15.29.   A binary Fresnel zone plate.

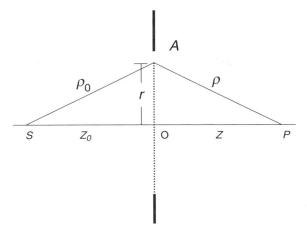

Figure 15.30. Side view of a FZP.

The path difference between two rays traveling along SOP and SAP is given by

$$\Delta(r) = (\rho_0 + \rho) - (z_0 + z) = \sqrt{r^2 + z_0^2} + \sqrt{r^2 + z^2} - (z_0 + z) \qquad (15.11\text{-}1)$$

The Fresnel-zone parameter, n, is defined such that the path difference is an integer multiple of half wavelengths [Hecht]:

$$n\frac{\lambda}{2} = \Delta(r) \qquad (15.11\text{-}2)$$

The nth *Fresnel zone* is the area between the circles with radii r_{n-1} and r_n. Note that the field at P coming from a point on the circle with radius r_n is half a wavelength out of phase with the field from a point on the circle with radius $r_{n\pm1}$. Using this fact, it is easy to show that the adjacent zones cancel each other out. Therefore, the total field at P will increase if either all even or all odd zones are blocked out, the remaining zones reinforce each other, thus creating a focal point at distance z. Assuming the zone plate is illuminated by a plane wave, $z_0 \approx \infty$, and setting $z = f_o$, Eq. (15.11-1) results in

$$\frac{r_n}{\lambda} = \sqrt{\frac{f_0}{\lambda}\left(n - \frac{1}{2}\right) + \frac{(2n-1)^2}{16}} \qquad (15.11\text{-}3)$$

f_0 is called the focal point. An FZP operates as a lens with focal length f_0. It can be shown that there exist other focal points at $f_0/3$, $f_0/5$, $f_0/7$, and so on [Hecht].

Let R_n be the radius of the nth circle on the FZP. It satisfies

$$R_n^2 = \left(f_o + \frac{n\lambda}{2}\right)^2 - f_o^2 \qquad (15.11\text{-}4)$$

Using Eq. (15.11-3), this can be written as

$$R_n^2 = f_o^2 \left[\frac{n\lambda}{f_o} + \frac{1}{4} \left(\frac{n\lambda}{f_o} \right)^2 \right] \tag{15.11-5}$$

When n is large, the second term on the right-hand side above can be neglected, yielding

$$R_n \simeq \sqrt{n\lambda f_o} \tag{15.11-6}$$

16

Diffractive Optics II

16.1 INTRODUCTION

This chapter is a continuation of Chapter 15, covering more methods on diffractive optics. It consists of seven sections. Section 16.2 is on the method of virtual holography which makes DOEs easier to handle by accommodating to the requirements of technologies of implementation. It also brings closer holography and classical optics.

Section 16.3 discusses the method of POCS discussed in Section 14.8 to design binary DOEs. Section 16.4 incorporates the POCS method in a novel approach called iterative interlacing technique. In this method, subholograms are created, and the iterative design of subholograms is based on minimizing the remaining reconstruction error at each iteration.

Section 16.5 combines IIT and the decimation-in-frequency property used in FFT algorithms to come up with a better and faster strategy to design the subholograms in a method called ODIFIIT. Section 16.6 further generalizes the constant amplitude Lohmann method discussed in Section 15.4 in quantization of amplitude and phase in each Lohmann cell. The POCS method is used for the optimization of the subholograms. Section 16.7 combines the Lohmann and ODIFIIT methods, resulting in considerably higher accuracy in terms of reduced mean square error in the desired image region.

16.2 VIRTUAL HOLOGRAPHY

It is known that most holograms are redundant, and only a small part of the hologram is sufficient to generate the desired information. If the hologram is to be viewed with the unaided eye, the resolution of image points is determined by the size of the eye lens aperture, which is of the order of several millimeters [Born and Wolf, 1969]. One desires to have as much information as possible in a hologram of this size.

When the hologram is recorded with a certain kind of wave, such as acoustical waves or microwaves, and reconstruction is done with another kind of wave, such as

---

Diffraction, Fourier Optics and Imaging, by Okan K. Ersoy
Copyright © 2007 John Wiley & Sons, Inc.

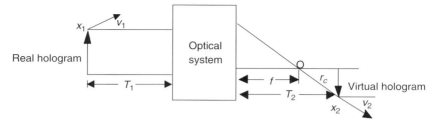

Figure 16.1.   System diagram for virtual holography.

visible light, it is necessary to change the hologram size at the ratio of the two wavelengths in order to prevent image distortions [Smith, 1975]. However, this is a time-consuming and error-prone operation.

A *virtual hologram* is defined as the hologram that is not recorded in a medium but exists in space as the image of another hologram that is recorded, and which is called the *real hologram* [Ersoy, August 79]. The information coming from the virtual hologram is the desired information whereas the information coming from the real hologram consists of transformed information. The transformation between the real hologram and the virtual hologram is achieved with an optical system as shown in Figure 16.1.

It can be said that neither the real hologram nor the virtual hologram is exactly like regular holograms. If one looks through the real hologram under reconstruction, one sees transformed information that is probably unrecognizable. The virtual hologram is more like a regular hologram, but it is not registered in a physical medium.

16.2.1   Determination of Phase

The rays coming from the real hologram parallel to the optical axis converge to point O as shown in Figure 16.1. Since the optical path lengths between a real hologram point and the corresponding virtual hologram point are the same, the phase at the virtual hologram point relative to the other virtual hologram points are determined by the radius vector length r_c between O and the virtual hologram point as shown in Figure 16.1.

The system transfer matrix $\mathbf{S}$ that connects the input and the output in the form [Gerrard and Burch, 1975]

$$\begin{bmatrix} x_2 \\ v_2 \end{bmatrix} = \mathbf{S} \begin{bmatrix} x_1 \\ v_1 \end{bmatrix} \tag{16.2-1}$$

can be determined as

$$\mathbf{S} = \begin{bmatrix} A' & B' \\ C' & D' \end{bmatrix} \tag{16.2-2}$$

where

$$A' = A + CT_2 \tag{16.2-3}$$

$$B' = (A + CT_2)T_1 + B + DT_2 \tag{16.2-4}$$

$$C' = C \tag{16.2-5}$$

$$D' = CT_1 + D \tag{16.2-6}$$

A, B, C, and D determine the optical system matrix; T_1 and T_2 are the distances shown in Figure 16.1. In order to have image generation, $B' = 0$ so that

$$M = \frac{x_2}{x_1} = A' = \frac{1}{D'} \tag{16.2-7}$$

A plane wave input to the optical system at an angle v_1 focuses to a point at a distance

$$f = -\frac{A}{C} \tag{16.2-8}$$

with the lateral coordinate x_2 given by

$$x_2 = (B + Df)v_1 \tag{16.2-9}$$

Using these equations, it is straightforward to calculate r_c shown in Figure 16.1.

r_c determines the type of the reference wave on the virtual hologram. The phase due to the reference wave on the real hologram is also to be transferred to the virtual hologram. Therefore, it makes sense to talk of the real reference wave and the virtual reference wave, respectively. This approach can be further extended by having several optical systems that give rise to several virtual holograms and reference waves. The end result would be to add all the reference waves on the last virtual hologram. However, the phase due to each reference wave would be determined by its position on the corresponding hologram.

Three examples will be considered. The first one is a single lens. This case corresponds to

$$A = 1, \quad B = 0, \quad C = -\frac{1}{f}, \quad D = 1 \tag{16.2-10}$$

so that

$$\frac{1}{T_1} + \frac{1}{T_2} = \frac{1}{f} \tag{16.2-11}$$

$$M = -\frac{T_2}{T_1} \tag{16.2-12}$$

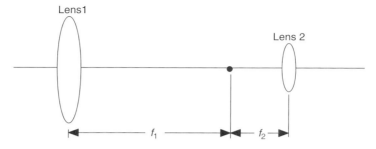

Figure 16.2. The telescopic system.

The second example is the telescopic system shown in Figure 16.2. Here f equals ∞ so that the virtual reference wave is planar. Using two lenses with focal lengths f_1 and f_2 yields

$$A' = -\frac{1}{F} = -\frac{f_2}{f_1} \qquad (16.2\text{-}13)$$

$$B' = f_1 + f_2 \qquad (16.2\text{-}14)$$

$$C' = 0 \qquad (16.2\text{-}15)$$

$$D' = -F \qquad (16.2\text{-}16)$$

such that

$$M = -\frac{1}{F} = -\frac{f_2}{f_1} \qquad (16.2\text{-}17)$$

$$T_2 = \frac{f_1 + f_2}{F} - \frac{T_1}{F^2} \qquad (16.2\text{-}18)$$

It is seen that M is independent of the real hologram coordinates, and, if F is large, T_2 is insensitive to T_1.

The third example is the lensless Fourier arrangement discussed in Section 13.3. It is obtained when the point O of Figure 16.1 lies on the image plane. If z_0 is the distance from the virtual hologram to the image plane, we have

$$\frac{A}{C} = -(T_2 + z_0) \qquad (16.2\text{-}19)$$

16.2.2   Aperture Effects

In order to obtain effective diffraction from the virtual hologram, it is desirable that virtual hologram apertures spread waves as much as possible. An interesting observation here is that aperture size may not be important in this context, since the virtual hologram is not to be recorded. In other words, even if virtual hologram apertures are overlapping, diffraction effects at a distance can be explained in terms of waves coming from point sources on the virtual hologram to interfere with each other in the volume of interest.

Spreading of waves can be discussed in terms of v_1 and v_2, the angles a ray makes at two reference planes. From Eq. (16.2-2) we find

$$v_2 = C'x_1 + D'v_1 \qquad (16.2\text{-}20)$$

In the imaging planes, the following is true:

$$C' = -\frac{1}{f} \qquad (16.2\text{-}21)$$

$$D' = \frac{1}{M} \qquad (16.2\text{-}22)$$

It follows that spreading of waves increases as f and M are reduced. In the telescopic system, $C' = 0$, and the input spreading of waves coming from an aperture is increased at the output by a factor of $1/M$.

In order to find the size of the virtual hologram apertures, both magnification and diffraction effects need to be considered. The size of a virtual hologram aperture d_v can be written as

$$d_v = Md_r + D \qquad (16.2\text{-}23)$$

where d_r is the size of the real hologram aperture, and D is the additional size obtained due to diffraction coming from the limited size of the optical system. For example, in the telescopic system, D can be approximated by [Gerrard and Burch, 1975]

$$D = \frac{2.44f\lambda}{d_A} \qquad (16.2\text{-}24)$$

where d_A is the diameter of the telescope objective, and f is its focal length.

16.2.3 Analysis of Image Formation

Analysis of image formation can be done in a similar way to the analysis of image formation in Section 15.8 on one-image-only holography. We assume that the contributions of various reference waves result in an effective reference wave coming from the point (x_c, y_c, z_c). This can always be done within the paraxial approximation. If $(x_i, y_i, 0)$ are the coordinates of a sample point on the virtual hologram, and (x_0, y_0, z_0) are the coordinates of an object point, the equation that determines the positions of various image harmonics is given by

$$-x_i\left(\frac{x_c}{r_c} + \frac{x_0}{r_0}\right) - y_i\left(\frac{y_c}{y_c} + \frac{y_0}{y_0}\right) + \left(\frac{x_i^2 + y_i^2}{2}\right)\left(\frac{1}{r_0} + \frac{1}{r_c}\right) = n\lambda + \frac{\phi\lambda}{2\pi} - F_{0c}$$

$$(16.2\text{-}25)$$

where ϕ is some constant phase, and

$$F_{0c} = r_0 + r_c \qquad (16.2\text{-}26)$$

$$r_0 = \pm(x_0^2 + y_0^2 + z_0^2)^{\frac{1}{2}} \qquad (16.2\text{-}27)$$

where $+(-)$ sign is used if the object point is real (virtual) with respect to the virtual hologram, and

$$r_c = \pm(x_c^2 + y_c^2 + z_c^2)^{\frac{1}{2}} \qquad (16.2\text{-}28)$$

where $+(-)$ sign is used if the focal point of the reference wave comes before (after) the virtual hologram.

If Eq. (16.2-25) is divided by M, we find that the real hologram is designed for the object coordinates

$$r_0' = \frac{r_0}{M} \qquad (16.2\text{-}29)$$

$$x_0' = \frac{x_0}{M} \qquad (16.2\text{-}30)$$

$$y_0' = \frac{y_0}{M} \qquad (16.2\text{-}31)$$

$$z_0' = (r_0'^2 - x_0'^2 - y_0'^2) \qquad (16.2\text{-}32)$$

and the reference wave originating from

$$r_c' = \frac{r_c}{M} \qquad (16.2\text{-}33)$$

$$x_c' = \frac{x_c}{M} \qquad (16.2\text{-}34)$$

$$y_c' = \frac{y_c}{M} \qquad (16.2\text{-}35)$$

$$z_c' = (r_c'^2 - x_c'^2 - y_c'^2) \qquad (16.2\text{-}36)$$

and the wavelength

$$\lambda' = \frac{\lambda}{M} \qquad (16.2\text{-}37)$$

If a reconstruction wave coming from (x_c'', y_c'', z_c'') and with wavelength λ is used, the mth harmonic without the optical system will be reconstructed at

$$\frac{1}{r_0''} = \frac{m}{M^2}\left[\frac{1}{r_0} + \frac{1}{r_c}\right] - \frac{1}{r_c''} \qquad (16.2\text{-}38)$$

$$x_0'' = r_0''\left\{mM\left[\frac{x_0}{r_0} + \frac{x_c}{r_c}\right] - \frac{x_c''}{r_c''}\right\} \qquad (16.2\text{-}39)$$

$$y_0'' = r_0''\left\{mM\left[\frac{y_0}{r_0} + \frac{y_c}{r_c}\right] - \frac{y_c''}{r_c''}\right\} \qquad (16.2\text{-}40)$$

Equations (16.2-38)–(16.2-40) indicate that there is also another object distorted with respect to the desired object. In analogy with the concept of virtual hologram, it will be called the *virtual object*. This means that we can start either from the virtual hologram to generate the real object or from the real hologram to generate the virtual object. It looks like it is easier to use the virtual hologram concept in most cases. However, there are occasions when it is more advantageous to use the virtual object concept.

Assuming plane perpendicular reconstruction wave $(x_c = y_c = 0)$, Eqs. (16.2-38)–(16.2-40) for the first harmonic reduce to

$$r_0'' = \frac{R}{M} r_0 \qquad (16.2.41)$$

$$x_0'' = R x_0 \qquad (16.2.42)$$

$$y_0'' = R y_0 \qquad (16.2.43)$$

$$R = \frac{r_c}{M(z_0 + r_c)} \qquad (16.2.44)$$

The last four equations indicate one computational advantage that can be achieved, namely, the use of convolution to determine the hologram. In digital holography, normally the hologram is much smaller than the image. Convolution requires the image and hologram sizes to be the same. With this method, one can obtain equal sizes by choosing R properly. For the case of a single lens, this is satisfied by

$$R = \frac{f}{r_0 + r_c} \qquad (16.2.45)$$

The appearance of various holograms and objects in this case is shown in Figure 16.3.

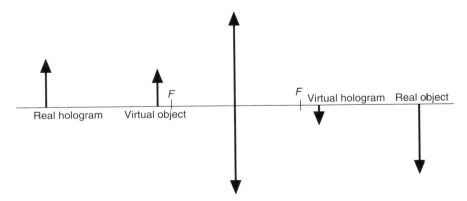

Figure 16.3. Virtual and real holograms and objects.

16.2.4   Information Capacity, Resolution, Bandwidth, and Redundancy

Considering the virtual hologram, the resolution requirements set the lower limit of its size. For a rectangular aperture, the minimum resolvable distance between object points, defined according to Rayleigh criterion, is given by [Born and Wolf]

$$h = \frac{z_0 \lambda}{L_v} \tag{16.2-46}$$

where L_v is the size of the virtual hologram. Accordingly, the virtual hologram size L_v should be large enough to give a desired h.

An equally important consideration is the spatial frequency limit of the recording medium. The distance between the fringes in the x direction is approximately given by

$$\Delta_F = \frac{\lambda}{|M|} \left[\frac{x_i - x_c}{r_c} + \frac{x_i - x_o}{r_o} \right]^{-1} \tag{16.2-47}$$

Equations (16.2-46) and (16.2-47) show that, if $|M|$ is made small enough, Δ_F can be increased to such a magnitude that almost any recording medium can be used to make a hologram. However, this is accompanied with an increase in h. Conversely, in order to increase image resolution, the virtual hologram can be made larger than the real hologram, provided that the frequency limit of the recording medium is not exceeded.

Given a certain hologram size, it is desirable to record information in such a way that redundancy is reduced to a desired degree, especially if multiplexing is to be used. It seems to be advantageous to discuss redundancy in terms of the number of fringes. If N fringes are needed to obtain a desired image resolution, the virtual hologram size can be chosen to cover N fringes. The real hologram will also have N fringes. In this way, the recording medium is used as efficiently as desired. If the recording medium has a space-bandwidth product SB given by [Caulfield and Lu, 1970]

$$\text{SB} = \frac{\text{hologram area}}{\text{area of minimum resolution element}} \tag{16.2-48}$$

and the average space-bandwidth product used per signal is SB_s, the number of signals that can be recorded is of the order of SB/SB_s times the capacity for linear addition of signals in the recording medium. Direct recording of the hologram would cause SB/SB_s to be of the order of 1. Thus, the information capacity of the real hologram can be used effectively via this method at the cost of reduced resolution. However, this may require a nonlinear recording technique such as hard clipping. If SB_s is reasonably small for a signal, coarse recording devices such as a laser printer can be used to make a digital hologram. Another advantage that can be cited is that it

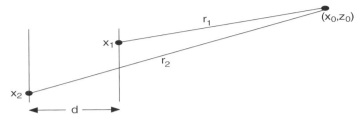

Figure 16.4. Interference of two spherical waves coming from points on two planes separated by distance d.

should be much faster to calculate the virtual hologram of reduced size because of a few numbers of fringes.

16.2.5 Volume Effects

Equations (16.2-17) and (16.2-18) describe the transformation obtained with the telescopic system. Lateral magnification is independent of real hologram coordinates, whereas T_2 versus T_1 varies as $1/F^2$. If F is reasonably large, and if several real hologram planes are separated by certain distances, these distances will be reduced by F^2 in the virtual hologram space. Thus, any errors made in positioning the real holograms would be reduced F times in the lateral direction and F^2 times in the vertical direction in the virtual hologram space. This means that it should be quite simple to obtain interference between various holograms and/or optical elements using the virtual holography concept and the telescopic system.

As a simple example, consider the interference at point O of two spherical waves coming from points at x_1 and x_2 on two planes separated by a plane d as shown in Figure 16.4.

Fraunhofer approximation will be assumed to be valid since the quadratic terms can always be removed with a lens if d is sufficiently small. Then, the difference between the two optical path lengths including the effect of a plane perpendicular reference wave can be approximated by

$$\Delta = \frac{x_0(x_2 - x_1)}{z_0} + \frac{d}{2}\left(\frac{x_0 - x_2}{z_0}\right)^2 \qquad (16.2\text{-}49)$$

If x_0 is much larger than x_2, this can be simplified to

$$\Delta = \frac{x_0(x_2 - x_1)}{z_0} + \frac{d}{2}\left(\frac{x_0}{z_0}\right)^2 \qquad (16.2\text{-}50)$$

If $d = 0$, the second term in Eq. (16.2-50) disappears, and the expression used by Lohmann to determine the position of a synthetic aperture is obtained [Lohmann, 1970]. For different object points, one needs to assume that x_0/z_0 remains approximately constant. However, this assumption can be relaxed to a large extent by finding the stationary point of Δ with respect to x_0/z_0. This is given by

$$x_2 - x_1 = -\frac{x_0}{z_0} d \tag{16.2-51}$$

$$\Delta = -\frac{d}{2}\left(\frac{x_0}{z_0}\right)^2 \tag{16.2-52}$$

It is possible to extend this approach further to make digital volume holograms. For example, the phase can be quantized by using several hologram planes separated by distance d, which is determined by Eq. (16.2-52). Then, the conjugate image problem would also disappear. The reason why it is practical now to do so is that d and $(x_2 - x_1)$ can be easily controlled by adjustments in the real hologram space where any errors are transformed to the virtual hologram space on a much reduced scale.

16.2.6 Distortions Due to Change of Wavelength and/or Hologram Size Between Construction and Reconstruction

It is well known in holographic microscopy and acoustical holography that lateral and vertical magnifications differ when wavelength and/or hologram size are changed between construction and reconstruction if these two changes are not done in the same ratio. Various techniques are proposed to get around this problem such as a phase plate [Firth, 1972]. With virtual holography, it is possible to match hologram size to wavelength change without actually matching the physical hologram size. However, distortions can be reduced without matching if a spherical reference wave other than the one used in construction is used on the virtual or the real hologram in reconstruction.

If N is the number of times the hologram size is changed between construction and reconstruction, k_1 and k_2 are the wave numbers during construction and reconstruction, respectively, the equation governing image formation can be written, similar to Eq. (16.2-25), as

$$k_1\left[-\frac{x_{01}}{r_{01}}Nx - \frac{y_{01}}{r_{01}}Ny + \frac{N^2(x^2+y^2)}{2r_{01}}\right]$$
$$= k_2\left[-x\left(\frac{x_{02}}{r_{02}} - \frac{x_c}{r_c}\right) - y\left(\frac{y_{02}}{r_{02}} - \frac{y_c}{r_c}\right) + \frac{x^2+y^2}{2}\left(\frac{1}{r_{02}} - \frac{1}{r_c}\right)\right] \tag{16.2-53}$$

where x_{01}, y_{01}, and z_{01} are the object coordinates, x_{02}, y_{02}, and z_{02} are the image coordinates, and

$$r_{01} = [x_{01}^2 + y_{01}^2 + z_{01}^2]^{\frac{1}{2}} \tag{16.2-54}$$
$$r_{02} = [x_{02}^2 + y_{02}^2 + z_{02}^2]^{\frac{1}{2}} \tag{16.2-55}$$

Equating corresponding terms, we obtain

$$\frac{1}{r_{02}} = \left(\frac{k_1 N^2}{r_{01}} + \frac{1}{r_c}\right)\frac{1}{k_2} \tag{16.2-56}$$

$$x_{02} = r_{02}\left(\frac{k_1 N x_{01}}{r_{01}} + \frac{x_c}{r_c}\right)\frac{1}{k_2} \tag{16.2-57}$$

$$y_{02} = r_{02}\left(\frac{k_1 N y_{01}}{r_{01}} + \frac{y_c}{r_c}\right)\frac{1}{k_2} \tag{16.2-58}$$

Various magnifications will be equal if

$$r = \frac{r_{01}}{N} \tag{16.2-59}$$

$$x_{02} = \frac{x_{01}}{N} \tag{16.2-60}$$

$$y_{02} = \frac{y_{01}}{N} \tag{16.2-61}$$

$$z_{02} = \frac{z_{01}}{N} \tag{16.2-62}$$

It is also possible to use an effective virtual reference wave which is a sum of a number of reference waves corresponding to different holograms, as discussed in Section 16.2.3, in order to reduce distortions and to scan over the different parts of the image field.

16.2.7 Experiments

The first digital holograms using this method were made with a scanning electron microscope system discussed previously in Chapter 15. All the calculations for encoding the hologram were done for the virtual hologram. The virtual hologram was then transformed in the computer using the optical system parameters into the real hologram that was physically generated. The one-image-only holography technique was used to encode the holograms. Here the position of each hologram aperture is chosen according to

$$\phi(x_i, y_i) + k r_{oi} = 2\pi n + \phi_0 \tag{16.2-63}$$

where $\phi(x_i, y_i)$ is the phase of the reference wave at the virtual aperture position $(x_i, y_i, 0)$, n is an integer, ϕ_0 is the desired phase at the object point with position coordinates (x_0, y_0, z_0), and

$$r_{oi} = \left[(x_0 - x_i)^2 + (y_0 - y_i)^2 + z_0^2\right]^{\frac{1}{2}} \tag{16.2-64}$$

Figure 16.5. Reconstruction with the first virtual hologram.

where $\phi(x_i, y_i, 0)$ is given by

$$\phi(x_i, y_i, 0) = kr_{ci} \tag{16.2-65}$$

$$r_{ci} = \left[(x_c - x_i)^2 + (y_c - y_i)^2 + z_c^2 \right]^{\frac{1}{2}} \tag{16.2-66}$$

where (x_c, y_c, z_c) are the coordinates of point O in Figure 16.1. For example, for an on-axis beam with a single lens, they can be chosen as

$$x_c = 0 \tag{16.2-67}$$

$$y_c = 0 \tag{16.2-68}$$

$$z_c = -fM \tag{16.2-69}$$

where M is the desired magnification, and f is the focal length of the lens system used. The aperture positions on the real hologram are x_i/M and y_i/M.

The reconstruction obtained from the first hologram generated in this way is shown in Figure 16.5. The object was chosen to be a circle slanted in the Z direction. This is why it looks slightly elliptic in the picture indicating the 3-D nature of the object points. The real hologram is 2 mm × 2 mm in size. This is reduced 4× with a 20 mm lens to give the virtual hologram. The distance $T_2 - f$ in Figure 16.1 was chosen to be 5 mm.

It was possible with the SEM system to generate holograms, 2 mm × 2 mm in size each, side by side up to a total size of 7.4 cm × 7.4 cm. At each shift there was an uncertainty of ±5 μ in positioning. Because each hologram is a window by itself, the light coming from each individual hologram will be directed in a different direction, which is similar to the problem of nondiffuse illumination. Even if the uncertainty in positioning was negligible and a large single hologram was made, this would not increase the information density, since only a small part of the total information coming from an area of the order of 2 mm × 2 mm is visible to the human eye at a time.

With this method it is possible to reduce the total virtual hologram set to a size of several millimeters on a side so that all the information coming from different

Figure 16.6. Reconstruction with the second virtual hologram generating part of a 3-D cube.

Figure 16.7. Reconstruction with the hologram set generating the 3-D cube.

holograms is made visible to the human eye, even though this is done at the expense of image resolution.

In order to show this, sixteen adjacent holograms arranged in a 4 × 4 matrix were generated. The virtual hologram set 2 mm × 2 mm in size was obtained by reducing the real hologram set 4× with a 50-mm objective. The distance $T_2 - f$ in Figure 16.1 was chosen as 12.5 mm. Each hologram generated one side or a diagonal, all the corner points, and the midpoint of a 3-D cube. The output from one such hologram is shown in Figure 16.6. When all the holograms were illuminated, the total image was obtained, as shown in Figure 16.7. Because all the holograms contribute to the corner points and midpoint, these object points appear much more intense than the other object points in Figure 16.7. Looking through the area where the virtual hologram set was supposed to be located, the whole cube in space could be seen.

16.3   THE METHOD OF POCS FOR THE DESIGN OF BINARY DOE

In the methods discussed in this section and the following sections of this chapter, the Fourier transform system shown in Figure 15.1 is used. The sampled hologram consists of an array of discrete points. The transmittance function of the hologram consisting of an $M \times N$ array of $\Delta v_x \times \Delta v_y$ sized pixels can be represented by the sum

$$G(v_x, v_y) = \sum_{k=0}^{M-1} \sum_{l=0}^{N-1} H(k,l) \mathrm{rect}\left(\frac{v_x - k\Delta v_x}{\Delta v_x}\right) \mathrm{rect}\left(\frac{v_y - l\Delta v_y}{\Delta v_y}\right) \qquad (16.3\text{-}1)$$

where $H(k,l)$ is the binary transmittance of the $(k,l)th$ point. The reconstructed image in the observation plane is given by the Fourier transform of the transmittance:

$$
\begin{aligned}
g(x, y) &= \iint G(v_x, v_y) e^{2\pi i [x v_x + y v_y]} dv_x dv_y \\
&= \Delta v_x \Delta v_y \sin c[\Delta v_x x] \sin c[\Delta v_y y] \sum \sum H(k, l) \exp[2\pi i (kx \Delta v_x + l y \Delta v_y)]
\end{aligned}
$$

$$(16.3\text{-}2)$$

By ignoring the two constants and the two $\sin c$ factors outside the sums, the reconstructed image is approximated by the two-dimensional inverse discrete Fourier transform (2D-IDFT) of the transmittance values.

The POCS method discussed in Section 14.8 is used to optimize the design of the hologram. Letting the dimensions of both the observation and CGH planes be $M \times N$, the relationship between the wave fronts at the observation plane $h(m, n)$ and the CGH plane $H(k, l)$ is given by the following discrete Fourier transform pair:

$$
h(m, n) = \frac{1}{MN} \sum_{k=0}^{M-1} \sum_{l=0}^{N-1} H(k, l) W_M^{mk} W_N^{nl}
$$

$$(16.3\text{-}3)$$

where $0 \leq m \leq M - 1, 0 \leq n \leq N - 1$, and

$$
H(k, l) = \sum_{m=0}^{M-1} \sum_{n=0}^{N-1} h(m, n) W_M^{-mk} W_N^{-nl}
$$

$$(16.3\text{-}4)$$

where $0 \leq k \leq M - 1, \leq l \leq N - 1$, and

$$
W_u = \exp(i2\pi/u)
$$

$$(16.3\text{-}5)$$

The goal of the POCS method is to generate the CGH whose reconstructed image most resembles the desired image.

Given a desired image $f(m, n)$ in a region R of the observation plane, the POCS method works as follows:

1. Using Eq. (16.3-4), compute $F(m, n)$ from $f(m, n)$.
2. Generate the binary transmittance values $H(k, l)$ from $F(k, l)$ as follows:

$$
H(k, l) = \begin{cases} 1 & \text{if } \mathrm{Re}[F(k, l)] \geq 0 \\ 0 & \text{otherwise} \end{cases}
$$

$$(16.3\text{-}6)$$

3. Using Eq. (16.3-2), find the reconstructed image $h(m, n)$. The accuracy of the reconstructed image is measured based on the mean square error (MSE)

between $f(m,n)$ and $h(m,n)$ within R, the region of the desired image. The MSE is defined as [Seldowitz]

$$\text{MSE} = \frac{1}{MN} \sum_{(m,n)\in R} \sum |f(m,n) - \lambda h(m,n)|^2 \qquad (16.3\text{-}7)$$

where λ is a scaling factor. The minimum MSE for $h(m,n)$ is achieved if,

$$\lambda = \frac{\displaystyle\sum_{(m,n)\in R} \sum f(m,n)h^*(m,n)}{\displaystyle\sum_{(m,n)\in R} \sum |h(m,n)|^2} \qquad (16.3\text{-}8)$$

4. Define a new input image $f'(m,n)$ such that
   (a) Outside R, $f'(m,n)$ equals $h(m,n)$.
   (b) Inside R, $f'(m,n)$ has the amplitude of the original image $f(m,n)$ and the phase of $h(m,n)$.
5. Letting $f(m,n) = f'(m,n)$, go to step 1.
6. Repeat steps 1–5 until the MSE converges or specified conditions are met.

16.4   ITERATIVE INTERLACING TECHNIQUE (IIT)

The IIT technique discussed in this section can be incorporated into any existing DOE synthesis method in order to improve its performance [Ersoy, Zhuang, Brede]. The interlacing technique (IT) will first be introduced, and then it will be generalized to the IIT. The IT divides the entire hologram plane into a set of subholograms. A subhologram consists of a set of cells, or points, referred to as a "block." All the subholograms are designed separately and then interlaced, or entangled, to create one hologram. Two examples of interlacing schemes with two subholograms are shown in Figures 16.8 and 16.9.

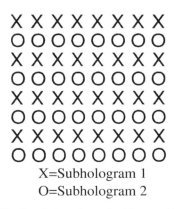

X=Subhologram 1
O=Subhologram 2

Figure 16.8.   Interlacing scheme 1 with two subholograms.

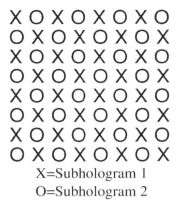

X=Subhologram 1
O=Subhologram 2

Figure 16.9. Interlacing scheme 2 with two subholograms.

In the IT method, once the entire hologram is divided into smaller subholograms, the first subhologram is designed to reconstruct the desired image $f(m,n)$. The reconstructed image due to the first subhologram is $h_1(m,n)$. Because the subhologram cannot perfectly reconstruct the desired image, there is an error image $e_1(m,n)$ defined as

$$e_1(m,n) = f(m,n) - \lambda_1 h_1(m,n) \qquad (16.4\text{-}1)$$

In order to eliminate this error, the second subhologram is designed with $e_1(m,n)/\lambda_1$ as the desired image. Since the Fourier transform is a linear operation, the total reconstruction due to both subholograms is simply the sum of the two individual reconstructions. If the second subhologram was perfect and its scaling factor matched λ_1, the sum of the two reconstructed images would produce $f(m,n)$. However, as with the first subhologram, there will be error. So, the third subhologram serves to reduce the left over error from the first two subholograms. Therefore, each subhologram is designed to reduce the error between the desired image and the sum of the reconstructed images of all the previous blocks. This procedure is repeated until each subhologram has been designed.

Each subhologram is generated suboptimally by the POCS algorithm (other methods can also be used). However, the total CGH may not yet reach the optimal result even after all the subholograms are utilized once. To overcome this problem, the method is generalized to the IIT.

The IIT is an iterative version of the IT method, which is designed to achieve the minimum MSE [Ersoy, Zhuang and Brede, 1992]. The reconstruction image of the ith subhologram at the jth iteration will be written as $h_i^j(m,n)$. After each subhologram has been designed using the IT method, the reconstruction due to the entire hologram $h_f(m,n)$ has a final error $e_f(m,n)$. To apply the iterative interlacing technique, a new sweep through the subholograms is generated. In the new sweep, the new desired image $f'(m, n)$ for the first subhologram is chosen as

$$f'(m, n) = h_1^1(m, n) - \frac{e_f(m, n)}{\lambda_f} \qquad (16.4\text{-}2)$$

where

$$e_f(m, n) = f(m, n) - \lambda_f h_f(m, n) \qquad (16.4\text{-}3)$$

λ_f is the scaling factor used after the last subhologram. Once the first subhologram is redesigned, the error image due to the entire hologram is calculated, including the new reconstruction created by the first subhologram. Similarly, the second subhologram is designed to reconstruct

$$h_2^1(m, n) - \frac{e_f'(m, n)}{\lambda_f'}, \quad e_f'(m, n)$$

being the updated error. This process is continued until convergence which is achieved when the absolute difference between successive reconstructed images

$$\Delta_j = \sum_{m=0}^{M-1} \sum_{n=0}^{N-1} |h_i^j(m, n) - h_b^{j-1}(m, n)| \qquad (16.4\text{-}4)$$

reaches a negligible value or remains steady for all the subholograms. By using the IIT method, the convergence tends to move away from the local-minimum MSE and moves towards the global-minimum MSE or at least a very deep minimum MSE.

16.4.1 Experiments with the IIT

In all the experiments carried out with the IIT, it was observed that the IIT can improve reconstruction results when used with another algorithm such as the POCS. A particular set of experiments were carried out with the image shown in Figure 16.10.

The associated iterative optimization algorithm used was the POCS. Table 16.1 shows the way the reconstruction MSE is reduced as the number of subholograms is increased.

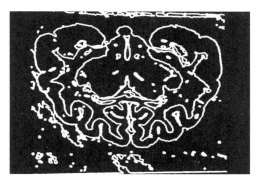

Figure 16.10. The cat brain image used in IIT experiments.

Table 16.1.   The MSE of reconstruction as a function the number of subholograms.

k	MSE	% Improvement
0	3230.70	0
1	2838.76	12.13
2	2064.35	36.10
3	1935.00	40.11

Figure 16.11.   The binary hologram generated with the IIT method for the cat brain image.

Figure 16.11 shows the binary hologram generated with the IIT method for the cat brain image. Figure 16.12 shows how the error is reduced as a function of the number of iterations. Figure 16.13 shows the corresponding He–Ne laser beam reconstruction.

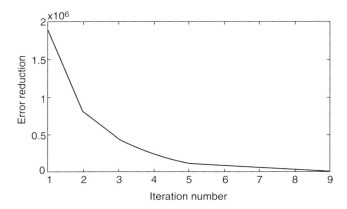

Figure 16.12.   Error reduction as a function of iteration number in IIT design.

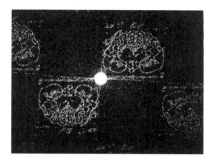

Figure 16.13. The He–Ne laser beam reconstruction of the cat brain image.

16.5   OPTIMAL DECIMATION-IN-FREQUENCY ITERATIVE INTERLACING TECHNIQUE (ODIFIIT)

The optimal decimation-in-frequency iterative interlacing technique (ODIFIIT) was developed as a result of trying to optimize the results of the IIT procedure [Zhuang and Ersoy, 1995]. The ODIFIIT exploits the decimation-in-frequency property of the Fast Fourier transform (FFT) when dividing the hologram into subholograms, and has two important advantages over IIT. It decreases computation time by reducing the dimensions of the Fourier transform and its inverse. The image inside the desired image region R is directly considered so that the design of each subhologram is more effectively processed because only contributions from the data of interest are taken into consideration.

The geometry of the reconstruction plane for the ODIFIIT method is shown in Figure 16.14. The desired amplitude $f_0(m, n)$ is a real-valued array of size $A \times B$. The desired image $f(m, n)$ at any time during optimization is $f_0(m, n)$ times a floating phase, which is determined by the phase of the current reconstruction inside region R. $f(m, n)$ is placed within region R, beginning at the point (M_1, N_1). The Hermitian

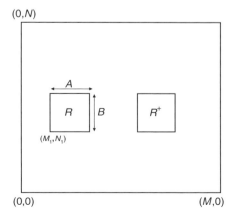

Figure 16.14. The reconstruction plane geometry in ODIFIIT.

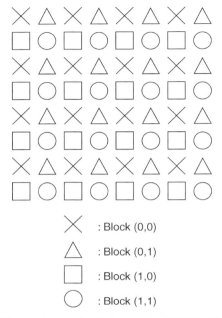

$\times$: Block (0,0)

$\triangle$: Block (0,1)

$\square$: Block (1,0)

$\bigcirc$: Block (1,1)

Figure 16.15. Interlacing of subholograms in ODIFIIT with $\mu = v = 2$.

conjugate of the reconstructed image exists in the region R^+ due to the real-valued CGH transmittance. Since the binary CGH has cell magnitude equal to unity, it is important that the desired image is scaled so that its DFT is normalized to allow a direct comparison to the reconstructed image $h(m,n)$.

The total CGH is divided into $\mu \times v$ subholograms, or blocks, where $\mu = M/A$ and $v = N/B$. μ and v are guaranteed to be integers if M, N, A, and B are all powers of two. Utilizing decimation-in-frequency [Brigham, 1974], the blocks are interlaced such that the $(\alpha, \beta)th$ block consists of the cells $(\mu k + \alpha, vl + \beta)$, where $0 \le k \le A - 1$, $0 \le l \le B - 1$, $0 \le \alpha \le \mu - 1$, and $0 \le \beta \le v - 1$. Figure 16.15 shows an example with $\mu = v = 2$.

Defining $H(k,l)$ as the sum of all the subholograms, the expression for the reconstructed image becomes

$$h(m,n) = \frac{1}{MN} \sum_{k=0}^{M-1} \sum_{l=0}^{N-1} H(k,l) W_M^{mk} W_N^{nl}$$

$$= \frac{1}{\mu v} \sum_{\alpha=0}^{\mu-1} \sum_{\beta=0}^{v-1} \left[\frac{1}{AB} \sum_{k=0}^{A-1} \sum_{l=0}^{B-1} H(\mu k + \alpha, vl + \beta) W_A^{mk} W_B^{nl} \right] W_M^{m\alpha} W_N^{n\beta} \qquad (16.5\text{-}1)$$

where $0 \le m < M$, $0 \le n < N$.

The reconstructed image in the region R is computed by replacing m and n by $m + M_1$ and $n + N_1$, respectively, and letting m and n span just the image region:

$$h(m + M_1, n + N_1)$$

$$= \frac{1}{\mu v} \sum_{\alpha=0}^{\mu-1} \sum_{\beta=0}^{v-1} \left[\frac{1}{AB} \sum_{k=0}^{A-1} \sum_{l=0}^{B-1} H(\mu k + \alpha, vl + \beta) W_A^{(m+M_1)k} W_B^{(n+N_1)l} \right] W_M^{(m+M_1)\alpha} W_N^{(n+N_1)\beta}$$

$$(16.5\text{-}2)$$

where $0 \leq m \leq A - 1$, $0 \leq n \leq B - 1$.

Let $h_{\alpha,\beta}(m, n)$ be the size $A \times B$ inverse discrete Fourier transform of the (α, β)th subhologram:

$$h_{\alpha,\beta}(m, n) = \text{IDFT}_{AB}[H(\mu k + \alpha, vl + \beta)]_{m,n}$$

$$= \frac{1}{AB} \sum_{k=0}^{A-1} \sum_{l=0}^{B-1} H(\mu k + \alpha, vl + \beta) W_A^{mk} W_B^{nl} \qquad (16.5\text{-}3)$$

where $0 \leq \alpha \leq \mu - 1$, $0 \leq \beta \leq v - 1$, $0 \leq m \leq A - 1$, $0 \leq n \leq B - 1$.

Using the IDFT of size $A \times B$, the reconstructed image inside the region R becomes

$$h(m + M_1, n + N_1) = \frac{1}{\mu v} \sum_{\alpha=0}^{\mu-1} \sum_{\beta=0}^{v-1} h_{\alpha,\beta}(m + M_1, n + N_1) W_M^{(m+M_1)\alpha} W_N^{(n+N_1)\beta}$$

$$(16.5\text{-}4)$$

where $0 \leq m \leq A - 1$, $0 \leq n \leq B - 1$.

The indices $(m + M_1)$ and $(n + N_1)$ of $h_{\alpha,\beta}(m + M_1, n + N_1)$ are implicitly assumed to be $(m + M_1)$ modulo A and $(n + N_1)$ modulo B, respectively. Equation (16.5.4) gives the reconstructed image in the region R in terms of the size $A \times B$ IDFTs of all the subholograms. From this equation, it can be seen that the reconstructed image in the region R due to the (α, β)th subhologram is given by

$$h'_{\alpha,\beta}(m + M_1, n + N_1) = \frac{1}{\mu v} h_{\alpha,\beta}(m + M_1, n + N_1) W_M^{(m+M_1)\alpha} W_N^{(n+N_1)\beta} \qquad (16.5\text{-}5)$$

which is the IDFT of the (α, β)th block times the appropriate phase factor, divided by μv.

An array, which will be useful later on, is defined as follows:

$$\tilde{h}_{\alpha,\beta}(m + M_1, n + N_1) = h(m + M_1, n + N_1) - h'_{\alpha,\beta}(m + M_1, n + N_1). \qquad (16.5\text{-}6)$$

This is the reconstructed image in the region R due to all the subholograms except the (α, β)th subhologram.

Conversely, given the desired image in the region R, the transmittance values can be obtained. From Eq. (16.5-2)

$$H(k,l) = \sum_{m=0}^{A-1} \sum_{n=0}^{B-1} h(m+M_1, n+N_1) W_M^{-(m+M_1)k} W_N^{-(n+N_1)l} \tag{16.5-7}$$

where $0 \leq k \leq M - 1$, $0 \leq l \leq N - 1$.

Dividing $H(k,l)$ into $u \times v$ blocks as before yields

$$H(\mu k + \alpha, vl + \beta)$$

$$= \sum_{\alpha=0}^{\mu-1} \sum_{\beta=0}^{v-1} \left[\sum_{m=0}^{A-1} \sum_{n=0}^{B-1} h(m+M_1, n+N_1) W_M^{-(m+M_1)\alpha} W_N^{-(n+N_1)\beta} W_A^{-(m+M_1)k} W_B^{-(n+N_1)l} \right]$$

$$= W_A^{-M_1 k} W_B^{-N_1 l} \sum_{\alpha=0}^{\mu-1} \sum_{\beta=0}^{v-1} \mathrm{DFT}_{AB} \left[h(m+M_1, n+N_1) W_M^{-(m+M_1)\alpha} W_N^{-(n+N_1)\beta} \right]_{k,l}$$

$$\tag{16.5-8}$$

where $0 \leq k \leq A - 1$, $0 \leq l \leq B - 1$, $0 \leq \alpha \leq \mu - 1$, $0 \leq \beta \leq v - 1$.

Therefore, the transmittance values of the subhologram (α, β) that create the image $h(m+M_1, n+N_1)$ in the region R are given by

$$H(\mu k + \alpha, vl + \beta) = W_A^{-M_1 k} W_B^{-N_1 l} \mathrm{DFT}_{AB} \left[h(m+M_1, n+N_1) W_M^{-(m+M_1)\alpha} W_N^{-(n+N_1)\beta} \right]_{k,l}$$

$$\tag{16.5-9}$$

where $0 \leq k \leq A - 1$, $0 \leq l \leq B - 1$.

Using Eqs. (16.5-5) and (16.5-9), we can compute the reconstructed image in the region R due to each individual subhologram, or, given a desired image in the region R, we can determine the transmittance values needed to reconstruct that desired image. Therefore, we can now utilize the IIT to design a CGH.

Letting $f_0(m+M_1, n+N_1)$, $0 \leq m \leq A - 1$, $0 \leq n \leq B - 1$, be the the desired image of size $A \times B$, the ODIFIIT algorithm can be summarized as follows:

1. Define the parameters M, N, A, B, M_1, and N_1, and determine μ and v. Then, divide the total CGH into $\mu \times v$ interlaced subholograms.

2. Create an initial $M \times N$ hologram with random transmittance values of 0 and 1.

3. Compute the $M \times N$ IDFT of the total hologram. The reconstructed image in the region R is the points inside the region R, namely, $h(m+M_1, n+N_1)$, $0 \leq m \leq A - 1$, $0 \leq n \leq B - 1$.

4. The desired image $f(m+M_1, n+N_1)$ is obtained by applying the phase of each point $h(m+M_1, n+N_1)$ to the amplitude $f_0(m+M_1, n+N_1)$ as in the POCS method. So,

$$f(m+M_1, n+N_1) = f_0(m+M_1, n+N_1) \exp(i\phi_{m+M_1, n+N_1}) \tag{16.5-10}$$

where $\phi_{m+M_1, n+N_1} = \arg\{h(m+M_1, n+N_1)\}$.

5. Find the optimization parameter λ using Eq. (16.3-5).
6. Using Eqs. (16.5-3), (16.5-5), and (16.5-6), find $\tilde{h}_{\alpha,\beta}(m + M_1, n + N_1)$. This is the reconstructed image in the region R due to all the subholograms except the (α, β)th subhologram.
7. Determine the error image that the (α, β)th subhologram uses to reconstruct (i.e., the error image) as

$$e(m + M_1, n + N_1) = \frac{f(m + M_1, n + N_1)}{\lambda} - \tilde{h}_{\alpha,\beta}(m + M_1, n + N_1) \qquad (16.5\text{-}11)$$

which is equivalent to the error image in the IIT method.
8. Using Eq. (16.5-9), find the transmittance values $E(\mu k + \alpha, \nu l + \beta)$ for the current block that reconstructs the error image.
9. Design the binary transmittance values of the current block as

$$H(\mu k + \alpha, \nu l + \beta) = \begin{cases} 1 & \text{if } \operatorname{Re}[E(\mu k + \alpha, \nu l + \beta)] \geq 0 \\ 0 & \text{otherwise} \end{cases} \qquad (16.5\text{-}12)$$

10. Find the new reconstructed image $h'_{\alpha,\beta}(m + M_1, n + N_1)$ in the region R due to the current block.
11. Determine the new total reconstructed image $h(m + M_1, n + N_1)$ by adding the new $h'_{\alpha,\beta}(m + M_1, n + N_1)$ to $\tilde{h}_{\alpha,\beta}(m + M_1, n + N_1)$.
12. With the new $h(m + M_1, n + N_1)$, use Eq. (16.5-10) to update $f(m + M_1, n + N_1)$.
13. Repeat steps 7–12 until the transmittance value at each point in the current block converges.
14. Update the total hologram with the newly designed transmittance values.
15. Keeping λ the same, repeat steps 3–14 (except step 5) for all the subholograms.
16. After all the blocks are designed, compute the MSE from Eq. (16.3-7).
17. Repeat steps 3–16 until the MSE converges. Convergence indicates that the optimal CGH has been designed for the current λ.

16.5.1   Experiments with ODIFIIT

The ODIFIIT method was used to design the DOEs of the same binary E and girl images that were used in testing Lohmann's method. There two images are shown in Figure 15.3 and Figure 15.8, respectively. A higher resolution 256×256 grayscale image shown in Figure 16.16 was also used.

The computer reconstructions from the ODIFIIT holograms are shown in Figures 16.17–16.19.

All the holograms designed using the ODIFIIT used the interlacing pattern as shown in Figure 16.15. There are many different ways in which the subholograms

Figure 16.16. The third image used in ODIFIIT investigations.

can be interlaced. However, it was observed experimentally that this interlacing method minimizes the MSE and thus produces the best results. In this scheme, each hologram was divided into 4×4 subholograms. So, the first image, the second image, and the third image were contained in 16×16, 128×128, and 256×256 desired image regions inside the entire image plane, respectively.

Figure 16.17. The reconstruction of the first image from its ODIFIIT hologram.

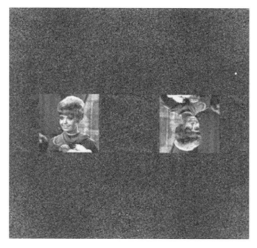

Figure 16.18.   The reconstruction of the second image from its ODIFIIT hologram.

These results showed very accurate reconstructions of the desired images. Compared to the results from the Lohmann method discussed in Section 15.8, there was consequently a significant increase in the reconstruction quality for high resolution gray-scale images.

Like the DOEs designed with the Lohmann method, the transmittance function designed with ODIFIIT can also be coded as $(-1,1)$ to make binary phase holograms. This can be further extended to code holograms with multiple levels of phase quantization. This corresponds to multilevel phase elements. For example, four levels of phase quantization can be implemented when encoding the hologram.

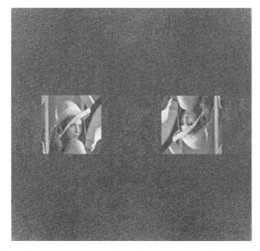

Figure 16.19.   The reconstruction of the third image from its ODIFIIT hologram.

So at each point in the hologram there would be a phase shift of 0, $\pi/2$, π, or $3\pi/2$. This generates a reconstruction with lower MSE because the hologram function has a closer approximation to the actual desired phase at each point.

16.6   COMBINED LOHMANN-ODIFIIT METHOD

Since the detour-phase method used in the binary Lohmann method cannot code amplitude and phase exactly in practise, there will always be some amount of inherent error in the reconstruction. Therefore, iterative optimization as in the IIT and ODIFIIT methods should be effective in reducing the final reconstruction error. For this reason, Lohmann's coding scheme was implemented together with the interlacing technique to create a better method for designing DOEs [Kuhl, Ersoy]. The combined method is called the *Lohmann-ODIFIIT method*, or *LM-ODIFIIT method*. In this approach, the desired amplitude and phase of each subhologram point is encoded using a Lohmann cell, but the hologram is divided into interlaced subholograms like in ODIFIIT.

It was discussed in Section 15.4 that three approximations are made for simplicity in designing a Lohmann hologram: (a) $\mathrm{sinc}[c\delta v(x + x_0)] \approx \mathrm{const}$, (b) $\mathrm{sinc}(yW_{nm}\delta v) \approx 1$, and (c) $\exp[2\pi i(xP_{nm}\delta_x)] \approx 1$. The effects of these approximations on the reconstructed image depend on several factors.

The sinc function $\mathrm{sinc}[c\delta v(x + x_0)]$ creates a drop-off in intensity in the x-direction proportional to the distance from the center of the image plane. Approximation (a) considers this sinc factor to be nearly constant inside the image region. A small aperture size c results in less drop-off in intensity. However, this also reduces the brightness of the image.

The sinc function $\mathrm{sinc}(yW_{nm}\delta v)$ indicates a drop-off in intensity similar to (a), but in the y-direction. This sinc function acts like a slight decrease in amplitude transmission by the factor $\mathrm{sinc}(yW_{nm}\delta v)$.

The phase shift $\exp[2\pi i(xP_{nm}\delta_x)]$ causes a phase error that varies with x location in the image plane. This phase error depends on x and P, and ranges from zero to $\pi/2M$. Ignoring the phase factor $\exp[2\pi i(xP_{nm}\delta_x)]$ leads to image deteriotation as if there is a phase error due to an improper aperture shift [Lohmann, 1970].

The solution used to account for the sinc roll-off in the x direction is to divide the desired image by $\mathrm{sinc}[c\delta v(x + x_0)]$. The desired image $f(x, y)$ becomes,

$$\frac{f(x, y)}{\mathrm{sinc}[c\delta v(x + x_0)]}.$$

The same thing cannot be done for the y direction because $\mathrm{sinc}(yW_{nm}\delta v)$ depends on the aperture parameters W_{nm} which are based on the desired image. In the LM-ODIFIIT and quantized LM-ODIFIIT, this $\sin c$ factor is accounted for as follows: the desired image is first divided by the $\sin c$ factor affecting the x direction, and the hologram is designed. Then, the y dependent $\sin c$ factors due to all the apertures are calculated and summed to determine the effect on the output image. The desired

Figure 16.20.   Reconstruction with the LM-ODIFIIT method.

image is next divided by the y dependent $\sin c$ factor just as it was for the x dependent factor. Next, the hologram is designed again, the effect on the output due to the new aperture heights is calculated, and the original desired image is divided by this new factor. This process is repeated until the reconstructed image does not change, or a convergence condition is met.

The phase shift $\exp[2\pi i(xP_{nm}\delta_x)]$ is accounted for by successive iterations while designing the hologram function.

16.6.1   Computer Experiments with the Lohmann-ODIFIIT Method

Figure 16.20 shows the computer reconstruction of the binary "E" image of Figure 15.3 from its LM-ODIFIIT hologram [Kuhl, Ersoy]. Figure 16.21 shows the computer reconstruction of the binary "E" image from its LM-ODIFIIT hologram with four levels of quantization. Figure 16.22 shows the computer reconstruction of

Figure 16.21.   Reconstruction with the LM-ODIFIIT method with four levels of phase quantization.

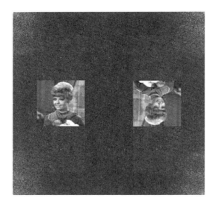

Figure 16.22.   Reconstruction with the LM-ODIFIIT method.

the girl image from its LM-ODIFIIT hologram. A comparison between the output of
the Lohmann method, the ODIFIIT, and the LM-ODIFIIT for a 16×16 gray-scale
image was also performed. The results are shown in Figure 16.23.

The tables to be discussed next are based on the results obtained with the "binary
E" image. Table 16.2 shows the mean square error and diffraction efficiency results
of the computer experiments. The MSE represents the difference between the
desired image and the reconstructed image inside the desired image region, and the
diffraction efficiency is a measure of how much of the incident wave in percent is
diffracted into the desired image region. The table includes comparative results for
the ODIFIIT, the Lohmann method (LM), the Lohmann method using a constant
amplitude for each cell (LMCA), ODIFIIT using Lohmann's coding method (LM-
ODIFIIT), and LM-ODIFIIT with constant cell amplitude (LMCA-ODIFIIT). All

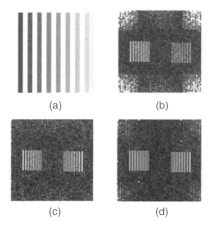

Figure 16.23.   (a), Original image, and reconstructions with (b) the Lohmann method, (c) the ODIFIIT,
(d) the LM-ODIFIIT.

Table 16.2.   MSE and diffraction efficiency results for binary DOEs.

Method	MSE	Efficiency (%)
LM	1	1.2
ODIFIIT	0.33	5.7
LMCA	2.03	5.9
LM-ODIFIIT	5.9×10^{-31}	1.2
LMCA-ODIFIIT	1	2.1

results are for binary amplitude holograms. Values of $c = 1/2$ and $M = 1$ in the Lohmann method were used throughout. For comparison purposes, the MSE occurring in the Lohmann method was normalized to one, so all other MSE values are relative to the Lohmann method.

Table 16.3 shows the results of the computer experiments in which the amplitude and phase of each cell took on quantized values, and the ODIFIIT was applied. Each Lohmann cell is divided into $N \times N$ smaller squares in N-level quantization. Quantization of amplitude and phase is done as discussed in Section 16.6. MSE is still relative to the MSE from the Lohmann method alone.

The experimental results indicate that the ODIFIIT is better than the Lohmann method in terms of both MSE and efficiency. In the constant amplitue Lohmann method, the error in the image region increased, but efficiency was significantly improved. Basically, this means reconstruction of a brighter image with less definition. This result seems reasonable for the following reasons: less information is used in LMCA, which increases the error, but the amplitude values throughout the hologram are greater due to aperture size always being as large as possible, which increases the brightness. The output displays a reconstruction of the desired image which is brighter, but has "fuzzy" edges. Less reconstruction occurs outside the desired image region further confirming the efficiency increase. In fact, the efficiency of LMCA surpassed the efficiency of ODIFIIT, but the significant increase in MSE is a disadvantage.

The results indicate that LM-ODIFIIT produces an output most resembling the desired image (lowest MSE), while its efficiency is the same as the original Lohmann's method (LM). Simulations show qualitatively, that LM-ODIFIIT

Table 16.3.   MSE and diffraction efficiency results as a function of number of quantization levels in LM-ODIFIIT and LMCA-ODIFIIT.

Levels of Quantization (N)	MSE	Efficiency (%)
2	2.33	0.6
2-CA	0.42	1.4
4	0.19	1.1
4-CA	0.93	1.1

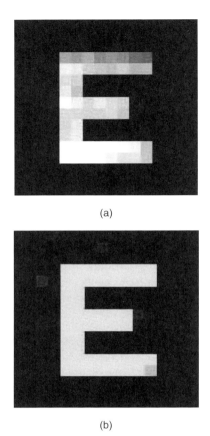

(a)

(b)

Figure 16.24. (a) The gray-scale image from its LM-ODIFIIT hologram, (b) the same image after the approximations are compensated for in the method.

generated an extremely sharp and uniform image inside the image region. The simulated reconstruction of the girl image especially demonstrates the accuracy of the LM-ODIFIIT, and should be compared to the LM and ODIFIIT results of the same image.

The MSE of the LMCA-ODIFIIT is equal to the MSE of the LM, greater than the MSE of the LM-ODIFIIT, and less than the MSE of the LMCA. Incorporating ODIFIIT into LMCA reduces the error, just as it did for LM. However, considering the extremely small MSE of LM-ODIFIIT, it is somewhat surprising that the MSE of the LMCA-ODIFIIT is not less. As expected, the efficiency of the LMCA-ODIFIIT is greater than the efficiency of the LM-ODIFIIT.

Quantized LM-ODIFIIT with $N = 2$ had the largest MSE and the lowest efficiency of all the methods. By using so few quantization levels, not enough information remains after encoding to allow for good reconstruction. Anytime the normalized amplitude is less than 1/2, the amplitude is quantized to zero, and phase information is lost. This explains high error and very poor efficiency. By implementing the constant amplitude technique, more phase information is retained. This results in better MSE and efficiency.

For $N = 4$, MSE dropped significantly, creating a very good reconstruction of the desired image. This method produced a low MSE, second only to the extremely low value of the original LM-ODIFIIT. Also, its efficiency is comparable to all other methods except ODIFIIT and LMCA. Making the amplitude a constant maintains the efficiency, but increases the error.

Figure 16.23 shows that the 128×128 LM and LM-ODIFIIT images display periodic arrays of dots. The horizontal array in the LM image occurs because of the periodic nature of the hologram. This is due to all the cells being symmetric relative to the direction in which the aperture height is adjusted. The horizontal array in the LM-ODIFIIT image occurs for the same reason. However, in the LM-ODIFIIT hologram, it was observed that the phase at each point of the subholograms converged to the same value. This was observed for every subhologram except the first. This gives a periodic nature to the other direction of the hologram, which results in the vertical array of dots.

Up to now, all simulations neglected the $\sin c$ roll-off that actually occurs in the physical reconstruction from a binary CGH. The compensating methods discussed in Section 16.7 were further used in simulations to correct for these errors. The computer reconstructions showed excellent homogeneity in the x direction indicating that the x dependent $\sin c$ factor was accurately accounted for. Slight variations were present in the y direction, but can be reduced to a desired level by continued iterations. Figure 16.24 shows the computer reconstruction of the gray-scale image E before and after the $\sin c$ roll-off corrections were factored into the output.

17

Computerized Imaging Techniques I: Synthetic Aperture Radar

17.1 INTRODUCTION

A number of modern imaging techniques rely on the Fourier transform and related computer algorithms for image reconstruction. For example, synthetic aperture radar (SAR), image reconstruction from projections, including computerized tomography, magnetic resonance imaging (MRI), and confocal microscopy discussed in Chapter 5 are among such techniques.

In this chapter, the SAR principles are discussed. The first SAR image was produced at the University of Michigan in 1958 with a coherent optical processing method since SAR is essentially similar to holography [Cutrona]. In the early 1970s digital signal processing methods were started to be utilized to computerize SAR image processing [Bennett], [Ulaby]. Currently, computerized SAR image processing is what is used in practice.

This chapter consists of ten sections. Section 17.2 describes SAR in general terms. Section 17.3 is on range resolution. In order to achieve high resolution of imaging, it is also necessary to carefully choose the pulse waveform used. This topic is introduced in Section 17.4. Matched filtering discussed in Section 17.5 is used both for maximizing signal-to-noise ratio (SNR) and achieving very sharp pulses. The way sharp pulses with sufficient energy are achieved with matched filtering is discussed in Section 17.6. Section 17.7 is on cross-range resolution.

The topics discussed so far lay the groundwork for SAR imaging. A simplified theory of SAR imaging is discussed in Section 17.8. Image reconstruction with the Fresnel approximation is highlighted in Section 17.9. Another digital SAR image reconstruction algorithm is discussed in Section 17.10.

17.2 SYNTHETIC APERTURE RADAR

The word *radar* is an acronym for *RAdio Detection And Ranging*, where the letters used are capitalized. *Synthetic aperture radar* is a technique to create

Diffraction, Fourier Optics and Imaging, by Okan K. Ersoy

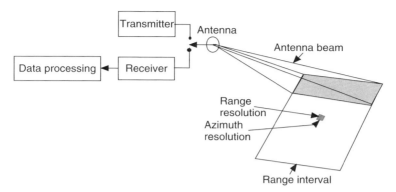

Figure 17.1. A visualization of a SAR system.

high-resolution images, which are useful in many areas such as geography, topographic and thematic mapping, oceanography, forestry, agriculture, urban planning, environmental sciences, prediction, and evaluation of natural disasters.

SAR is a coherent imaging technique which makes use of the motion of a radar system mounted on a vehicle such as an aircraft (airborne SAR) or a satellite (spaceborne SAR). A target area is illuminated with the radar's radiation pattern, resulting in a collection of echoed signals while the vehicle is in motion. These echoed signals are integrated together, using special algorithms, yielding a high-resolution image of the target area.

A simple block diagram for SAR imaging is shown in Figure 17.1. An EM pulse at a microwave frequency is sent by the transmitter and received by the receiver after it gets reflected by the target area. In order to synthesize a large artificial antenna, the flight path in a straight direction is used. In this way, the effective antenna length can be several kilometers. Azimuth (or cross-range) is the direction along the flight path. Range (or slant range) is the direction perpendicular to the azimuth.

One return or echo of the pulse generates information to be used to obtain the image of the target area. High resolution (small pixel sizes) and large signal-to-noise ratio are obtained by illuminating the target area many times with periodically emitted pulses. This is equivalent to signals, each of which is emitted from a single element of a large antenna array. Hence, the name synthetic aperture radar.

The azimuth resolution at range R from an antenna of dimension L is approximately $\lambda R/L$ as shown in Section 17.7. Without the synthetic aperture, making L large at microwave frequencies is impractical. The synthetic aperture corresponds to an antenna array with antenna elements distributed over the flight path.

SAR imaging is often carried out with a satellite system. For example, the Seasat satellite system put into orbit by JPL/NASA in June 1978 had the task of creating SAR images of large regions of the earth's surface. The Shuttle Imaging Radar (SIR)

Figure 17.2. A SAR image generated in September, 1995 [Courtesy of Center for Remote Imaging, Sensing and Processing, National University of Singapore (CRISP)].

systems carried on the space shuttles are similar SAR imaging systems. Examples of airborne SAR systems carried out with airplanes are the E-3 AWACS (Airborne Warning and Control System), used in the Persian Gulf region to detect and track maritime and airborne targets, and the E-8C Joint STARS (Surveillance Target Attack Radar System), which was used during the Gulf War to detect and locate ground targets.

An example of a SAR image is shown in Figure 17.2. This image of South Greenland was acquired on February 16, 2006 by Envisat's Medium Resolution Imaging Spectrometer (MERIS) [ESA].

17.3 RANGE RESOLUTION

In SAR as well as other types of radar, range resolution is obtained by using a pulse of EM wave. The range resolution has to do with ambiguity of the received signal due to overlap of the received pulse from closely spaced objects.

In addition to nearby objects, there are noise problems, such as random fluctuations due to interfering EM signals, atmospheric effects, and thermal variations in electronic components. Hence, it is necessary to increase signal-to-noise ratio (SNR) as well as to achieve large range resolution.

The distance R to a single object reflecting the pulse is $tc/2$, where t is the interval of time between sending and receiving the pulse, and c is the speed of light,

3×10^8 m/sec. Suppose the pulse duration is T seconds. Then, the delay between two objects must be at least T seconds so that there is no overlap between the two pulse echoes. This means the objects must be separated by $cT/2$ meters (if MKS units are used). Reducing T results in better range resolution. However, high pulse energy is also required for good detection, and short pulses mean lower energy in practice. In order to avoid this problem, matched filtering discussed in Section 17.5 is often used to convert a pulse of long duration to a pulse of short duration at the receiver. In this way, the received echoes are sharpened, and the overall system possesses the range resolution of a short pulse. The peak transmitter power is also greatly reduced for a constant average power. In such systems, matched filtering is used both for pulse compression as well as detection by SNR optimization. This is further discussed in Section 17.5.

17.4 CHOICE OF PULSE WAVEFORM

The shape of a pulse is significant in order to differentiate nearby objects. Suppose that $p(t)$ is the pulse signal of duration T, which is nonzero for $0tT$. The returned pulse from one object can be written as

$$p_1(t) = \sigma_1 p(t - \tau_1) \tag{17.4-1}$$

where σ_1 is the attenuation constant, and τ_1 is the time delay. The returned pulse from a second object can be written as

$$p_2(t) = \sigma_2 p(t - \tau_2) \tag{17.4-2}$$

The shape of the pulse should be optimized such that $p_1(t)$ is as dissimilar from $p_2(t)$ for $\tau_1 \neq \tau_2$ as possible.

The most often used measure of similarity between two waveforms $p_1(t)$ and $p_2(t)$ is the Euclidian distance given by

$$D^2 = \int [p_1(t) - p_2(t)]^2 dt \tag{17.4-3}$$

D^2 can be written as

$$D^2 = \sigma_1^2 \int p^2(t - \tau_1) dt + \sigma_2^2 \int p^2(t - \tau_2) dt - 2\sigma_1\sigma_2 \int p(t - \tau_1) p(t - \tau_2) dt \tag{17.4-4}$$

The first two terms on the right-hand side above are proportional to the pulse energy, which can be separately controlled by scaling. Hence, only the last term is significant for optimization. It should be minimized for $\tau_1 \neq \tau_2$ in order to maximize D^2. The integral in the last term is rewritten as

$$R(\tau_1, \tau_2) = \int p(t - \tau_1)p(t - \tau_2)dt \qquad (17.4\text{-}5)$$

which is the same as

$$R(\tau) = \int p(t)p(t + \tau)dt \qquad (17.4\text{-}6)$$

where τ equals $\tau_1 - \tau_2$ or $\tau_2 - \tau_1$. It is observed that $R(\tau)$ is the autocorrelation of $p(t)$.

A *linear frequency modulated* (*linear fm*) *signal*, also called a *chirp signal* has the property of very sharp autocorrelation which is close to zero for $\tau \neq 0$. It can be written as

$$x(t) = A\cos(2\pi(ft + \gamma t^2)) \qquad (17.4\text{-}7)$$

or more generally as

$$x(t) = e^{j2\pi(ft+\gamma t^2)} \qquad (17.4\text{-}8)$$

The larger γ signifies larger variation of instantaneous frequency f_i, which is the derivative of the phase:

$$f_i = f + 2\gamma t \qquad (17.4\text{-}9)$$

It is observed that f_i varies linearly with t. The autocorrelation function of $x(t)$ can be shown to be

$$R(\tau) = e^{j2\pi\tau(f+\gamma\Sigma)} \int e^{j4\pi\gamma\tau t}dt \qquad (17.4\text{-}10)$$

Suppose that a pulse centered at T_0 and of duration T is expressed as

$$p(t) = \text{rect}\left(\frac{t - T_0}{T}\right)x(t) \qquad (17.4\text{-}11)$$

The autocorrelation function of $p(t)$ is given by

$$R\left(\sum\right) = e^{j2\pi\int\left[f+\frac{\gamma}{2}\left(T_0+\frac{\tau}{2}\right)\right]}\text{tri}\left((T_-|\tau|)\ \text{sin}\,c\left(\frac{\pi\gamma\tau}{2}(T - |\tau|)\right)\right)$$

The width of the main lobe of this function is approximately equal to $1/\gamma T$. The rectangular windowing function is often replaced by a more smooth function such as a Gaussian function, as discussed in Section 14.2. Because of the properties discussed above, a chirp signal within a finite duration window is often the pulse waveform chosen for good range resolution.

17.5   THE MATCHED FILTER

One basic problem addressed by matched filtering is to decide whether a signal of a given form exists in the presence of noise. In the classical case, the filter is also constrained to be a LTI system. Matched filters are used in many other applications as well, such as pulse compression as discussed in the next section, and image reconstruction as discussed in Sections 17.8–17.10.

Suppose the input consists of the deterministic signal $x(t)$ plus noise $N(t)$. Let the corresponding outputs from the linear system used be $x_0(t)$ and $N_0(t)$, respectively. This is shown in Figure 17.3. The criterion of optimality to be used to determine whether $x(t)$ is present is the maximum SNR at the system output. It is shown below that the LTI filter that maximizes the SNR is the matched filter.

If $n_0(t)$ is assumed to be a sample function of a wide-sense stationary (WSS) process, the SNR at time T_0 can be defined as

$$\text{SNR} = \frac{|x_0(T_0)|^2}{E[N_0^2(T_0)]} \tag{17.5-1}$$

The output signal $x_0(t)$ is given by

$$x_0(T) = \int_{-\infty}^{\infty} X(f)H(f)e^{j2\pi fT_0}\,df \tag{17.5-2}$$

The output average noise power is given by

$$E[N_0^2(t)] = \int_{-\infty}^{\infty} |H(f)|^2 S_N(f)\,df \tag{17.5-3}$$

where $S_N(f)$ is the spectral density of $N(t)$.

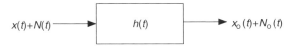

Figure 17.3.   The matched filter as a LTI filter to optimally reduce noise.

Now, the SNR can be written as

$$\text{SNR} = \frac{\left| \int\limits_{-\infty}^{\infty} X(f)H(f)e^{j2\pi f T_0} \right|^2}{\int\limits_{-\infty}^{\infty} |H(f)|^2 S_N(f)\, df}$$

(17.5-4)

To optimize the SNR, the Schwarz inequality can be used. If $A(f)$ and $B(f)$ are two possibly complex functions of f, the Schwarz inequality is given by

$$\left| \int\limits_{-\infty}^{\infty} A(f)B(f)\, df \right|^2 \leq \int\limits_{-\infty}^{\infty} |A(f)|^2\, df \int\limits_{-\infty}^{\infty} |B(f)|^2\, df$$

(17.5-5)

with equality iff

$$A(f) = CB^*(f)$$

(17.5-6)

C being an arbitrary real constant. Let

$$A(f) = \sqrt{S_N(f)}\, H(f)$$

$$B(f) = \frac{X(f)e^{j2\pi f T_0}}{\sqrt{S_N(f)}}$$

(17.5-7)

Then, the Schwarz inequality gives

$$\left| \int\limits_{-\infty}^{\infty} X(f)H(f)e^{j2\pi f T}\, df \right|^2 \leq \left[\int\limits_{-\infty}^{\infty} S_N(f)|H(f)|^2\, df \right] \left[\int\limits_{-\infty}^{\infty} \frac{|X(f)|^2}{S_N(f)}\, df \right]$$

(17.5-8)

or

$$\text{SNR} \leq \int\limits_{-\infty}^{\infty} \frac{|X(f)|^2}{S_N(f)}\, df$$

(17.5-9)

The SNR is maximized and becomes equal to the right-hand side of Eq. (17.5-9) when the equality holds according to Eq. (17.5-6), in other words, when

$$H(f) = H_{\text{opt}}(f) = C\frac{X^*(f)}{S_N(f)}e^{-j2\pi f T_0}$$

(17.5-10)

The filter whose transfer function is given by Eq. (17.5-10) is called the *matched filter*. It is observed that $H_{\text{opt}}(f)$ is proportional to the complex conjugate of the FT of the input signal, and inversely proportional to the spectral density of the input noise. The factor

$$e^{-j2\pi f T_0}$$

serves to adjust the time T_0 at which the maximum SNR occurs.

EXAMPLE 17.1 If the input noise is white with spectral density equal to N_0, find the matched filter transfer function and impulse response. Also show the convolution operation in the time-domain with the matched filter.
Solution: Substituting N for $S_N(f)$ in Eq. (17.5-10) yields

$$H_{\text{opt}}(f) = KX^*(f)e^{-j2\pi f T_0}$$

where K is an arbitrary constant. The impulse response is the inverse FT of $H_{\text{opt}}(f)$, and is given by

$$h_{\text{opt}}(t) = Kx(T_0 - t) \tag{17.5-11}$$

The input signal is convolved with $H_{\text{opt}}(f)$ to yield the output. Thus,

$$y(t) = \int_{-\infty}^{\infty} x(\tau)h_{\text{opt}}(t - \tau)\mathrm{d}\tau$$

$$= K \int_{-\infty}^{\infty} x(\tau)x(T_0 - t + \tau)\mathrm{d}\tau$$

The peak value occurs at $t = T_0$, and is given by

$$y(T) = K \int_{-\infty}^{\infty} x^2(\tau)\mathrm{d}\tau$$

which is proportional to the energy of the input signal.

17.6 PULSE COMPRESSION BY MATCHED FILTERING

In applications such as pulse radar and sonar, it is important to have pulses of very short duration to obtain good range resolution. However, high pulse energy is also required for good detection and short pulses mean lower energy in practice. In order to avoid this problem, matched filtering is often used to convert a pulse of long duration to a pulse of short duration at the receiver. In this way, the received echos are sharpened, and the overall system possesses the range resolution of a short pulse. The peak transmitter power is also greatly reduced for a constant average power. In such systems, matched filtering is used both for pulse compression as well as detection by SNR optimization.

The input pulse waveform is chosen such that the output pulse is narrow. For example, the input can be chosen as a chirp pulse in the form

$$x(t) = e^{t^2/T^2} e^{j(2\pi f_0 t + \gamma t^2)} \qquad (17.6\text{-}1)$$

where T is called the pulse duration. In practice, $|\gamma|T$ is much less than $2\pi f_0$. The spectrum of the pulse is given by

$$X(f) = F_B \sqrt{m/\pi} \, e^{-4\pi^2 (f-f_0)^2 / F_B^2} e^{j\left[\gamma T^2 (f-f_0^2) / F_B^2 - \frac{1}{2}\tan^{-1}\gamma T^2\right]} \qquad (17.6\text{-}2)$$

where

$$m = [1 + \gamma^2 T^4]^{\frac{1}{2}} \qquad (17.6\text{-}3)$$

$$F_B = \frac{m}{\pi T} \qquad (17.6\text{-}4)$$

F_B is the effective bandwidth of the spectrum since the spectrum is also Gaussian with center at f_0.

Disregarding constant terms, the matched filter for the signal $x(t)$ (assuming noise is white and $T_0 = 1$) is given by

$$H(f) = e^{-4\pi^2 (f-f_0)^2 / F_B^2} e^{-j\left[\gamma T^2 (f-f_0)^2 / F_B^2 - \frac{1}{2}\tan^{-1}\gamma T^2\right]} \qquad (17.6\text{-}5)$$

In practice, the amplitude of $H(f)$ in Eq. (17.6-5) actually reduces the amplitude of the final result. This can be prevented by using the phase-only filter given by

$$H(f) = e^{-j\left[\gamma T^2 (f-f_0)^2 / F_B^2 - \frac{1}{2}\tan^{-1}\gamma T^2\right]} \qquad (17.6\text{-}6)$$

Then, the output of the matched filter is

$$y(t) = \int_{-\infty}^{\infty} X(f)H(f)e^{j2\pi ft}\,df$$

$$= \sqrt{m}\, e^{-t^2/(T/m)^2} e^{j2\pi f_0 t} \qquad (17.6\text{-}7)$$

It is observed that the output signal is again a Gaussian pulse with the frequency modulation removed, and the pulse duration compressed by the factor m. In addition, the pulse amplitude is increased by the factor $\sqrt{m}$ so that the energy of the signal is unchanged.

If Eq. (17.6-5) is used instead of Eq. (17.6-6), the same results are valid with the replacement of m by $m/\sqrt{2}$.

In pulse radar and sonar, range accuracy, and resolution are a function of the pulse duration. Long duration signals reflected from near targets blend together, lowering the resolution. The maximum range is a function of the SNR, and thereby the energy in the pulse. Thus, with the technique described above, both high accuracy, resolution, and long range are achieved.

Pulse compression discussed above is a general property rather than being dependent on the particular signal. The discussion below is for a general pulse signal with $T_0 = 0$. It is observed from Eq. (6.11.7) that the matched filter generates a spectrum which has zero phase, and an amplitude which is $|X(f)|^2$ when Eq. (17.11-5) is used. The output signal at $t = 0$ is given by

$$y(0) = \int_{-\infty}^{\infty} |X(f)|^2 df = E$$

which is large.

In order to define the degree of compression, it is necessary to define practical measures for time and frequency duration. Let the pulse have the energy E and a maximum amplitude A_{max}. The input pulse duration can be defined as

$$T_x = \frac{E}{A_{max}^2} \qquad (17.6\text{-}8)$$

Similarly, the spectral width F is defined by

$$F = \frac{E}{B_{max}^2} \qquad (17.6\text{-}9)$$

where B_{max} is the maximum amplitude of the spectrum.

The input signal energy is the same as $y(0)$. The output signal energy is

$$E_y = \int_{-\infty}^{\infty} |X(f)|^4 df \qquad (17.6\text{-}10)$$

The compression ratio m is given by

$$m = \frac{T_x}{T_y} \qquad (17.6\text{-}11)$$

where T_y is the pulse width of the output, which is also given by

$$C_{max}^2 T_y = E_y$$

C_{max} is the maximum amplitude of the output which is $y(0)$. Since $y(0)$ is the same as E, Eq. (17.6-11) can be written as

$$m = \frac{T_x E^2}{E_y} = \alpha T_x F \qquad (17.6\text{-}12)$$

where

$$\alpha = \frac{E}{E_y} B_{max}^2 = B_{max}^2 \frac{\displaystyle\int_{-\infty}^{\infty} |X(f)|^2 df}{\displaystyle\int_{-\infty}^{\infty} |X(f)|^4 df} \qquad (17.6\text{-}13)$$

α can also be written as

$$\alpha = \frac{\displaystyle\int_{-\infty}^{\infty} D^2(f) df}{\displaystyle\int_{-\infty}^{\infty} D^4(f) df} \qquad (17.6\text{-}14)$$

where

$$D(f) = \frac{|X(f)|}{B_{max}} \qquad (17.6\text{-}15)$$

Since $D(f)$ is less than or equal to 1, α is greater than or equal to 1. For example, three types of spectra and corresponding α are the following:

Spectrum	Rectangular	Triangular	Gaussian
α	1	1.67	2.22

It is seen that the Gaussian spectrum has the best pulse compression property. Equation (17.6-12) shows that the compression ratio is proportional to the *product of the signal duration T_x and the spectral width F*. This is often called the *time-bandwidth product*.

17.7   CROSS-RANGE RESOLUTION

In order to understand the cross-range (azimuth) resolution properties of an antenna of length L, consider the geometry shown in Figure 17.4. It is assumed that the target is so far away that the echo signal impinges on the antenna at an angle ϕ at all positions on the antenna.

Assuming that the antenna continuously integrates incident energy, the integrated antenna response can be written as

$$E(\phi) = A \int_{-L/2}^{L/2} e^{j\psi(y)} dy \qquad (17.7\text{-}1)$$

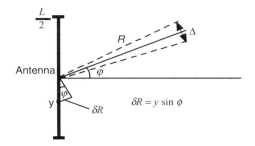

Figure 17.4. Geometry for estimating cross-range resolution.

where A is the incident amplitude assumed constant, and

$$\psi(y) = \frac{2\pi}{\lambda} y \sin \phi \qquad (17.7\text{-}2)$$

is the phase shift due to a distance $\delta = y \sin \phi$. The imaginary part of $E(\phi)$ integrates to zero, and the real part gives

$$E(\phi) = \frac{\sin\left(\frac{\pi}{\lambda} L \sin \phi\right)}{\left(\frac{\pi}{\lambda} L \sin \phi\right)} L \qquad (17.7\text{-}3)$$

Normalizing to unity at $\phi = 0$, the antenna power gain becomes

$$G(\phi) = \left|\frac{E(\phi)}{E(0)}\right|^2 = \frac{\sin^2\left(\frac{\pi}{\lambda} L \sin \phi\right)}{\left(\frac{\pi}{\lambda} L \sin \phi\right)^2} \qquad (17.7\text{-}4)$$

Setting $G(\phi) = \frac{1}{2}$ at half-power points yields, after some simplified computations,

$$\phi\Big|_{3dB} \simeq \pm 0.44 \frac{\lambda}{L} \qquad (17.7\text{-}5)$$

For a target at a range R, this translates to the following cross-range resolution:

$$\Delta \simeq 0.88 \frac{R\lambda}{L} \qquad (17.7\text{-}6)$$

Thus, improved cross-range resolution occurs for short wavelengths and large antenna lengths.

17.8 A SIMPLIFIED THEORY OF SAR IMAGING

The imaging geometry is shown in Figure 17.5. A 2-D geometry is used for the sake of simplicity. A 3-D real-world geometry would be obtained by replacing x by $\sqrt{x^2 + z^2}$, z being the height.

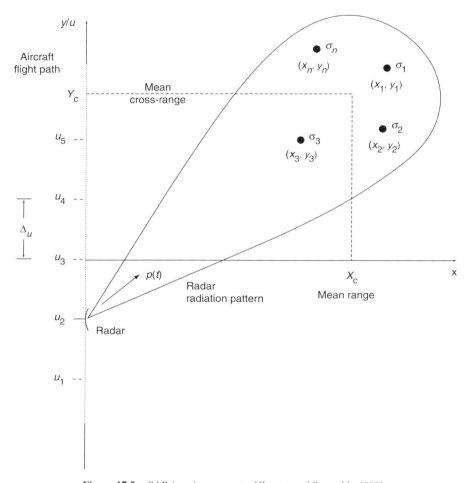

Figure 17.5.   SAR imaging geometry [Courtesy of Soumekh, 1999].

SAR imaging is usually discussed in one of two modes. In the *squint mode*, the center of the target area is at (x_c, y_c) where y_c is nonzero. This is what is shown in Figure 17.5. In the *spotlight mode*, y_c equals zero such that the target region appears perpendicular to the direction of flight.

Assume that the target region consists of stationary point reflectors with reflectivity σ_n at coordinates (x_n, y_n), $n = 1, 2, 3, \ldots$. A pulse signal $p(\bullet)$ is used to illuminate the target area. A radar receiver at $(0, y)$ receives the echo signal $s(t, y)$ reflected back from the targets as

$$s(t, y) = \sum_n \sigma_n p(t - t_n) \qquad (17.8\text{-}1)$$

where t_n is the round-trip delay from the radar to the nth target, given by

$$t_n = \frac{2r_n}{c} \tag{17.8-2}$$

and r_n is the distance to the nth target, namely,

$$r_n = [x_n^2 + (y_n - y)^2]^{\frac{1}{2}} \tag{17.8-3}$$

The desired image to be reconstructed can be represented as

$$u(x, y) = \sum_n \sigma_n \delta(x - x_n, y - y_n) \tag{17.8-4}$$

So the problem is how to obtain an approximation of $u(x, y)$ from the measured signal $s(t, y)$. The 2-D Fourier transform of $u(x, y)$ is given by

$$U(f_x, f_y) = \sum_n \sigma_n \exp[-j2\pi(f_x x_n + f_y y_n)] \tag{17.8-5}$$

The 1-D Fourier transform of $s(t, y)$ with respect to the time variable t is given by

$$S(f, y) = P(f) \sum_n = \sigma_n \exp[-j2\pi f t_n] = P(f) \sum_n \sigma_n \exp[-j2kr_n] \tag{17.8-6}$$

where $k = 2\pi f/c$ is the wave number, and $P(f)$ is the Fourier transform of $p(t)$.

The 2-D Fourier transform of $s(t, y)$ with respect to both t and y can now be computed as the 1-D Fourier transform of $S(f, y)$ with respect to y, and is given by

$$S(f, f_y) = P(f) \sum_n = \sigma_n \exp[-j(k_x x_n + k_y y_n)] \tag{17.8-7}$$

where

$$k_y = 2\pi f_y$$
$$k_x = [4k^2 - k_y^2]^{\frac{1}{2}} = 2\pi f_x \tag{17.8-8}$$

Note that $U(f_x, f_y)$ can be obtained from $S(f, f_y)$ by

$$U(f_x, f_y) = \frac{S(f, f_y)}{P(f)} \tag{17.8-9}$$

This operation is called *source deconvolution* where the source is $P(f)$. However, using Eq. (17.8-9) usually yields erroneous results since it is an ill-conditioned operation, especially in the presence of noise. Instead matched filtering is used in the form

$$U'(f_x, f_y) = S(f, f_y)P^*(f)$$
$$= |P(f)|^2 \sum_n \sigma_n \exp[-j2\pi(f_x x + f_y y)] \qquad (17.8\text{-}10)$$

In the space-time domain, it can be shown that this operation corresponds to the convolution of $s(t, y)$ with $p * (-t)$:

$$u'(t, y) = s(t, y) * p^*(-t)$$
$$= \sum_n \sigma_n h\left(t - \frac{2r_n}{c}\right) \qquad (17.8\text{-}11)$$

where $h(\cdot)$ is the point spread function given by

$$h(t) = F^{-1}[|P(f)|^2] \qquad (17.8\text{-}12)$$

Equation (17.8-10) as compared to Eq. (17.8-4) shows that the spectrum of the reconstructed signal is different from the spectrum of the desired signal by the factor of $|P(f)|^2$. Since this is typically a slowly varying amplitude function of f with zero phase, reconstruction by matched filtering usually gives much better results than source deconvolution in the presence of noise.

17.9   IMAGE RECONSTRUCTION WITH FRESNEL APPROXIMATION

It is possible to simplify the reconstruction algorithm by using the Fresnel approximation. Suppose that the center of the target area is at $(x_c, 0)$. The Taylor series expansion of r_n around $(x_c, 0)$ with $x_n = x_c + x_n$ yields

$$r_n = x_c + x_n + \frac{(y_n - y)^2}{2x_c} + \cdots \qquad (17.9\text{-}1)$$

Fresnel approximation corresponds to keeping the terms shown. $S(f, y)$ of Eq. (17.8-6) can now be written as

$$S(f, y) \simeq P(f)\exp(-j2kx_c) \sum_n \sigma_n \exp\left[-j2kx_n + \frac{(y_n - y)^2}{x_c}\right] \qquad (17.9\text{-}2)$$

Consider the 1-D Fourier transform of the ideal target function given by Eq. (17.8-4) with respect to the x-coordinate:

$$U(k_x, y) = \sum_n \sigma_n \exp(-jk_x x_n)\delta(y - y_n) \qquad (17.9\text{-}3)$$

where $k_x = 2\pi f_x$ and $\delta(x - x_n, y - y_n) = \delta(x - x_n)\delta(y - y_n)$ are used. Letting $k_x = 2k = \frac{2w}{c}$, comparison of Eqs. (17.9-2) and (17.9-3) shows that $S(f, y)$ is approximately equal to the convolution of $P(f)U(2k, y)$ with $\exp(-jky^2/x_c)$ in the y-coordinate:

$$S(f, y) \simeq P(f)U(2k, y) * \exp\left(\frac{-jky^2}{x_c}\right) \qquad (17.9\text{-}4)$$

In order to recover $u(x, y)$, matched filtering is performed with respect to both $P(f)$ and the chirp signal $\exp(-jky^2/x_c)$. The reconstruction signal spectrum can be written as

$$U'(f_x, y) = P * (f)S(f, y) * \exp\left[j\frac{ky^2}{x_c}\right] \qquad (17.9\text{-}5)$$

In general, the radar bandwidth is much smaller than its carrier frequency k_c so that $|k - k_c| \ll k_c$ is usually true. This is referred to as *narrow bandwidth*. In addition, the y-values are typically much smaller than the target range. This is referred to as *narrow beamwidth*. Then, $\exp(jky^2/x_c)$ can be approximated by $\exp(jk_cy^2/x_c)$, and Eq. (17.9-5) becomes

$$U'(f_x, y) = P * (f)S(f, y) * \exp\left[\frac{jk_cy^2}{x_c}\right] \qquad (17.9\text{-}6)$$

where

$$f_x = \frac{k}{\pi} = \frac{w}{\pi c} \qquad (17.9\text{-}7)$$

Note that there are two filtering operations above. The first one is with the filter transfer function $P^*(f)$, which corresponds to $p^*(-t)$ in the time domain. The second one is with the filter impulse function $\exp(jk_cy^2/x_c)$.

We recall that the correspondence between the time variable t and the range variable x with the stated approximations is given by

$$x = \frac{ct}{2} \qquad (17.9\text{-}8)$$

In summary, when narrow bandwidth and narrow beamwidth assumptions are valid, image reconstruction is carried out as follows:

1. Filter with the impulse response $p^*(-t)$ in the time direction.
2. Filter with the impulse response $\exp(jk_cy^2/x_c)$ in the y-direction.
3. Identify x as $ct/2$ or t as $2x/c$.

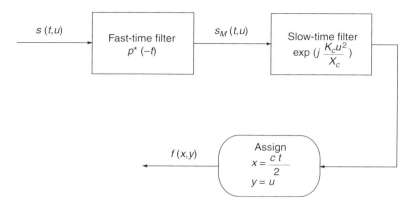

Figure 17.6.   Block diagram of SAR reconstruction with Fresnel approximation [Courtesy of Soumekh, 1999].

The block diagram for the algorithm is shown in Figure 17.6. When the approximations discussed above are not valid, more sophisticated image reconstruction algorithms are used, as discussed in the next section.

17.10   ALGORITHMS FOR DIGITAL IMAGE RECONSTRUCTION

The goal of SAR imaging is to obtain the reflectivity function $u(x, y)$ of the target area from the measured signal $s(t, y)$. There have been developed several digital reconstruction algorithms for this purpose, namely, the *spatial frequency interpolation algorithm*, the *range stacking algorithm*, the *time domain correlation algorithm*, and the *backprojection algorithm* [Soumekh]. The spatial frequency interpolation algorithm is briefly discussed below.

17.10.1   Spatial Frequency Interpolation

The 2-D Fourier transform of $s(t, y)$ yields $S(f, f_y)$, which can be written as $S(k, k_y)$ where $k = 2\pi f/c$ and $k_y = 2\pi f_y$.

After matched filtering, it is necessary to map k to k_x so that the reconstructed image can be obtained after 2-D inverse Fourier transform. The relationship between k_x, k, and k_y is given by

$$k_x^2 = 4k^2 - k_y^2 \qquad (17.10\text{-}1)$$

When k and k_y are sampled in equal intervals in order to use the DFT, k_x is sampled in nonequal intervals. This is shown in Figure 17.7. Then, it is necessary to use interpolation in order to sample both k_x and k_y in a rectangle lattice. Interpolation

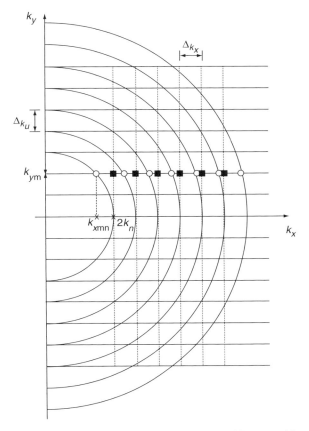

Figure 17.7. SAR spatial frequency sampling for discrete data [Courtesy of Soumekh, 1999].

is not necessary only when $k_x \simeq 2k$, as discussed in the case of narrow bandwidth and narrow beamwidth approximation in Section 17.9.

Interpolation should be done with a lowpass signal. Note that the radar range swath is $x \in [x_c - x_0, x_c + x_0]$, $y \in [y_c - y_0, y_c + y_0]$ where (x_c, y_c) is the center of the target region. Since (x_c, y_c) is not $(0, 0)$, $S(k_x, k_y)$ is a bandpass signal with fast variations. In order to convert it to a lowpass signal, the following computation is performed:

$$S'(k_x, k_y) = S(k_x, k_y)e^{j(k_x x_c + k_y y_c)} \tag{17.10-2}$$

Next $S'(k_x, k_y)$ is interpolated so that the sampled values of k_x lie in regular intervals. The interpolated $S'(k_x, k_y)$ is inverse Fourier transformed in order to obtain the reconstructed image. The block diagram of the algorithm is shown in Figure 17.8.

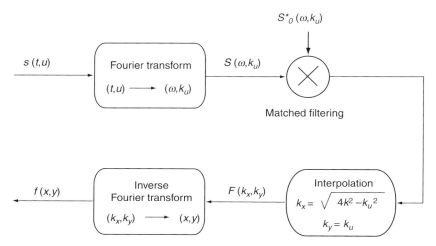

Figure 17.8. Block diagram of SAR digital reconstruction algorithm with spatial frequency domain interpolation [Courtesy of Soumekh, 1999].

Interpolation is to be done with unevenly spaced data in the k_x direction. Let k_y and k be sampled as

$$k_y = m\Delta k_y = k_{ym}$$
$$k = n\Delta k = k_n$$

$$(17.10\text{-}3)$$

Then, the sampled values of k_x are given by

$$k_{xmn} = [4n^2(\Delta k)^2 - m^2(\Delta k_y)^2]^{\frac{1}{2}}$$

$$(17.10\text{-}4)$$

The function $S'(k_x, k_y)$ is to be interpolated for regularly sampled values of k_x. This is done as follows [Soumekh, 1988]:

$$S'(k_x, k_{ym}) = \sum_n J_{mn} S'(k_{xmn}, k_{ym}) h(k_x - k_{xmn}) |k_x - k_{xmn}| \leq N\Delta k_x$$

$$(17.10\text{-}5)$$

where N is the number of sampled points along both positive and negative k_x directions, Δk_x is the desired sampling interval in the k_x direction, $h(\bullet)$ is the interpolating function to be discussed below, and J_m is the Jacobian given by

$$J_{mn} = \frac{\mathrm{d}}{\mathrm{d}w}[4k^2 - k_{ym}^2]^{\frac{1}{2}} = \frac{4k_n}{c[4k_n^2 - k_{ym}^2]}$$

$$(17.10\text{-}6)$$

where $k = w/c$. k_x is chosen as

$$k_x = i\Delta k_x = k_{xi}$$

$$(17.10\text{-}7)$$

The choice of the function $h(\bullet)$ is dictated by the sampling theorem in digital signal processing and communications. If a signal $V(k_x, k_y)$ is sampled as $V(n\Delta k_x, m\Delta k_y)$ it can be interpolated as

$$V(k_x, k_y) = \sum_n V(n\Delta k_x, n\Delta k_y) \sin c \left(\frac{k_x}{\Delta k_x} - n \right) \qquad (17.10\text{-}8)$$

where

$$\sin c(v) = \frac{\sin \pi v}{\pi v} \qquad (17.10\text{-}9)$$

The $\sin c$ function has infinitely long tails. In order to avoid this problem in practise, it is replaced by its truncated version by using a window function $w(k_x)$ in the form

$$h(k_x) = \sin c \left(\frac{k_x}{\Delta k_x} \right) w(k_x) \qquad (17.10\text{-}10)$$

$w(k_x)$ can be chosen in a number of ways as discussed in Section 14.2. For example, the *Hamming window* is given by

$$w(k_x) = \begin{cases} 0.54 + 0.46 \cos\left(\dfrac{\pi k_x}{N\Delta k_x} \right) & |k_x| \leq N\Delta k_x \\ 0 & \text{otherwise} \end{cases} \qquad (17.10\text{-}11)$$

18

Computerized Imaging II: Image Reconstruction from Projections

18.1 INTRODUCTION

In the second part of computerized imaging involving Fourier-related transforms, image reconstruction from projections including tomography is discussed. The fundamental transform for this purpose is the *Radon transform*. In this chapter, the Radon transform and its inverse are first described in detail, followed by imaging algorithms used in tomography and other related areas.

Computed tomography (CT) is mostly used as a medical imaging technique in which an area of the subject's body that is not externally visible is investigated. A 3-D image of the object is obtained from a large series of 2-D x-ray measurements. An example of CT image is shown in Figure 18.1. CT is also used in other fields such as nondestructive materials testing.

This chapter consists of eight sections. The Radon transform is introduced in Section 18.2. The *projection slice theorem* that shows how the 1-D Fourier transforms of projections are equivalent to the slices in the 2-D Fourier transform plane of the image is discussed in Section 18.3. The inverse Radon transform (IRT) is covered in Section 18.4. The properties of the Radon transform are described in Section 18.5.

The remaining sections are on the reconstruction algorithms. Section 18.6 illustrates the sampling issues involved for the reconstruction of a 2-D signal from its projections. Section 18.7 covers the *Fourier reconstruction algorithm*, and Section 18.8 describes the *filtered backprojection algorithm*.

18.2 THE RADON TRANSFORM

Consider Figure 18.2. The $x - y$ axis are rotated by θ to give the axis $u - v$. The relationship between (x, y) and (u, v) is given by a plane rotation of

Diffraction, Fourier Optics and Imaging,   by Okan K. Ersoy
Copyright © 2007 John Wiley & Sons, Inc.

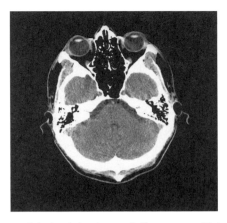

Figure 18.1. The CT image of the head showing cerebellum, temporal lobe and sinuses [Courtesy of Wikipedia].

θ degrees:

$$\begin{bmatrix} x \\ y \end{bmatrix} = \begin{bmatrix} \cos\theta & -\sin\theta \\ \sin\theta & \cos\theta \end{bmatrix} \begin{bmatrix} u \\ v \end{bmatrix} \tag{18.2-1}$$

The Radon transform $p(u, \theta)$ of a signal $g(x, y)$ shown in Figure 18.1 is the line integral of $g(x, y)$ parallel to the v-axis at the distance u on the u-axis which makes the angle $0 \le \theta < \pi$ with the x-axis:

$$p(u, \theta) = \int_{-\infty}^{\infty} g(x, y)\mathrm{d}v$$

$$= \int_{-\infty}^{\infty} g(u\cos\theta - v\sin\theta, u\sin\theta + v\cos\theta)\mathrm{d}v \tag{18.2-2}$$

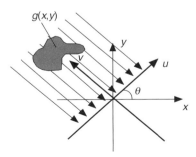

Figure 18.2. Rotation of Cartesian coordinates for line integration along the v-axis.

$p(u, \theta)$ is also called the *ray-sum* or the *ray integral* of $g(x, y)$ at the angle θ. A set of ray integrals forms a *projection*. Below θ will be implicitly treated by writing $p(u, \theta)$ as $p(u)$.

A projection taken along a set of parallel rays as shown in Figure 18.2 is called a *parallel projection*. If the set of rays emanates from a point source, the resulting projection is called a *fan-beam projection*. Only the case of parallel projections is considered in this chapter.

The Radon transform can also be expressed in terms of the Dirac-delta function. u can be written as $\alpha^t r$ where $\alpha = [\cos \theta \ \sin \theta]^t$, $r = [x \ y]^t$. Then, Eq. (18.2-1) is the same as

$$
\begin{aligned}
p(u, \theta) &= \int g(r)\delta(u - \alpha^t r)\mathrm{d}r \\
&= \iint_{-\infty}^{\infty} g(x, y)\delta(u - \cos \theta\, x - \sin \theta\, y)\mathrm{d}x\mathrm{d}y
\end{aligned}
\tag{18.2-3}
$$

The Radon transform has many applications in areas such as computerized tomography, geophysical and multidimensional signal processing. The most common problem is to reconstruct $g(x, y)$ from its projections at a finite number of the values of θ. The reconstruction is exact if projections for all θ are known. This is due to the theorem given in the next section.

18.3   THE PROJECTION SLICE THEOREM

The 1-D FT of $p(u)$ denoted by $P(f)$ is equal to the central slice at angle θ of the 2-D FT of $g(x, y)$ denoted by $G(f_x, f_y)$:

$$
P(f) = G(f \cos \theta, f \sin \theta)
\tag{18.3-1}
$$

The slice of frequency components in the 2-D frequency plane is visualized in Figure 18.3.

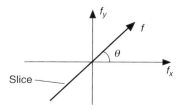

Figure 18.3.   The slice of frequencies used in the projection slice theorem.

Proof:

$$P(f) = \int\limits_{-\infty}^{\infty} p(u)e^{-j2\pi f t u}\,du$$

$$= \iint\limits_{-\infty}^{\infty} g(u\cos\theta - v\sin\theta, u\sin\theta + v\cos\theta)e^{-j2\pi f u}\,dv\,du$$

In the unrotated coordinate system, this is the same as

$$P(f) = \iint\limits_{-\infty}^{\infty} g(x,y)e^{-j2\pi(fx\cos\theta + fy\sin\theta)}\,dx\,dy$$

$$= G(f\cos\theta, f\sin\theta)$$

EXAMPLE 18.1 Find the Radon transform of $g(x,y) = e^{-x^2-y^2}$.
Solution:

$$p(u,\theta) = \iint\limits_{-\infty}^{\infty} e^{-x^2-y^2}\delta(u - x\cos\theta - y\sin\theta)\,dx\,dy$$

Since the transformation $(x,y) \rightarrow (u,v)$ as given by Eq. (18.2-1) is orthonormal, $x^2 + y^2 = u^2 + v^2$. Hence,

$$p(u,\theta) = \int\limits_{-\infty}^{\infty} e^{-u^2}e^{-v^2}\,dv$$

$$= \sqrt{\pi}e^{-u^2}$$

EXAMPLE 18.2 Find the Radon transform of $g(x,y)$ if

$$g(x,y) = \begin{cases} (1 - x^2 - y^2)^{\lambda-1} & x^2 + y^2 < 1 \\ 0 & \text{otherwise} \end{cases}$$

Solution: $g(x,dy)$ is nonzero inside the unit circle. Since $x^2 + y^2 = u^2 + v^2$, the limits of integration with respect to u are $\pm\sqrt{1-u^2}$. Thus,

$$p(u,\theta) = \int\limits_{-\sqrt{1-u^2}}^{\sqrt{1-u^2}} [1 - u^2 - v^2]^{\lambda-1}\,dv$$

The following definite integral can be found in [Erdelyi]:

$$\int_{-a}^{a} (a^2 - t^2)^{\lambda-1} dt = \frac{a^{2\lambda-1}\sqrt{\pi}\,\Gamma(\lambda)}{\Gamma\left(\lambda + \dfrac{1}{2}\right)}$$

Utilizing this result above yields

$$p(u, \theta) = \begin{cases} \dfrac{\sqrt{\pi}\,\Gamma(\lambda)}{\Gamma\left(\lambda + \dfrac{1}{2}\right)}(1 - u^2)^{\lambda-1/2} & -1 < u < 1 \\ 0 & \text{otherwise} \end{cases}$$

18.4   THE INVERSE RADON TRANSFORM

The 2-D FT representation of $g(x, y)$ is given by

$$g(x, y) = \iint_{-\infty}^{\infty} G(f_x, f_y) e^{j2\pi(f_x x + f_y y)} df_x df_y \qquad (18.4\text{-}1)$$

Converting (f_x, f_y) to polar coordinates (f, θ) gives

$$g(x, y) = \int_{0}^{\pi} \int_{-\infty}^{\infty} G(f \cos\theta, f \sin\theta) e^{j2\pi f(x\cos\theta + y\sin\theta)} |f| df d\theta \qquad (18.4\text{-}2)$$

Since $P(f) = G(f \cos\theta, f \sin\theta)$, this is the same as

$$g(x, y) = \int_{0}^{\pi} \int_{-\infty}^{\infty} fP(f)\operatorname{sgn}(f) e^{j2\pi f(x\cos\theta + y\sin\theta)} df d\theta \qquad (18.4\text{-}3)$$

By convolution theorem, with $u = x\cos\theta + y\sin\theta$, the first integral is given by

$$\int_{-\infty}^{\infty} fP(f)\operatorname{sgn}(f) e^{j2\pi f u} df = \frac{1}{2j\pi}\frac{\partial}{\partial u}p(u) * \left(\frac{-1}{j\pi u}\right)$$

$$= \frac{1}{2\pi^2}\int_{-\infty}^{\infty} \frac{\partial p(\tau)}{\partial\tau}\frac{1}{u - \tau} d\tau \qquad (18.4\text{-}4)$$

The last result is the Hilbert transform of $\frac{1}{2\pi}\frac{\partial}{\partial u}p(u)$. Let

$$\hat{p}(u) = \frac{1}{2\pi^2} \int\limits_{\infty}^{\infty} \frac{\partial p(\tau)}{\partial \tau} \frac{1}{u - \tau} d\tau \qquad (18.4\text{-}5)$$

$\hat{p}(u)$ is called the *filtered projection*. The inverse Radon transform becomes

$$\begin{aligned} g(x,y) &= \int\limits_0^\pi \hat{p}(u)d\theta \\ &= \int\limits_0^\pi \hat{p}(x\cos\theta + y\sin\theta)d\theta \end{aligned} \qquad (18.4\text{-}6)$$

This is equivalent to the back-projection of $\hat{p}(u)$ along the angular direction θ. This means the projections for all θ are needed for perfect reconstruction if $g(x,y)$ is not band-limited.

18.5   PROPERTIES OF THE RADON TRANSFORM

The Radon transform has a number of properties that are very useful in applications. The most important ones are summarized below. When needed, $p(u)$ will be written as $p_g(u)$ to show that the corresponding signal is $g(x,y)$.

Property 1: Linearity
   If $h(x,y) = g_1(x,y) + g_2(x,y)$,

$$p_h(u) = p_{g_1}(u) + p_{g_2}(u) \qquad (18.5\text{-}1)$$

Property 2: Periodicity

$$p(u,\theta) = p(u,\theta + 2\pi k), k \text{ an integer} \qquad (18.5\text{-}2)$$

Property 3: Mass conservation

$$\int\limits_{-\infty}^{\infty}\!\!\int\limits_{-\infty}^{\infty} g(x,y)dxdy = \int\limits_{-\infty}^{\infty} p(u)du \qquad (18.5\text{-}3)$$

Property 4: Symmetry

$$p(u,\theta) = p(-u,\theta \pm \pi) \qquad (18.5\text{-}4)$$

Property 5: Bounded signal

If $g(x, y) = 0$ for $|x|, |y| > \frac{D}{2}$,

$$p(u) = 0 \quad \text{for} \quad |u| > \frac{D}{\sqrt{2}} \tag{18.5-5}$$

Property 6: Shift

If $h(x, y) = x(x - x_0, y - y_0)$,

$$p_h(u, \theta) = p_g(u - u_0, \theta) \tag{18.5-6}$$

where $u_0 = x_0\cos\theta + y_0\sin\theta$.

Property 7: Rotation

$g(x, y)$ in polar coordinates is $g(r, \phi)$. If $h(r, \phi) = g(r, \phi + \phi_0)$,

$$p_h(u, \theta) = p_g(u, \theta + \phi_0) \tag{18.5-6}$$

Property 8: Scaling

If $h(x, y) = g(ax, ay)$, $a \neq 0$,

$$p_h(u) = \frac{1}{|a|} p_g(au) \tag{18.5-7}$$

Property 9: Convolution

If $h(x, y) = g_1(x, y) * g_2(x, y)$ (where $*$ represents 2-D convolution),

$$
\begin{aligned}
p_h(u) &= p_{g_1}(u) * p_{g_2}(u) \\
&= \int_{-\infty}^{\infty} p_{g_1}(\tau) p_{g_2}(u - \tau) d\tau \\
&= \int_{-\infty}^{\infty} p_{g_1}(u - \tau) p_{g_2}(\tau) d\tau
\end{aligned}
\tag{18.5-8}
$$

This property is useful to implement 2-D linear filters by 1-D linear filters.

18.6   RECONSTRUCTION OF A SIGNAL FROM ITS PROJECTIONS

The signal is given by its inverse Radon transform. For computer reconstruction, it is necessary to discretize both θ and u. A major consideration is how many samples are needed.

Assume that projections are available at angles $\theta_0, \theta_1, \theta_2 \ldots \theta_{N-1}$. In order to estimate N, the number of projections, the signal is assumed to be both space-limited

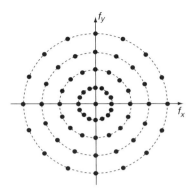

Figure 18.4. Samples in the Fourier domain when all projections are sampled at the same sampling rate.

and band-limited (even though this is not theoretically possible, it is a good approximation in practice). Thus, $g(x, y)$ will be assumed to be band-limited such that $G(f_x, f_y) = 0$ for $f_x^2 + f_y^2 \geq F_0^2$. By the 2-D sampling theorem, $g(x, y)$ can be sampled as $g(\Delta x m, \Delta y n)$ without loss of information if $\Delta x \leq 1/2F_0$ and $\Delta y \leq 1/2F_0$.

$g(x, y)$ will also be assumed to be space-limited such that $g(x, y) = 0$ for $x^2 + y^2 \geq T_0^2$. By the 2-D sampling theorem with the domains reversed, $G(f_x, f_y)$ can be sampled as $G(\Delta f_x m, \Delta f_y n)$ without loss of information if $\Delta f_x \leq 1/2T_0$, $\Delta f_y \leq 1/2T_0$.

Consider Figure 18.4 in which a polar raster of samples in the Fourier domain is shown [Dudgeon]. On the outermost circle with N projections, the distance between the samples can be estimated as $\pi F_0/N$. This distance is assumed to satisfy the same criterion as Δf_x and Δf_y. Thus,

$$\frac{\pi F_0}{N} \leq \frac{1}{2T_0} \tag{18.6-1}$$

With the sampling interval in the signal domain approximated as $\Delta = \Delta x = \Delta y \sim 1/2F_0$, the number of projections N should satisfy

$$N \geq \frac{\pi T_0}{\Delta} \tag{18.6-2}$$

Equation (18.6-2) should be considered as a rule of thumb.

18.7   THE FOURIER RECONSTRUCTION METHOD

We will assume that there are N projections, each projection is sampled, and the number of samples per projection is M. A DFT of each sampled projection is

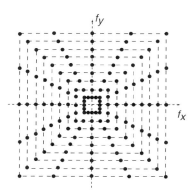

Figure 18.5. The sampling points on a polar lattice to be interpolated.

computed. These DFT values are interpreted as samples of the FT of $g(x, y)$ on a regular polar lattice, as shown in Figure 18.5 [Dudgeon]. These samples are interpolated to yield estimates on a rectangular lattice. Finally, the 2-D inverse DFT is computed to obtain $g(m, n)$. Usually, the size of the inverse DFT is chosen as two to three times that of each dimension of $g(m, n)$.

Appropriate windowing as discussed in Section 14.2 before the inverse DFT allows the minimization of the effects of Fourier domain truncation.

The simplest approach to interpolation is *zeroth order* or *linear interpolation*. Each desired rectangular sample point (f_x, f_y) is surrounded by four polar samples. In zeroth order interpolation, the rectangular sample is assigned the value of the nearest polar sample. In linear interpolation, a weighted average of the four nearest polar samples is used. The weighting can be taken inversely proportional to the Euclidian distance between the points.

A simple method is to use *bilinear interpolation* in terms of the polar coordinates: let (f, θ) be the coordinates of the point at which the interpolated sample is desired; let (f_i, θ_j), $i, j = 1, 2$ denote the coordinates of the four nearest polar samples. Then, the bilinear interpolation can be written as

$$G(f, \theta) = \sum_{i=1}^{2} \sum_{j=1}^{2} G(f_i, \theta_j) h_1(f - f_i) h_2(\theta - \theta_j) \qquad (18.7\text{-}1)$$

where

$$h_1(f') = 1 - \frac{|f'|}{\Delta f} \quad |f'| \leq \Delta f$$

$$\qquad (18.7\text{-}2)$$

$$h_2(\theta') = 1 - \frac{|\theta'|}{\Delta \theta} \quad |\theta'| \leq \Delta \theta$$

and Δf, $\Delta \theta$ are the sampling intervals in polar coordinates.

Since the density of the polar samples is less dense at high frequencies than at low frequencies, the interpolation results are also less accurate at high frequencies. This causes some degradation in the reconstructed image.

18.8   THE FILTERED-BACKPROJECTION ALGORITHM

The inverse Radon transform can be written as

$$g(x, y) = \int_0^\pi \hat{p}(x \cos \theta + y \sin \theta) d\theta \qquad (18.8\text{-}1)$$

where

$$\hat{p}(u) = \int_{-\infty}^{\infty} |f| P(f) e^{j2\pi fu} df \qquad (18.8\text{-}2)$$

$P(f)$ is zero outside $[-F_0, F_0]$. Its samples can be approximated at intervals of $\Delta f = 2F_0/N$. $p(u)$ is also sampled at intervals of $\Delta \le 1/2F_0$. By choosing $\Delta f \Delta = 1/N$, $P(\Delta f k)$ can be approximated by a DFT of size N as follows:

$$P(k) = P(\Delta f k) = \Delta \sum_{n=-N/2+1}^{N/2} p(\Delta n) e^{-j2\pi nk/N}, \quad -\frac{N}{2} < k \le \frac{N}{2} \qquad (18.8\text{-}3)$$

or

$$P(k) = \frac{1}{2F_0} \sum_{n=0}^{N-1} p(n) e^{-j2\pi nk/N}, \quad k = 0, 1 \ldots, (N-1) \qquad (18.8\text{-}4)$$

since $p(-|\ell|)$ and $P(-|m|)$ can be chosen equal to $p(N - |\ell|)$ and $P(N - |m|)$, respectively.

Equation (18.8-2) can be written as

$$\hat{p}(u) = \int_{-F_0}^{F_0} |f| P(f) e^{j2\pi fu} df \qquad (18.8\text{-}5)$$

$\hat{p}(u)$ can be approximated by an inverse DFT by using the same sampling interval Δ as follows:

$$\hat{p}(n) = \hat{p}(n\Delta) = \Delta f \sum_{k=-N/2+1}^{N/2} P(\Delta f k) |k \Delta f| e^{j2\pi nk/N} \qquad (18.8\text{-}6)$$

or

$$\hat{p}(n) = \Delta f \sum_{k=0}^{N-1} P'(k) e^{j2\pi nk/N}, \quad n = 0, 1, \ldots, (N-1) \tag{18.8-7}$$

where

$$P'(k) = \begin{cases} P(k)|k\Delta f| & 0 \leq k \leq \dfrac{N}{2} \\ P(k)|(N-k)\Delta f| & \dfrac{N}{2} < k < N \end{cases} \tag{18.8-8}$$

Better reconstructions are obtained if $P(k)$ is multiplied by a windowing function as discussed in Section 14.2. The windowing function deemphasizes the high frequencies at which measurement noise is more serious.

Finally, the reconstructed image is obtained by the numerical integration of Eq. (18.8-1). If there are M projections, $g(x, y)$ can be approximated as $g'(x, y)$ given by

$$g'(x, y) = \frac{\pi}{M} \sum_{k=1}^{M} \hat{p}(x \cos \theta_k + y \sin \theta_k) \tag{18.8-9}$$

$u = (x \cos \theta_k + y \sin \theta_k)$ may not correspond to $p(\Delta n)$ of Eq. (18.8-4). Then, interpolation is necessary again to find $p(u)$ given $p(n\Delta)$, $n = 0, 1, 2 \ldots, (N-1)$. Linear interpolation is usually sufficient for this purpose.

The algorithm as described above has certain drawbacks. The use of the DFT with N points means that linear convolution of Eq. (18.4-5) is replaced by a "partial" circular convolution. The word "partial" is used since the discretization of $|f|$ is exact. If the size N DFTs are replaced by size $2N$ DFTs by zero-padding, the reconstruction improves since circular convolution becomes linear in the desired output terms.

However, best results are achieved if the convolution expressed in the transform domain by Eq. (18.4-5) is first written in the time-domain, and then the whole processing is done by DFTs of zero-padded sequences.

The transfer function in Eq. (18.4-5) is given by

$$H(f) = \begin{cases} |f| & |f| \leq F_0 \\ 0 & \text{otherwise} \end{cases} \tag{18.8-10}$$

The corresponding impulse response is given by

$$h(t) = \int_{-F_0}^{F_0} H(f) e^{j2\pi ft} \mathrm{d}f$$

$$= 2F_0^2 \frac{\sin 2\pi t F_0}{2\pi x F_0} - F_0^2 \left(\frac{\sin \pi t F_0}{\pi x F_0} \right)^2 \tag{18.8-11}$$

The discretized impulse response at $t = n\Delta t$ is given by

$$h(n\Delta t) = \begin{cases} F_0^2 & n = 0 \\ 0 & n \text{ even} \\ -\dfrac{4F_0^2}{n^2\pi^2} & n \text{ odd} \end{cases} \qquad (18.8\text{-}12)$$

Equation (18.4-5) as a convolution in the time-domain is given by

$$\hat{p}(u) = \int_{-\infty}^{\infty} p(\tau)h(u - \tau)\mathrm{d}\tau \qquad (18.8\text{-}13)$$

Its discrete version is given by

$$\hat{p}(n\Delta) \simeq \Delta \sum_{m=-\infty}^{\infty} p(\Delta m)h((n - m)\Delta) \qquad (18.8\text{-}14)$$

Since $p(\Delta tm)$ is zero for $|m| > \frac{N}{2}$,

$$\hat{p}(n\Delta t) \simeq \Delta t \sum_{m=-N/2}^{N/2} p(\Delta tm)h((n - m)\Delta t) \qquad (18.8\text{-}15)$$

This linear convolution can now be computed with DFTs of size $2N$ after zero-padding both $p(\bullet)$ and $h(\bullet)$. Further improvement is possible by properly windowing the frequency domain results. The whole procedure can be written as follows:

1. Zero-pad $p(\bullet)$ and $h(\bullet)$ in the form

$$p(n) = \begin{cases} p(n) & 0 \leq n \leq \dfrac{N}{2} \\ 0 & \dfrac{N}{2} < n < \dfrac{3N}{2} \\ p(n - 2N) & n > \dfrac{3N}{2} \end{cases} \qquad (18.8\text{-}16)$$

   and similarly for $h(\bullet)$.
2. Compute the size $2N$ DFTs of $p(\bullet)$ and $h(\bullet)$.
3. Do the transform domain operations.
4. Window the results of step 3 by a proper window.
5. Compute the size $2N$ inverse DFT.

19

Dense Wavelength Division Multiplexing

19.1 INTRODUCTION

In recent years, optical Fourier techniques have found major applications in optical communications and networking. One such application area is *arrayed waveguide grating* (AWG) technology used in *dense wavelength division multiplexing* (DWDM) systems. DWDM provides a new direction for solving capacity and flexibility problems in optical communications and networking. It offers a very large transmission capacity and novel network architectures [Brackett, 90], [Brackett, 93]. Major components in DWDM systems are the wavelength multiplexers and demultiplexers. Commercially available components are based on fiber-optic or microoptic techniques [Pennings, 1995], [Pennings, 1996].

Research on integrated-optic (de)multiplexers has increasingly been focused on grating-based and phased-array-based (PHASAR) devices (also called arrayed waveguide gratings) [Laude, 1993], [Smit, 1988]. Both are imaging devices, that is, they image the field of an input waveguide onto an array of output waveguides in a dispersive way. In grating-based devices, a vertically etched reflection grating provides the focusing and dispersive properties required for demultiplexing. In phased-array-based devices, these properties are provided by an array of waveguides, whose lengths are chosen to satisfy the required imaging and dispersive properties.

As phased-array-based devices are realized in conventional waveguide technology and do not require the vertical etching step needed in grating-based devices, they appear to be more robust and fabrication tolerant. The first devices operating at short wavelengths were reported by Vellekoop and Smit [1989] [Verbeek and Smit, 1995]. Takahashi *et al.* [1990] reported the first devices operating in the long wavelength window around 1.6 micron. Dragone [1991] extended the phased-array concept from $1 \times N$ to $N \times N$ devices.

This chapter consists of six sections. The principle of arrayed waveguide grating is discussed in Section 19.2. A major issue is that the number of available channels is limited due to the fact that each focused beam at a particular wavelength repeats at periodic locations. Section 19.3 discusses a method called method of irregularly sampled zero-crossings (MISZC) developed to significantly reduce this problem.

Section 4 provides detailed computer simulations. Section 19.4 provides an analysis of the properties of the MISZC. Computer simulations with the method in 2-D and 3-D are described in Section 19.5. Implementational issues in 2-D and 3-D are covered in Section 19.6.

19.2   ARRAY WAVEGUIDE GRATING

The AWG-based multiplexers and demultiplexers are essentially the same. Depending on the direction of light wave propagation, the device can be used as either a multiplexer or a demultiplexer due to the reciprocity principle. For the sake of simplicity, the demultiplexer operation is discussed here.

The AWG consists of two arrays of input/output waveguides, two focusing slab regions, and the array grating waveguides. This is illustrated in Figure 19.1. A single fiber containing the multiwavelength input is connected to the array of input waveguides. The input multiwavelength signal is evenly split among the input waveguides, and the signal propagates through the input waveguides to reach the input focusing slab region. The light wave travels through the focusing slab and is coupled into the array grating waveguides. A linear phase shift occurs in the light wave traveling through each array grating waveguide due to the path length differences between the array grating waveguides.

The light wave is subsequently coupled into the output focusing slab, in which the multiwavelength input signal is split into different beams according to their wavelengths due to diffraction and wavelength dispersion. The length of the array waveguides and the path length difference ΔL between two adjacent waveguides are chosen in such a way that the phase retardation for the light wave of the center wavelength

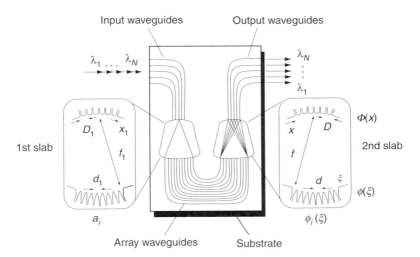

Figure 19.1.   Schematic for the arrayed waveguide grating device [Courtesy of Okamoto].

passing through every array waveguide is $2\pi m$, m being the *diffraction order* equal to an integer. The phase retardations of the light waves of wavelengths other than the center wavelength are different from the phase retardation of the center wavelength. As a result, a unique phase front is created for each wavelength component which is then focused to a different position on the output side of the output focusing slab region. Each wavelength component is then fed into an output waveguide.

An approximate analysis of the demultiplexing operation is discussed below. The definitions in the input and output focusing slab regions are as follows:

Input focusing slab region:

D_1: the spacing between the ends of adjacent waveguides,

d_1: the spacing between the ends of adjacent waveguides on the output side,

x_1: distance measured from the center of the input side,

f_1: the radius of the output curvature.

Output focusing slab region:

d: the spacing between the ends of adjacent waveguides connected to the array waveguides,

D: the spacing between the ends of adjacent waveguides connected to the output waveguides,

f: the radius of the output curvature.

We reiterate that the path difference between two adjacent waveguides is ΔL, and the corresponding phase retardation is $2\pi m$ with respect to the center wavelength.

Consider the light beams passing through the ith and (i-1)th array waveguides. In order for the two light beams to interfere constructively, their phase difference should be a multiple of 2π as they reach the output side of the focusing slab region. The condition for constructive interference is then given by

$$
\begin{aligned}
&\beta_s(\lambda_0)\left[f_1 - \frac{d_1 x_1}{2f_1}\right] + \beta_c(\lambda_0)[L_c + (i-1)\Delta L] + \beta_s(\lambda_0)\left[f + \frac{dx}{2f}\right] \\
&= \beta_s(\lambda_0)\left[f_1 + \frac{d_1 x_1}{2f_1}\right] + \beta_c(\lambda_0)[L_c + i\Delta L] + \beta_s(\lambda_0)\left[f - \frac{dx}{2f}\right] - 2\pi m
\end{aligned}
\tag{19.2-1}
$$

where β_s and β_c denote the propagation constants (wave numbers) in the slab region and the array waveguide, respectively, m is the diffraction order, λ_0 is the center wavelength of the multiple wavelength input, and L_c is the minimum array waveguide length. Subtracting common terms from Eq. (19.2-1), we obtain

$$
\beta_s(\lambda_0)\frac{d_1 x_1}{f_1} - \beta_s(\lambda_0)\frac{dx}{f} + \beta_c(\lambda_0)\Delta L = 2\pi m
\tag{19.2-2}
$$

When the condition

$$
\beta_c(\lambda_0)\Delta L = 2\pi m
\tag{19.2-3}
$$

is satisfied for λ_0, the light input position x_1 and the output position x satisfy the condition

$$\frac{d_1 x_1}{f_1} = \frac{dx}{f} \tag{19.2-4}$$

The above equation means that when light is coupled into the input position x_1, the output position x is determined by Eq. (19.2-4).

The path length difference ΔL can be shown to be

$$\Delta L = \frac{n_s d D \lambda_0}{n_c f \Delta \lambda} \tag{19.2-5}$$

The spatial separation (free spectral range) of the mth and $(m + l)$th focused beams for the same wavelength can be derived from Eq. (19.2-2) as [Okamoto]

$$X_{\mathrm{FSR}} = x_m - x_{m+1} = \frac{\lambda_0 f}{n_s d} \tag{19.2-6}$$

The number of available wavelength channels N_{ch} is obtained by dividing X_{FSR} by the output waveguide separation D as

$$N_{\mathrm{ch}} = \frac{X_{\mathrm{FSR}}}{D} = \frac{\lambda_0 f}{n_s d D} \tag{19.2-7}$$

In practice, achieving the layout of the waveguides in a planar geometry such that the length difference between two waveguides is $m\lambda$ is not a trivial task. Professional computer-aided design programs are usually used for this purpose. An example is shown in Figure 19.2 in which the BeamPROP software package by Rsoft Inc. was used to carry out the design [Lu, Ersoy].

Two examples of the results obtained with PHASAR simulation at a center wavelength of 1.55 μ and channel spacing of 0.8 μ are shown in Figures 19.3 and 19.4 to illustrate how the number of channels are limited [Lu and Ersoy et al., 2003]. In Figure 19.3, there are 16 channels, and the second order channels on either side of the central channels do not overlap with the central channels. On the other hand, in Figure 19.4, there are 64 channels, and the second order channels on either side of the central channels start overlapping with the central channels. In this particular case, the number of channels could not be increased any further. Currently, PHASAR devices being marketed have of the order of 40 channels.

19.3   METHOD OF IRREGULARLY SAMPLED ZERO-CROSSINGS (MISZC)

In this method, the design of a DWDM device is undertaken such that there is only one image per wavelength so that the number of channels is not restricted due to the distance between successive orders as discussed with Eq. (19.2-6) above

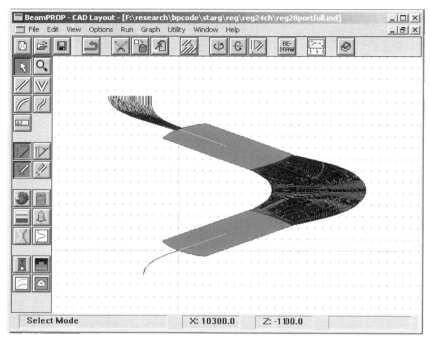

Figure 19.2. A 180 channel regular AWG layout designed with BeamPROP [Lu and Ersoy *et al.*, 2003].

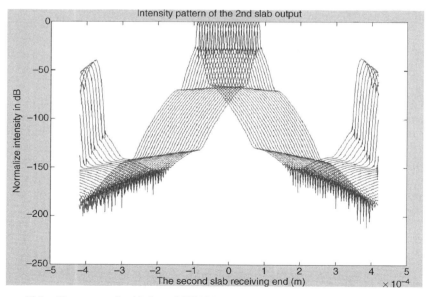

Figure 19.3. The output of a 16-channel PHASAR at a center wavelength of 1.550 µ, and channel spacing of 0.8 µ [Lu, Ersoy].

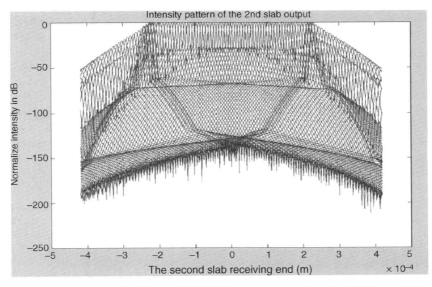

Figure 19.4. The output of a 64-channel PHASAR at a center wavelength of $1.550\,\mu$, and channel spacing of $0.8\,\mu$ [Lu, Ersoy].

[Lu, Ersoy], [Ersoy, 2005]. The beams at their focal points will be referred to as images. We will in particular discuss how to achieve the design in the presence of phase modulation corresponding to a combination of a linear and a spherical reference wave. Figure 19.5 shows a visualization of the reference waves and geometry involved with two wavelengths.

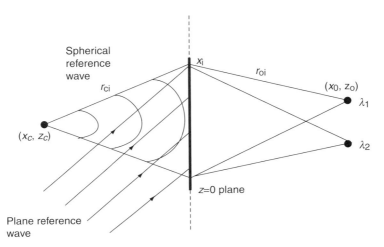

Figure 19.5. A visualization of the reference waves and geometry involved with two wavelengths in MISZC.

Once the required phase is computed, its implementation in the case of a AWG can be done by choosing the length of each waveguide to yield the required phase. This is the way it is already done with the regular PHASAR devices with only the linear grating phase modulation in confocal or Rowland geometries. The method is first discussed below for illustration purposes with respect to a planar geometry, meaning that the phased array apertures are placed on a plane (line for the 2-D case). Generalizations of the results to confocal, Rowland and 3-D geometries are given in the subsequent sections.

The method is based on first randomly choosing the locations of the centers of radiating apertures and then either by creating the negative phase of the phasefront (possibly plus a constant) at the chosen locations so that the overall phase is zero (or a constant), or slightly adjusting locations of the centers of radiating apertures such that the total phase shift from such a center to the desired image point equals a constant value, say, zero modulo 2π. In both approaches, such points will be referred to as *zero-crossings*. In the second approach, they will be referred to as *automatic zero-crossings*. In practice, the sampling points are chosen semi-irregularly as discussed below.

The total number of zero-crossings can be a very large number, especially in the presence of linear and spherical phase modulation. Practical implementations allow only a small number of apertures, for example, 300 being a typicalnumber in the case of PHASARS. In order to avoid the problem of too many apertures, and to avoid harmonics generated due to regular sampling [Ishimaru, 1962], [Lao, 1964], we choose irregularly sampled sparse number of apertures. One way to determine zero-crossing locations is given below as a procedure.

Step 1: The aperture points are initialized by choosing one point at a time randomly along the phased array surface. In order to achieve this systematically on the total surface of the phased array, the following approach can be used:

Initial point locations = uniformly spaced point locations + small random shifts

Step 2(a): If the method of creating the negative phase of the phasefront at the chosen locations is used, the said phase is created physically, for example, by correctly choosing the lengths of the waveguides in the case of PHASAR devices.

Step 2(b): If the method of automatic zero-crossings is used, correction values are calculated for each of the initial points generated in step 1 to find the nearest zero-crossing points as

Final locations of zero-crossings = Initial point locations from step 1 + correction terms

The two approaches work similarly. Below one algorithm to calculate the correction terms to generate the automatic zero-crossings is discussed.

19.3.1 Computational Method for Calculating the Correction Terms

The phased array equation including the linear grating term for image formation is given by

$$\delta x_i + \varphi_i(x_i) + kr_{oi} = 2\pi n + \phi_0 \qquad (19.3\text{-}1)$$

where

$$\varphi_i(x_i) = kr_{ci} \qquad (19.3\text{-}2)$$

$$r_{oi} = \sqrt{(x_0 - x_i)^2 + z_0^2} \qquad (19.3\text{-}4)$$

$$r_{ci} = \sqrt{(x_c - x_i)^2 + z_c^2} \qquad (19.3\text{-}5)$$

Equation (19.3-1) can be written as

$$\delta x_i + r_{ci} + r_{oi} = n\lambda + \phi_0 \lambda / 2\pi \qquad (19.3\text{-}6)$$

For an arbitrary position x_i on the phased array, Eq. (19.3-6) becomes

$$\delta x_i + r_{ci} + r_{oi} = n\lambda + \phi_0 \lambda / 2\pi + B \qquad (19.3\text{-}7)$$

where B represents error. Let us assume that the position of the aperture is to be moved a distance Δ in the positive x-direction such that the phase array Eq. (19.3-6) is satisfied. Then, the following is true:

$$\delta x_i' + r_{ci}' + r_{oi}' = \delta x_i + r_{ci} + r_{oi} - B \qquad (19.3\text{-}8)$$

where r_{oi} and r_{ci} are given by Eqs. (19.3-4) and (19.3-5). Since $x_i' = x_i + \Delta$, the following can be written:

$$r_{oi}' = \sqrt{r_{oi}^2 - 2\Delta(x_0 - x_i) + \Delta^2} \qquad (19.3\text{-}9)$$

$$r_{ci}' = \sqrt{r_{ci}^2 - 2\Delta(x_c - x_i) + \Delta^2} \qquad (19.3\text{-}10)$$

Using these equations leads to a fourth order polynomial equation for Δ as

$$\Delta^4 F_4 + \Delta^3 F_3 + \Delta^2 F_2 + \Delta F_1 + F_0 = 0 \qquad (19.3\text{-}11)$$

where

$$F_4 = G_1^2 - 1 \tag{19.3-12}$$

$$F_3 = 2G_1 G_2 + 2X_c + 2X_o \tag{19.3-13}$$

$$F_2 = G_2^2 + 2G_1 G_3 - 4X_c X_o - r_{oi}^2 - r_{ci}^2 \tag{19.3-14}$$

$$F_1 = 2G_2 G_3 + 2X_c r_{oi}^2 + 2X_o r_{ci}^2 \tag{19.3-15}$$

$$F_0 = G_3^2 - r_{oi}^2 r_{ci}^2 \tag{19.3-16}$$

$$G_1 = \frac{\delta^2}{2} - 1 \tag{19.3-17}$$

$$G_2 = -(r_{ci} + r_{oi} - B)\delta + X_c + X_o \tag{19.3-18}$$

$$G_3 = r_{ci} r_{oi} + \frac{B^2}{2} - B(r_{ci} + r_{oi}) \tag{19.3-19}$$

Δ is obtained as the root of the fourth order polynomial in Eq. (19.3-11). It is interesting to observe that this equation reduces to a second order polynomial equation when there is no linear phase modulation due to a grating, as discussed in Section 15.9. The locations of the chosen zero-crossing sampling points correspond to the positions of the waveguide apertures on the phased array surface in the case of PHASAR devices.

19.3.2 Extension of MISZC to 3-D Geometry

Extension to 3-D geometry is useful because other technologies can also be used. For example, the arrayed waveguides can be arranged in a 2-D plane or a 2-D curvature, instead of along a 1-D line discussed above. Other technologies such as scanning electron beam microscopy [Ersoy, 1979] and reactive ion etching which are used for manufacturing diffractive optical elements could also be potentially used. The basic method in 3-D is conceptually the same as before. In other words, the locations of the centers of radiating apertures are first (semi)randomly chosen; then either the negative phase of the phasefront (possibly plus a constant) at the chosen locations is physically generated so that the overall phase is zero (or a constant), or the locations of the centers of radiating apertures are slightly adjusted so that the total phase shift from such a center to the desired image point equals a constant value, say, zero modulo 2π.

In the case of choosing automatic zero-crossings, Eq. (19.3-11) is still valid if adjustment is done only along the x-direction, and the following replacements are made:

$$\begin{aligned} z_0^2 &\rightarrow z_0^2 + (y_0 - y_i)^2 \\ z_c^2 &\rightarrow z_c^2 + (y_c - y_i)^2 \end{aligned} \tag{19.3-20}$$

19.4 ANALYSIS OF MISZC

In the method discussed above, the problem of higher order harmonic images is minimized. In this section, an analysis in 3-D is provided to explain why this is the case. In planar devices such as optical PHASARS, two dimensions are used. The 2-D analysis needed in planar devices such as optical PHASARS is achieved simply by skipping one dimension, say, the y variable from the equations.

The MISZC is a nonlinear encoding method. In general, with such an encoding technique, the harmonic images are generated due to two mechanisms: (1) regular sampling and (2) nonlinear encoding. In MISZC, harmonic images due to regular sampling are converted into tolerable background noise by irregular sampling [Doles, 1988]. The analysis of why the harmonic images due to nonlinear encoding with zero-crossings are also eliminated in the presence of phase modulation is given below.

Equation (19.3-1) can be written more generally as

$$\varphi(x_i, y_i) + \theta(x_i, y_i) + kr_{oi} = 2n\pi + \varphi_0 \tag{19.4-1}$$

where $\varphi(x_i, y_i)$ is the phase shift caused by the wave propagation from the origin of the spherical reference wave (x_c, y_c, z_c) to the ith coupling aperture (x_i, y_i) on the surface of the phased array; $\theta(x_i, y_i)$ is another phase shift, for example, the linear phase shift in Eq. (19.3-1); kr_{oi} is the phase shift caused by the wave propagation from the aperture (x_i, y_i) on the surface of the phased array to the image point (object point) located at (x_o, y_o, z_o). In a PHASAR device, $\theta(x_i, y_i)$ can be expressed as $n_c \alpha k x_i$, where n_c is the effective index of refraction inside the waveguide.

For the center wavelength λ, Eq. (19.4-1) is written as

$$kr_{ci} + nkx_i\alpha + kr_{oi} = 2n\pi + \varphi_0 \tag{19.4-2}$$

Based on paraxial approximation, we write

$$r_{oi} = \sqrt{(x_0 - x_i)^2 + (y_0 - y_i)^2 + z_0^2} \cong z_0 + \frac{x_0^2 + y_0^2 + x_i^2 + y_i^2}{2z_0} - \frac{x_0 x_i + y_0 y_i}{z_0}$$

$$r_{ci} = \sqrt{(x_c - x_i)^2 + (y_c - y_i)^2 + z_c^2} \cong z_c + \frac{x_c^2 + y_c^2 + x_i^2 + y_i^2}{2z_c} - \frac{x_c x_i + y_c y_i}{z_c}$$

$$\tag{19.4-3}$$

Substituting Eq. (19.4-3) into Eq. (19.4-2) and neglecting constant phase terms results in

$$x_i\delta - x_i\left(\frac{x_c}{z_c} + \frac{x_o}{z_o}\right) + \frac{x_i^2}{2}\left(\frac{1}{z_0} + \frac{1}{z_c}\right) = n\lambda \tag{19.4-4}$$

where $\delta = n_c\alpha$.

Suppose that the wavelength is changed from λ to λ'. Equation (19.4-4) remains valid at another image point (x_0', z_0'). Taking the ratio of the two equations at λ and λ' yields

$$\frac{x_i\delta - x_i\left(\dfrac{x_c}{z_c} + \dfrac{x_o}{z_o}\right) + \dfrac{x_i^2}{2}\left(\dfrac{1}{z_0} + \dfrac{1}{z_c}\right)}{x_i\delta - x_i\left(\dfrac{x_c}{z_c} + \dfrac{x_o'}{z_o'}\right) + \dfrac{x_i^2}{2}\left(\dfrac{1}{z_o'} + \dfrac{1}{z_c}\right)} = \frac{\lambda}{\lambda'} = R \qquad (19.4\text{-}5)$$

Equating the coefficients of the terms with x_i, the new focal point (x_0', z_0') is obtained as

$$z_0' = \frac{R}{\dfrac{1-R}{z_c} + \dfrac{1}{z_0}} \approx Rz_0 \qquad (19.4\text{-}6)$$

$$x_0' = \frac{\dfrac{x_0}{z_0} - (1-R)\left(\delta - \dfrac{x_c}{z_c}\right)}{\dfrac{1-R}{z_c} + \dfrac{1}{z_0}} \approx x_0 - z_0(1-R)\left(\delta - \dfrac{x_c}{z_c}\right) \qquad (19.4\text{-}7)$$

where the approximations are based on $1\text{-}R \ll 1$ and $z_c \ll z_0$.

From the above derivation, it is observed that the focal point location z_0' is very close to the original z_0. Along the x-direction, the dispersion relationship is given as

$$\Delta x_0 = x_0' - x_0 = -z_0(1-R)\left(\delta - \frac{x_c}{z_c}\right) \qquad (19.4\text{-}8)$$

$$\left|\frac{\Delta x_0}{\Delta \lambda}\right| \approx \frac{z_0}{\lambda}\left|\delta - \frac{x_c}{z_c}\right| \qquad (19.4\text{-}9)$$

The image points of higher harmonics due to nonlinear encoding with zero-crossings occur when the imaging equation satisfies

$$x_i\delta - x_i\left(\frac{x_c}{z_c} + \frac{x_0'}{z_0'}\right) + \frac{x_i^2}{2}\left(\frac{1}{z_0'} + \frac{1}{z_c}\right) = nm\lambda' \qquad (19.4\text{-}10)$$

Taking the ratio of Eqs. (19.4-2) and (19.4-10) within the paraxial approximation yields

$$\frac{x_i\delta - x_i\left(\dfrac{x_c}{z_c} + \dfrac{x_0}{z_0}\right) + \dfrac{x_i^2}{2}\left(\dfrac{1}{z_0'} + \dfrac{1}{z_c}\right)}{x_i\delta - x_i\left(\dfrac{x_c}{z_c} + \dfrac{x_0'}{z_0'}\right) + \dfrac{x_i^2}{2}\left(\dfrac{1}{z_0'} + \dfrac{1}{z_c}\right)} = \frac{\lambda}{m\lambda'} = \frac{R}{m} \qquad (19.4\text{-}11)$$

Solving for x_0' and z_0' in the same way, the higher order harmonic image point locations are obtained as

$$z_0' = \frac{R}{\dfrac{m-R}{z_c} + \dfrac{m}{z_0}} \tag{19.4-12}$$

$$x_0' = \frac{\dfrac{mx_0}{z_0} - (m-R)\left(\delta - \dfrac{x_c}{z_c}\right)}{\dfrac{m-R}{z_c} + \dfrac{m}{z_0}} \tag{19.4-13}$$

From the above equations, we observe that a significant move of imaging position in the z-direction occurs as z_0' shrinks with increasing harmonic order. This means that the higher harmonics are forced to move towards locations very near the phased array. However, at such close distances to the phased array, the paraxial approximation is not valid. Hence, there is no longer any valid imaging equation. Consequently, the higher harmonics turn into noise. It can be argued that there may still be some imaging equation even if the paraxial approximation is not valid. However, the simulation results discussed in Section 19.4 indicate that there is no such valid imaging equation, and the conclusion that the higher harmonic images turn into noise is believed to be valid. Even if they are imaged very close to the phased array, they would appear as background noise at the relatively distant locations where the image points are. Simulations of Section 19.4 indicate that the signal-to-noise ratio in the presence of such noise is satisfactory, and remains satisfactory as the number of channels are increased.

19.4.1 Dispersion Analysis

The analysis in this subsection is based on the simulation results from Eqs. (19.4-6), (19.4-7), (19.4-12), and (19.4-13) in the previous subsection.

Case 1: Spherical wave case ($0.1 < z_c/z_0 < 10$)
For the first order harmonics ($m = 1$), the positions of the desired focal point for λ', that is, x_0' and z_0' have linear relationship with the wavelength λ'. The slope of this relationship decreases as the ratio z_c/z_0 decreases. For the higher order harmonics ($m \geq 2$), x_0' is much greater than $x_0 = 0$ and z_0' is much less than z_0. Therefore, we conclude that the higher order harmonics turn into background noise as discussed in the previous subsection.

Case 2: Plane wave case ($z_c/z_0 \gg 1$)
In this case, Eqs. (19.4-6), (19.4-7), (19.4-12), and (19.4-13) can be simplified as

$$z_0' = \frac{R}{\dfrac{m-R}{z_c} + \dfrac{m}{z_0}} \approx \frac{R}{m}z_0 \approx \frac{1}{m}z_0 \tag{19.4-14}$$

$$x_0' = \frac{\dfrac{mx_0}{z_0} - (m-R)\left(\delta - \dfrac{x_c}{z_c}\right)}{\dfrac{m-R}{z_c} + \dfrac{m}{z_0}} \approx x_0 - z_0\left(1 - \frac{R}{m}\right)\left(\delta - \frac{x_c}{z_c}\right) \tag{19.3-15}$$

Then, the dispersion relations for the first order ($m = 1$) are derived as

$$\left|\frac{\Delta z}{\Delta \lambda}\right| \approx \frac{z_0}{\lambda} \tag{19.4-16}$$

$$\left|\frac{\Delta x}{\Delta \lambda}\right| \approx \frac{z_0}{\lambda}\left|\delta - \frac{x_c}{z_c}\right| \tag{19.4-17}$$

19.4.1.1 3-D Dispersion. The mathematical derivation for the 3-D case is very much similar to that for the 2-D case discussed before [Hu, Ersoy]. However, instead of viewing the y variables as constants, thus neglecting them in the derivation, we investigate the y variables along with the x variables, and then obtain independent equations that lead to dispersion relations in both the x-direction and the y-direction. It is concluded that if the x-coordinates and y-coordinates of the points are chosen independently, the dispersion relations are given by

$$\left|\frac{\Delta x}{\Delta \lambda}\right| \approx \frac{z_0}{\lambda}\left|\delta_x - \frac{x_c}{z_c}\right| \tag{19.4-18}$$

$$\left|\frac{\Delta y}{\Delta \lambda}\right| \approx \frac{z_0}{\lambda}\left|\delta_y - \frac{y_c}{z_c}\right| \tag{19.4-19}$$

19.4.2 Finite-Sized Apertures

So far in the theoretical discussions, the apertures of the phased array are assumed to be point sources. In general, this assumption works well provided that the phase does not vary much within each aperture. In addition, since the zero-crossings are chosen to be the centers of the apertures, there is maximal tolerance to phase variations, for example, in the range $[-\pi/2, \pi/2]$. In this section, PHASAR types of devices are considered such that phase modulation is controlled by waveguides truncated at the surface of the phased array.

We use a cylindrical coordinate system (r, ϕ, z) to denote points on an aperture, and a spherical coordinate system (R, Θ, Φ) for points outside the aperture. In terms of these variables, the Fresnel–Kirchhoff diffraction formula for radiation fields in the Fraunhofer region is given by [Lu, Ersoy, 1993]

$$E_{\text{FF}}(R, \Theta, \Phi) = jk \frac{e^{-jkR}}{2\pi R} \frac{1 + \cos \Theta}{2} \int_S E(r, \phi, 0) e^{jkr\sin\Theta\cos(\Phi-\phi)} r\,dr\,d\phi \tag{19.4-20}$$

The transverse electric field of the LP_{01} mode may be accurately approximated as a Gaussian function:

$$E_{\text{GB}}(r, \phi, 0) = E_0 e^{-r^2/w^2} \tag{19.4-21}$$

where w is the waist radius of the gaussian beam. The field in the Fraunhofer region radiated by such a Gaussian field is obtained by substituting Eq. (19.4-21) into

Eq. (19.4-20). The result is given by

$$E_{\mathrm{GBFF}}(R, \Theta, \Phi) = jkE_0 \frac{e^{-jkR}}{R} \frac{w^2}{2} e^{-(kw \sin \Theta)^2/4} \qquad (19.4\text{-}22)$$

The far field approximation is valid with the very small sizes of the apertures. Equation (19.4-22) is what is utilized in the simulation of designed phased arrays with finite aperture sizes in Section 19.5.2.

19.5   COMPUTER EXPERIMENTS

We first define the parameters used to illustrate the results as follows:

- M: the number of phased array apertures (equal to the number of waveguides used in the case of PHASARS)
- L: The number of channels (wavelengths to be demultiplexed)
- $\Delta\lambda$: The wavelength separation between channels
- r: random coefficient in the range of [0,1] defined as the fraction of the uniform spacing length Δ (hence the random shift is in the range $[-r\Delta, r\Delta]$.

The results are shown in Figures 19.6–19.12. The title of each figure also contains the values of the parameters used. Unless otherwise specified, r is assumed to be 1. In Section 19.5.1, the apertures of the phased array are assumed to be point sources. In Section 19.5.2, the case of finite-sized apertures are considered.

19.5.1   Point-Source Apertures

Figure 19.6 shows the intensity distribution on the image plane and the zero-crossing locations of the phased array with 16 channels when the central wavelength is 1550 nm, and the wavelength separation is 0.4 nm between adjacent channels. There are no harmonic images observed on the output plane which is in agreement with the claims of Section 19.3.

In order to verify the dispersion relation given by Eq. (19.4.17), the linear relationship of Δx with respect to $\Delta\lambda$, δ, and different values of z_0 were investigated, respectively. The simulation results shown in Figure 19.7 give the slope of each straight line as $1.18, 0.78, 0.40\ (\times 10^6)$, which are in excellent agreement with the theoretically calculated values using Eq. (19.4-17) with $\delta = 30, \lambda_0 = 1550\,\mathrm{nm}$, namely, 1.16, 0.77, and $0.39(\times 10^6)$.

In MISZC, both random sampling and implemention of zero-crossings are crucial to achieve good results. In the following, comparitive results are given to discuss the importance of less than random sampling. Figures 19.8 and 19.9 show the results in cases where total random sampling is not used. All the parameters are the same as in Figure 19.5, except that the parameter r is fixed as 0, 1/4 and 1/2,

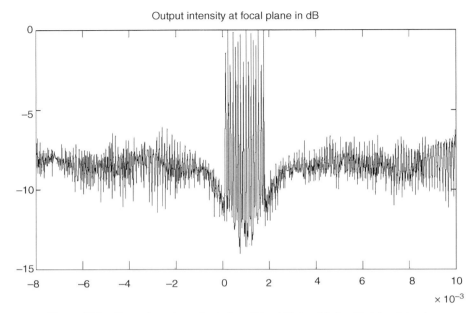

Figure 19.6.   Output intensity at focal plane ($M = 100, L = 16, \delta = 15, \Delta\lambda = 0.4\,\text{nm}$).

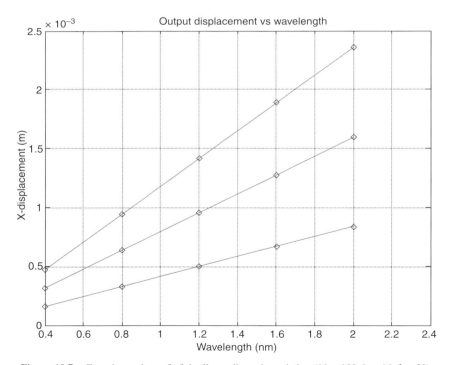

Figure 19.7.   Experimental proof of the linear dispersion relation ($M = 100, L = 16, \delta = 30$).

Output intensity at focal plane in dB

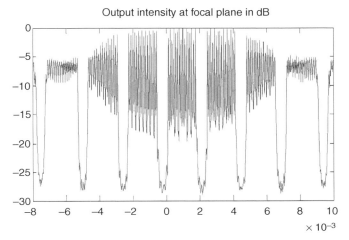

Figure 19.8. Harmonics with nonrandom sampling ($M = 100, L = 16, \delta = 15, \Delta\lambda = 0.4\,\text{nm}, r = 0$).

respectively. It is observed that the harmonics of different orders start showing up when r is less than 1, that is, with less than total randomness. In comparison, Figure 19.5 shows the case with $r=1$, and no harmonics appear since total random sampling is used in this case.

19.5.2 Large Number of Channels

The major benefit of the removal of the harmonic images is the ability to increase the possible number of channels. A number of cases with 64, 128, and 256 channels were designed to study large number of channels. In the figure below, the number of

Output intensity at focal plane in dB

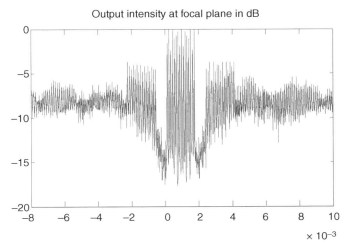

Figure 19.9. Harmonics with partial random sampling ($M = 100, L = 16, \delta = 15, \Delta\lambda = 0.4\,\text{nm}, r = 0.5$).

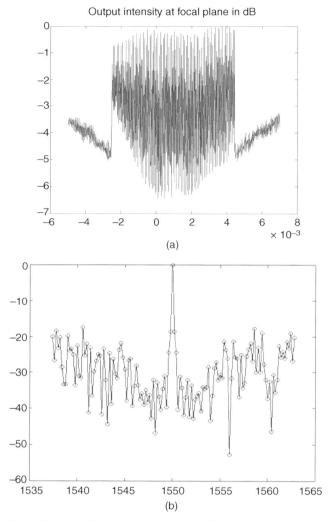

Figure 19.10.   The case of large number of channels ($M = 200, L = 128, \Delta\lambda = 0.2\,\mathrm{nm}$).

phased arrayed apertures, the number of channels, and the wavelength separation are represented by $M, L,$ and $\Delta\lambda$, respectively. Figure 19.10 shows the results for $M = 200, L = 128,$ and $\Delta\lambda = 0.2\,\mathrm{nm}$. The figure consists of two parts. The top figure shows the demultiplexing properties under simultaneous multichannel operation. In this figure, we observe that the nonuniformity among all the channels are in the range of $\sim 2\,\mathrm{dB}$. It is also usual in the literature on WDM devices to characterize the cross talk performance by specifying the single channel cross talk figure under the worst case. The bottom figure is the normalized transmission spectrum with respect to the applied wavelengths in the central output port. The

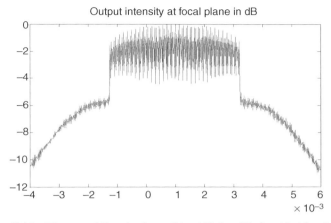

Figure 19.11. The case of Gaussian beam ($M = 150, L = 128, \delta; = 10, \Delta\lambda = 0.2\,\text{nm}$).

cross talk value is estimated to be 20 dB. It was observed that cross talk value improves when more apertures (waveguides in the case of PHASARS) are used.

19.5.3 Finite-Sized Apertures

The theory for the case of finite-sized apertures yielding beams with Gaussian profile was discussed in Section 19.4.2. Using Eq. (19.4-22), a number of simulations were conducted. The results with 128 channels are shown in Figure 19.11. It is observed that the results are quite acceptable.

19.5.4 The Method of Creating the Negative Phase

The experimental results up to this point are for the method of automatic zero-crossings. Figure 19.12 shows an example with the method of creating the negative of the phase of the total phasefront with 16 channels [Hu, Ersoy]. It is observed that the results are equally valid as in the previous cases.

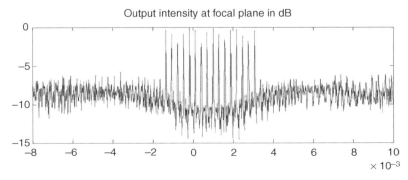

Figure 19.12. Sixteen-channel design with the method of creating the negative phase of the wave front.

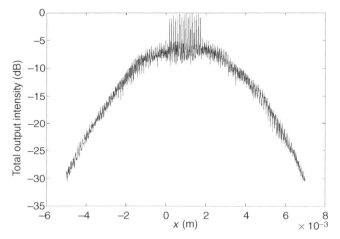

Figure 19.13. Sixteen-channel design with phase errors (ERR = 0.25π).

19.5.5 Error Tolerances

Phase errors are expected to be produced during fabrication. The phase error tolerance was investigated by applying random phase error to each array aperture. The random phase errors were approximated by uniform distribution in the range of [-ERR, ERR] where ERR is the specified maximum error. As long as the maximum error satisfies

$$|\text{ERR}| \leq \pi/2 \qquad (19.5\text{-}1)$$

the phasors point in similar direction so that there is positive contribution from each aperture. Hence, satisfactory results are expected. This was confirmed by simulation experiments. An example is shown in Figure 19.13, corresponding to ERR = 0.25π.

19.5.6 3-D Simulations

The 3-D method was investigated through simulations in a similar fashion [Hu and Ersoy, 2002]. Figure 19.14 shows one example of focusing and demultiplexing on the image plane (x-y plane at $z = z_0$). The four wavelengths used were 1549.2, 1549.6, 1550, and 1550.4 nm, spaced by 0.4 nm (50 GHz). The array was generated with 50×50 apertures on a 2×2 mm square plane. The diffraction order δ_x in the x-direction was set to 5, while that in the y-direction, δ_y, was set to zero.

In Figure 19.14, part (a) shows demultiplexing on the image plane, and part (b) shows the corresponding insertion loss on the output line (x-direction) on the same plane.

It is observed that a reasonably small value of diffraction order ($\delta_x \sim 5$) is sufficient to generate satisfactory results. This is significant since it indicates that manufacturing in 3-D can indeed be achievable with current technology. A major advantage in 3-D is that the number of apertures can be much larger as compared to 2-D.

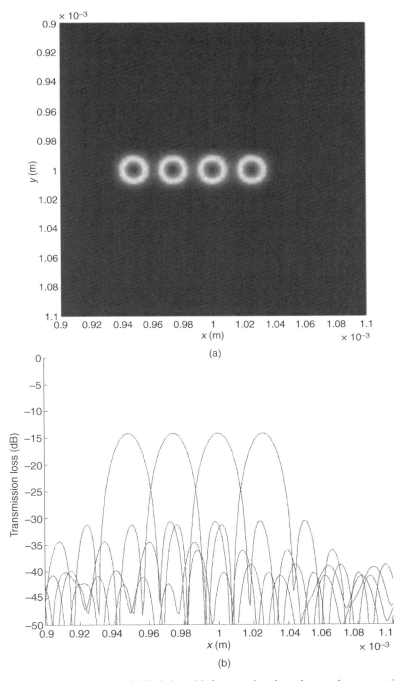

Figure 19.14. An example of 3-D design with four wavelengths and exact phase generation.

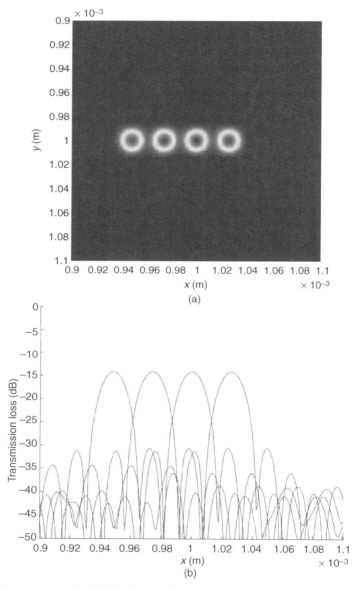

Figure 19.15. An example of 3-D design with four wavelengths and four-level phase quantization.

19.5.7   Phase Quantization

In actual fabrication, phase is often quantized. The technology used decides the number of quantization levels. Figure 19.15 shows the demultiplexing results with four quantization levels, and otherwise the same parameters as in Figure 19.14. The results are satisfactory.

19.6 IMPLEMENTATIONAL ISSUES

It is interesting to compare different approaches in terms of implementational issues. For example, in PHASAR type of devices with waveguides on a plane, the lengths of the waveguides should be chosen according to the following equations:

(a) Plane reference wave only, automatic zero-crossings:

$$L_i = \alpha x_i \qquad (19.6\text{-}1)$$

(b) Plane reference wave only, negative phase implementation

$$L_i = \alpha x_i - \theta_i/k \qquad (19.6\text{-}2)$$

where θ_i is the desired phase, and k is the wave number.

(c) Plane and spherical reference waves, automatic zero-crossings

$$L_i = \alpha(x_i + r_{ci}) \qquad (19.6\text{-}3)$$

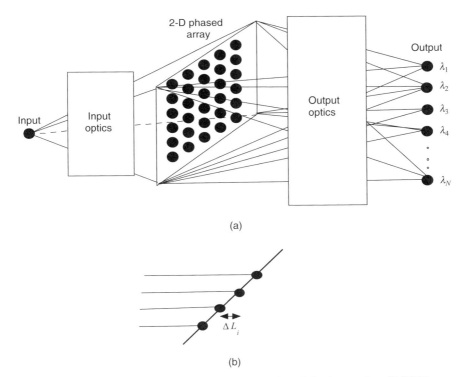

Figure 19.16. The visualization of a possible setup for 3-D implementation of MISZC.

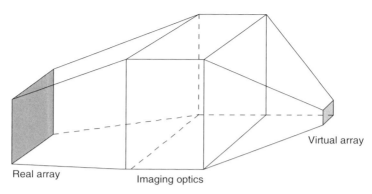

Figure 19.17. The visualization of a virtual holography setup in connection with Figure 19.16 to achieve desired phase modulation and size.

(d) Plane and spherical reference waves, negative phase implementation

$$L_i = \alpha(x_i + r_{ci}) - \theta_i/k \qquad (19.6\text{-}4)$$

The equations above show that the method of physical generation of the negative phase appears to be more difficult to implement than the method of automatic zero-crossings in terms of waveguide length control. However, in the method of automatic zero-crossings, the positions of the apertures have to be carefully adjusted. Since the initial positions of apertures are randomly chosen, this is not expected to generate additional difficulties since the result is another random number after adjustment.

In 3-D, the disadvantage is that it may be more difficult to achieve large δ. The big advantage is that there are technologies for diffractive optical element design with many apertures, which can also be used for phased array devices for DWDM. In our simulations, it was observed that δ of the order of 5 is sufficient to achieve satisfactory resolution.

This can be achieved in a number of ways. One possible method is by using a setup as in Figure 19.16, together with the method of virtual holography discussed in Section 16.2. In order to achieve large δ, the array can be manufactured, say, five times larger than normal, and arranged tilted as shown in Figure 19.16(b) so that ΔL_i shown in the figure is large. Then, the array (now called the real array) has the necessary phase modulation, and is imaged to the virtual array as shown in Figure 19.17, following the method of virtual holography. If M is the demagnification used in the lateral direction, the demagnification in the z-direction is M^2. As a result, the tilt at the virtual array in the z-direction can be neglected. The virtual array has the necessary size and phase modulation in order to operate as desired to focus different wavelengths at different positions as discussed above.

20

Numerical Methods for Rigorous Diffraction Theory

20.1 INTRODUCTION

Scalar diffraction theory is valid when the sizes of diffractive apertures are large as compared to the wavelength of the incident wave. Geometries considered are also usually simple. With the increasing trend to produce smaller and smaller devices where refractive optical components become impractical, and increasingly complex geometries and material properties, there is a demand for diffractive elements with sizes and characteristics of the order of a wavelength or smaller. However, in this size range, scalar diffraction theory may not be sufficiently accurate, and analytical results are usually very difficult to generate. Complex geometries may also require new computational approaches.

A practical approach is to solve Maxwell's equations by using numerical methods. There are several emerging approaches for this purpose. One approach is based on numerical methods which utilize finite differences. Another approach is based on Fourier modal analysis using the Fourier series. A third approach is the use of the method of finite elements for the solution of Maxwell's equations with boundary conditions.

This chapter consists of seven sections. Section 20.2 describes the formulation of the paraxial BPM method discussed in Section 12.4 in terms of finite differences using the *Crank–Nicholson method*. Section 20.3 discusses the *wide angle BPM* using the *Pâdé approximation*. This method is commonly used when there are large angular changes as in waveguide problems. Section 20.4 covers the method of finite differences as a preparation for the next section. Section 20.5 introduces the *finite difference time domain* (FDTD) *method*, which is currently a popular method in many applications. Section 20.6 describes some experiments with the FDTD method. Section 20.7 concludes the chapter with another competitive approach called the *Fourier modal method* (FMM).

Diffraction, Fourier Optics and Imaging, by Okan K. Ersoy
Copyright © 2007 John Wiley & Sons, Inc.

20.2 BPM BASED ON FINITE DIFFERENCES

Consider the Helmholtz equation for inhomogeneous media given by Eq. (12.2-4) repeated below for convenience:

$$(\nabla^2 + k^2(x, y, z))U'(x, y, z) = 0 \tag{20.2-1}$$

where

$$U'(x, y, z) = u(x, y, z, t)e^{-jwt} \tag{20.2-2}$$

and the position-dependent wave number $k(x, y, z)$ is given by

$$k(x, y, z) = n(x, y, z)k_0 \tag{20.2-3}$$

in which k_0 is the free space wave number, and $n(x, y, z)$ is the inhomogeneous index of refraction.

As in Section 12.3, the variation of $n(x, y, z)$ is written as

$$n(x, y, z) = \bar{n} + \Delta n(x, y, z) \tag{20.2-4}$$

where $\bar{n}$ is the average index of refraction. The Helmholtz equation (20.2-1) becomes

$$[\nabla^2 + \bar{n}^2 k_0^2 + 2\bar{n}\Delta n k_0^2]U = 0 \tag{20.2-5}$$

where the $(\Delta n)^2 k^2$ term has been neglected. Next the field is assumed to vary as

$$U'(x, y, z) = U(x, y, z)e^{j\bar{k}z} \tag{20.2-6}$$

in which $U(x, y, z)$ is assumed to be a slowly varying function of z, and $\bar{k}$ equals $\bar{n}k_0$. Substituting Eq. (20.2-6) in the Helmholtz equation yields

$$\nabla^2 U + 2j\bar{k}\frac{\delta}{\delta z}U + (k^2 - \bar{k}^2)U = 0 \tag{20.2-7}$$

Equation (20.2-7) is the Helmholtz equation used in various finite difference formulations of the BPM.

When the paraxial Helmholtz equation is valid as discussed in Section 12.3, Eq. (20.2-7) becomes

$$\frac{\delta}{\delta z}U = \frac{j}{2\bar{k}}\left[\frac{\delta^2 U}{\delta x^2} + \frac{\delta^2 U}{\delta y^2} + (k^2 - \bar{k}^2)U\right] \tag{20.2-8}$$

This is the basic paraxial equation used in BPM in 3-D; the 2-D paraxial equation is obtained by omitting the y-dependent terms. The paraxial approximation in this form allows two advantages. First, since the rapid phase variation with respect to z is factored out, the slowly varying field can be represented numerically along the longitudinal grid, with spacing which can be many orders of magnitude larger than the wavelength. This effect makes the BPM much more efficient than purely finite difference based techniques, which would require grid spacing of the order of one tenth the wavelength. Secondly, eliminating the second order derivative term in z enables the problem to be treated as a first order initial value problem instead of a second order boundary value problem. A second order boundary value problem usually requires iteration or eigenvalue analysis whereas the first order initial value problem can be solved by simple integration in the z-direction. This effect similarly decreases the computational complexity to a large extent as compared to full numerical solution of the Helmholtz equation.

However, there are also prices paid for the reduction of computational complexity. The slow envelope approximation assumes the field propagation is primarily along the z-axis (i.e., the paraxial direction), and it also limits the size of refractive index variations. Removing the second derivative in the approximation also eliminates the backward traveling wave solutions. So reflection-based devices are not covered by this approach. However, these issues can be resolved by reformulating the approximations. Extensions such as wide-angle BPM and bidirectional BPM for this purpose will be discussed later.

The FFT-based numerical method for the solution of Eq. (12.3-5) was discussed in Section 12.4. Another approach called FD-BPM is based on the method of finite differences, especially using the Crank–Nicholson method [Yevick, 1989]. Sometimes the finite difference method gives more accurate results [Yevick, 1989], [Yevick, 1990]. It can also use larger longitudinal step size to ease the computational complexity without compromising accuracy [Scarmozzino].

In the finite-difference method, the field is represented by discrete points on a grid in transverse planes, perpendicular to the longitudinal or z-direction in equal intervals along the z-direction. Once the input field is known, say, at $z = 0$, the field on the next transverse plane is calculated. In this way, wave propagation is calculated one step at a time along the z-direction through the domain of interest. The method will be illustrated in 2-D. Extension to 3-D is straightforward.

Let u_i^n denote the field at a transverse grid point i and a longitudinal point indexed by n. Also assume the grid points and planes are equally spaced by Δx and Δz apart, respectively. In the Crank–Nicholson (C-N) method, Eq. (12.3-5) is represented at a fictitious midplane between the known plane n and the unknown plane $n + 1$. This is shown in Figure 20.1.

The representation of the first order and second order derivatives on the midplane in the C-N method is as follows:

$$\left. \frac{\delta u_i}{\delta z} \right|_{z = \Delta z(\,n+1/2)} = \frac{u_i^{n+1} - u_i^n}{\Delta z} \tag{20.2-9}$$

$$\left. \frac{\delta^2 u_i}{\delta x^2} \right|_{z = \Delta z(n+1/2)} = \frac{\delta^2 u_i^{n+1} - \delta^2 u_i^n}{2\Delta z^2} \tag{20.2-10}$$

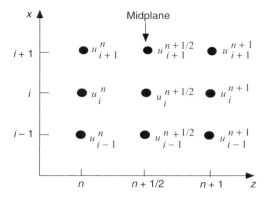

Figure 20.1.   The sampling points in the Crank–Nicholson method.

where $z_{n+1/2} = z_n + \Delta z/2$, and δ^2 represents the second order difference operator given by

$$\delta^2 u_i = u_{i+1} + u_{i-1} - 2u_i \qquad (20.2\text{-}11)$$

With these approximations, Eq. (20.2-8) becomes

$$\frac{u_i^{n+1} - u_i^n}{\Delta z} = \frac{j}{2\bar{k}} \left(\frac{\delta^2}{\Delta x^2} + (k(x_i, z_{n+1/2}) - \bar{k}^2) \right) \frac{u_i^{n+1} + u_i^n}{2} \qquad (20.2\text{-}12)$$

Equation (20.2-12) can be written as a tridiagonal matrix equation as [Scarmozzino]

$$a_i u_{i-1}^{n+1} + b_i u_i^{n+1} + c_i u_{i+1}^{n+1} = d_i \qquad (20.2\text{-}13)$$

This equation is solved for the unknowns u_i^{n+1} at the $(n+1)$th layer along the z-direction. Because of its tridiagonal nature, the solution can be found rapidly in $O(N)$ operations, N being the number of grid points along the x-direction.

At the boundary points $i = 0$ and N, appropriate boundary conditions must be used. The transparent boundary condition is commonly preferred [Hadley, 90], [Hadley, 92].

20.3   WIDE ANGLE BPM

The paraxial approximation used above allows BPM to be used in applications with paraxial waves and small variations of index of refraction. In order to extend the method, it is necessary to reintroduce the skipped second derivative term into Eq. (20.2-8) and thereby start with Eq. (20.2-7), which can be written as

$$\frac{\partial^2 U}{\partial z^2} + 2i\bar{k}\frac{\partial U}{\partial z} + \frac{\partial^2 U}{\partial x^2} + \frac{\partial^2 U}{\partial y^2} + (k^2 - \bar{k}^2)U = 0 \qquad (20.3\text{-}1)$$

Let D represent $\partial/\partial z$, and D^2 represent $\partial^2/\partial^2 z$. Equation (20.3-1) is written as

$$D^2 U + 2i\bar{k}DU + \left(\frac{\partial^2}{\partial x^2} + \frac{\partial^2}{\partial y^2} + (k^2 - \bar{k}^2)\right)U = 0 \qquad (20.3\text{-}2)$$

It is valid to consider this equation as an algebraic equation in D. Solving it for D yields

$$D = j\bar{k}[\sqrt{1 + P} - 1] \qquad (20.3\text{-}3)$$

where

$$P = \frac{1}{\bar{k}^2}\left(\frac{\partial^2}{\partial x^2} + \frac{\partial^2}{\partial y^2} + (k^2 - \bar{k}^2)\right) \qquad (20.3\text{-}4)$$

P and D represent operators on U. Thus, Eq. (20.3-2) becomes [Hadley]

$$\frac{\partial U}{\partial z} = j\bar{k}(\sqrt{1 + P} - 1)U \qquad (20.3\text{-}5)$$

This equation only gives the forward propagation solution. The backward solution can be found by changing the sign of the square root term. Either solution is exact.

To be able to solve for U in Eq. (20.3-5), it is necessary to expand the term $\sqrt{1 + P}$. The Pâdé expansion yields the best accuracy with the fewest terms of expansion for this purpose [Hadley]. The Pâdé approximation with order (m,n) is a rational function of the form

$$R(x) = \frac{\sum\limits_{k=0}^{M} a_k x^k}{1 + \sum\limits_{k=1}^{N} b_k x^k} \qquad (20.3\text{-}6)$$

Table 20.1 shows the low-order Pâdé approximants for the term $\bar{k}[\sqrt{1 + P} - 1]$.

Using the Pâdé approximation, Eq. (20.3-5) is written as

$$\frac{\partial U}{\partial z} = j\bar{k}\frac{N_m(P)}{D_n(P)}U \qquad (20.3\text{-}7)$$

where $N_m(P)$ and $D_n(P)$ are polynomials in P corresponding to the Pâdé approximant of order (m,n).

As the Pâdé order increases, the accuracy of the solution improves. The $(1,1)$ order Pâdé approximation is sufficiently accurate up to 30 degrees while the $(3,3)$ Pâdé approximation is equivalent to the 15th order Taylor expansion of the exact square root term in Eq. (20.3-5).

The wide angle BPM is commonly used in applications in which wide angles are necessary such as in the design of AWGs discussed in Chapter 19. For example,

Table 20.1. Low-order Pâdé approximants for the term $\bar{k}[\sqrt{1+P}-1]$.

Order	$\bar{k}[\sqrt{1+P}-1]$
(1,0)	$\dfrac{P}{2\bar{k}}$
(1,1)	$\dfrac{\dfrac{P}{2\bar{k}}}{1+\dfrac{P}{4\bar{k}^2}}$
(2,2)	$\dfrac{\dfrac{P}{2\bar{k}}+\dfrac{P^2}{4\bar{k}^3}}{1+\dfrac{3P}{4\bar{k}^2}+\dfrac{P^2}{16\bar{k}^4}}$
(3,3)	$\dfrac{\dfrac{P}{2\bar{k}}+\dfrac{P^2}{2\bar{k}^3}+\dfrac{3P^2}{32\bar{k}^5}}{1+\dfrac{5P}{4\bar{k}^2}+\dfrac{3P^2}{8\bar{k}^4}+\dfrac{P^3}{64\bar{k}^6}}$

Figure 19.2 shows the layout of the waveguides with wide angles of bending to accommodate the length variations. Figure 20.2 shows the output intensities in a 200-channel AWG design computed with the wide-angle BPM using the RSoft software called BeamPROP.

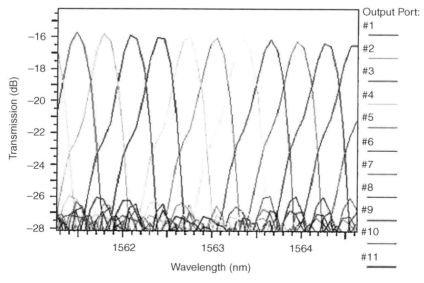

Figure 20.2. A close-up view of the output channel intensities in a 200-channel phasar design using wide-angle BPM [Lu, *et al.*, 2003].

20.4   FINITE DIFFERENCES

In the next section, the finite difference time domain technique is discussed. This method is based on the finite difference approximations of the first and second derivatives. They are discussed below.

The Taylor series expansion of $u(x,t_n)$ at $x_i + \Delta x$ about the point x_i for a fixed time t_n is given by [Kuhl, Ersoy]

$$u(x_i + \Delta x, t_n)\big|_{t_n} = u\big|_{x_i,t_n} + \Delta x \frac{\partial u}{\partial x}\bigg|_{x_i,t_n} + \frac{\Delta x^2}{2}\frac{\partial^2 u}{\partial x^2}\bigg|_{x_i,t_n} + \frac{\Delta x^3}{6}\frac{\partial^3 u}{\partial x^3}\bigg|_{x_i,t_n} + \frac{\Delta x^4}{24}\frac{\partial^4 u}{\partial x^4}\bigg|_{\xi_1,t_n}$$

$$(20.4\text{-}1)$$

The last term is an error term, with ξ_1 being a point in the interval $(x_i, x_i + \Delta x)$. Expansion at the point $x_i - \Delta x$ for fixed time t_n is similarly given by

$$u(x_i - \Delta x, t_n)\big|_{t_n} = u\big|_{x_i,t_n} - \Delta x \frac{\partial u}{\partial x}\bigg|_{x_i,t_n} + \frac{\Delta x^2}{2}\frac{\partial^2 u}{\partial x^2}\bigg|_{x_i,t_n} - \frac{\Delta x^3}{6}\frac{\partial^3 u}{\partial x^3}\bigg|_{x_i,t_n} + \frac{\Delta x^4}{24}\frac{\partial^4 u}{\partial x^4}\bigg|_{\xi_2,t_n}$$

$$(20.4\text{-}2)$$

where ξ_2 is in the interval $(x_i - \Delta x, x_i)$. Adding the two expansions gives

$$u(x_i + \Delta x) + u(x_i - \Delta x) = 2u\big|_{x_i,t_n} + \Delta x^2 \frac{\partial^2 u}{\partial x^2}\bigg|_{x_i,t_n} + \frac{\Delta x^4}{12}\frac{\partial^4 u}{\partial x^4}\bigg|_{\xi_3,t_n} \qquad (20.4\text{-}3)$$

where ξ_3 lies in the interval $(x_i - \Delta x, x_i + \Delta x)$. Rearranging the above expression yields

$$\frac{\partial^2 u}{\partial x^2}\bigg|_{x_i,t_n} = \left[\frac{u(x_i + \Delta x) - 2u(x_i) + u(x_i - \Delta x)}{(\Delta x)^2}\right]_{t_n} + O\big[(\Delta x)^2\big] \qquad (20.4\text{-}4)$$

Eq. (20.4-4) can be written as

$$\frac{\partial^2 u}{\partial x^2}\bigg|_{x_i,t_n} = \frac{u_{i+1}^n - 2u_i^n + u_{i-1}^n}{(\Delta x)^2} + O\big[(\Delta x)^2\big] \qquad (20.4\text{-}5)$$

The second partial derivative of u with respect to time is similarly given by

$$\frac{\partial^2 u}{\partial t^2}\bigg|_{x_i,t_n} = \frac{u_i^{n+1} - 2u_i^n + u_i^{n-1}}{(\Delta t)^2} + O\big[(\Delta t)^2\big] \qquad (20.4\text{-}6)$$

20.5   FINITE DIFFERENCE TIME DOMAIN METHOD

The FDTD method is an effective numerical method increasingly used in applications. In this method, the Maxwell's equations are represented in terms of central-difference equations. The resulting equations are solved in a leapfrog manner. In other words, the electric field is solved at a given instant in time, then the magnetic field is solved at the next instant in time, and the process is repeated iteratively many times.

Consider Eqs. (3.3-14) and (3.3-15) of Chapter 3 for a source-free region (no electric or magnetic current sources) rewritten below for convenience:

$$\nabla \times H = \frac{\partial D}{\partial t} = \varepsilon \frac{\partial E}{\partial t} \tag{3.3-14}$$

$$\nabla \times E = -\frac{\partial B}{\partial t} = -\mu \frac{\partial H}{\partial t} \tag{3.3-15}$$

Eq. (3.3-14) shows that the time derivative of the E field is proportional to the Curl of the H field. This means that the new value of the E field can be obtained from the previous value of the E field and the difference in the old value of the H field on either side of the E field point in space.

Equations (3.3-14) and (3.3-15) can be written in terms of the vector components as follows:

$$\frac{\partial H_x}{\partial t} = \frac{1}{\mu}\left(\frac{\partial E_y}{\partial z} - \frac{\partial E_z}{\partial y} - \rho' H_x\right) \tag{20.5-1}$$

$$\frac{\partial H_y}{\partial t} = \frac{1}{\mu}\left(\frac{\partial E_z}{\partial x} - \frac{\partial E_x}{\partial z} - \rho' H_y\right) \tag{20.5-2}$$

$$\frac{\partial H_z}{\partial t} = \frac{1}{\mu}\left(\frac{\partial E_x}{\partial y} - \frac{\partial E_y}{\partial x} - \rho' H_z\right) \tag{20.5-3}$$

$$\frac{\partial E_x}{\partial t} = \frac{1}{\varepsilon}\left(\frac{\partial H_z}{\partial y} - \frac{\partial H_y}{\partial z} - \sigma E_x\right) \tag{20.5-4}$$

$$\frac{\partial E_y}{\partial t} = \frac{1}{\varepsilon}\left(\frac{\partial H_x}{\partial z} - \frac{\partial H_z}{\partial x} - \sigma E_y\right) \tag{20.5-5}$$

$$\frac{\partial E_z}{\partial t} = \frac{1}{\varepsilon}\left(\frac{\partial H_y}{\partial x} - \frac{\partial H_x}{\partial y} - \sigma E_z\right) \tag{20.5-6}$$

These equations are next represented in terms of finite differences, using the Yee algorithm discussed below.

20.5.1   Yee's Algorithm

Yee's algorithm solves Maxwell's curl equations by using a set of finite-difference equations [Yee]. Using finite differences, each electric field component in 3-D is

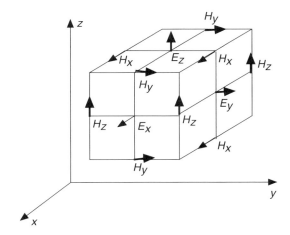

Figure 20.3. Electric and magnetic field vectors in a Yee cell [Yee].

surrounded by four circulating magnetic field components, and every magnetic field component is surrounded by four circulating electric field components as shown in Figure 20.3. The electric and magnetic field components are also centered in time in a leapfrog arrangement. This means all the electric components are computed and stored for a particular time using previously stored magnetic components. Then, all the magnetic components are determined using the previous electric field components [Kuhl, Ersoy].

With respect to the Yee cell, discretization is done as follows:

- $[i, j, k] = (i\Delta x, j\Delta y, k\Delta z)$ where $\Delta x, \Delta y$, and Δz are the space increments in the x, y, and z directions, respectively, and i, j, k are integers.
- $t_n = n\Delta t$
- $u(i\Delta x, j\Delta y, k\Delta z, n\Delta t) = u_{i,j,k}^n$

Yee's centered finite-difference expression for the first partial space derivative of u in the x-direction, evaluated at time $t_n = n\Delta t$ is given by [Yee]

$$\frac{\partial u}{\partial x}(i\Delta x, j\Delta y, k\Delta z, n\Delta t) = \frac{u_{i+1/2,j,k}^n - u_{i-1/2,j,k}^n}{\Delta x} + O\left[(\Delta x)^2\right] \qquad (20.5\text{-}7)$$

It is observed that the data a distance $\Delta x/2$ away from the point in question is used, similarly to the Crank–Nicholson method. The first partial derivative of u with respect to time for a particular space point is also given by

$$\frac{\partial u}{\partial t}(i\Delta x, j\Delta y, k\Delta z, n\Delta t) = \frac{u_{i,j,k}^{n+1/2} - u_{i,j,k}^{n-1/2}}{\Delta t} + O\left[(\Delta t)^2\right] \qquad (20.5\text{-}8)$$

Substituting in the space and time derivatives for the point (i, j, k) at time n in Eq. (20.5-4) yields

$$\frac{E_x|_{i,j,k}^{n+1/2} + E_x|_{i,j,k}^{n-1/2}}{\Delta t} = \frac{1}{\varepsilon_{i,j,k}} \left(\begin{array}{c} \dfrac{H_z|_{i,j+1/2,k}^{n} - H_z|_{i,j-1/2,k}^{n}}{\Delta y} \\ -\dfrac{H_y|_{i,j,k+1/2}^{n} - H_y|_{i,j,k-1/2}^{n}}{\Delta z} - \sigma_{i,j,k} E_x|_{i,j,k}^{n} \end{array} \right) \quad (20.5\text{-}9)$$

All the terms on the right side of this equation are evaluated at time step n. All the magnetic field components needed at time n are known. Only the values of E_x up to time $n - 1/2$ are stored. For a region without loss, this term is zero. If $\sigma_{i,j,k}$ is nonzero, then E_x can be estimated as [Yee]

$$E_x|_{i,j,k}^{n} = \frac{E_x|_{i,j,k}^{n+1/2} + E_x|_{i,j,k}^{n-1/2}}{2} \quad (20.5\text{-}10)$$

which is the average of the known value of E_x at time $n - 1/2$ and the unknown value at time $n + 1/2$. Using Eq. (20.5-10) in Eq. (20.5-9) yields

$$E_x|_{i,j,k}^{n+1/2} + E_x|_{i,j,k}^{n-1/2} = \frac{\Delta t}{\varepsilon_{i,j,k}} \left(\begin{array}{c} \dfrac{H_z|_{i,j+1/2,k}^{n} - H_z|_{i,j-1/2,k}^{n}}{\Delta y} - \dfrac{H_y|_{i,j,k+1/2}^{n} - H_y|_{i,j,k-1/2}^{n}}{\Delta z} \\ -\sigma_{i,j,k} \dfrac{E_x|_{i,j,k}^{n+1/2} + E_x|_{i,j,k}^{n-1/2}}{2} \end{array} \right)$$

$$(20.5\text{-}11)$$

Collecting like terms and solving for $E_x\big|_{i,j,k}^{n+1/2}$ results in

$$E_x|_{i,j,k}^{n+1/2} = \left(\frac{1 - \dfrac{\sigma_{i,j,k}\Delta t}{2\varepsilon_{i,j,k}}}{1 + \dfrac{\sigma_{i,j,k}\Delta t}{2\varepsilon_{i,j,k}}} \right) E_x|_{i,j,k}^{n-1/2} + \left(\frac{\dfrac{\Delta t}{\varepsilon_{i,j,k}}}{1 + \dfrac{\sigma_{i,j,k}\Delta t}{2\varepsilon_{i,j,k}}} \right) \left(\begin{array}{c} \dfrac{H_z|_{i,j+1/2,k}^{n} - H_z|_{i,j-1/2,k}^{n}}{\Delta y} \\ -\dfrac{H_y|_{i,j,k+1/2}^{n} - H_y|_{i,j,k-1/2}^{n}}{\Delta z} \end{array} \right)$$

$$(20.5\text{-}12)$$

All the other unknown field components can be similarly evaluated as

$$E_y|_{i,j,k}^{n+1/2} = \left(\frac{1 - \dfrac{\sigma_{i,j,k}\Delta t}{2\varepsilon_{i,j,k}}}{1 + \dfrac{\sigma_{i,j,k}\Delta t}{2\varepsilon_{i,j,k}}} \right) E_y|_{i,j,k}^{n-1/2} + \left(\frac{\dfrac{\Delta t}{\varepsilon_{i,j,k}}}{1 + \dfrac{\sigma_{i,j,k}\Delta t}{2\varepsilon_{i,j,k}}} \right) \left(\begin{array}{c} \dfrac{H_x|_{i,j,k+1/2}^{n} - H_x|_{i,j,k-1/2}^{n}}{\Delta z} \\ -\dfrac{H_z|_{i+1/2,j,k}^{n} - H_z|_{i-1/2,j,k}^{n}}{\Delta x} \end{array} \right)$$

$$(20.5\text{-}13)$$

$$E_z\big|_{i,j,k}^{n+1/2} = \left(\frac{1 - \dfrac{\sigma_{i,j,k}\Delta t}{2\varepsilon_{i,j,k}}}{1 + \dfrac{\sigma_{i,j,k}\Delta t}{2\varepsilon_{i,j,k}}}\right) E_z\big|_{i,j,k}^{n-1/2} + \left(\frac{\dfrac{\Delta t}{\varepsilon_{i,j,k}}}{1 + \dfrac{\sigma_{i,j,k}\Delta t}{2\varepsilon_{i,j,k}}}\right) \left(\frac{H_y\big|_{i+1/2,j,k}^{n} - H_y\big|_{i-1/2,j,k}^{n}}{\Delta x} - \frac{H_x\big|_{i,j+1/2,k}^{n} - H_x\big|_{i,j-1/2,k}^{n}}{\Delta y}\right)$$

$$(20.5\text{-}14)$$

$$H_x\big|_{i,j,k}^{n+1} = \left(\frac{1 - \dfrac{\rho'_{i,j,k}\Delta t}{2\mu_{i,j,k}}}{1 + \dfrac{\rho'_{i,j,k}\Delta t}{2\mu_{i,j,k}}}\right) H_x\big|_{i,j,k}^{n} + \left(\frac{\dfrac{\Delta t}{\mu_{i,j,k}}}{1 + \dfrac{\rho'_{i,j,k}\Delta t}{2\mu_{i,j,k}}}\right) \left(\frac{E_y\big|_{i,j,k+1/2}^{n+1/2} - E_y\big|_{i,j,k-1/2}^{n+1/2}}{\Delta z} - \frac{E_z\big|_{i,j+1/2,k}^{n+1/2} - E_z\big|_{i,j-1/2,k}^{n+1/2}}{\Delta y}\right)$$

$$(20.5\text{-}15)$$

$$H_y\big|_{i,j,k}^{n+1} = \left(\frac{1 - \dfrac{\rho'_{i,j,k}\Delta t}{2\mu_{i,j,k}}}{1 + \dfrac{\rho'_{i,j,k}\Delta t}{2\mu_{i,j,k}}}\right) H_y\big|_{i,j,k}^{n} + \left(\frac{\dfrac{\Delta t}{\mu_{i,j,k}}}{1 + \dfrac{\rho'_{i,j,k}\Delta t}{2\mu_{i,j,k}}}\right) \left(\frac{E_z\big|_{i+1/2,j,k}^{n+1/2} - E_z\big|_{i-1/2,j,k}^{n+1/2}}{\Delta x} - \frac{E_x\big|_{i,j,k+1/2}^{n+1/2} - E_x\big|_{i,j,k-1/2}^{n+1/2}}{\Delta z}\right)$$

$$(20.5\text{-}16)$$

$$H_z\big|_{i,j,k}^{n+1} = \left(\frac{1 - \dfrac{\rho'_{i,j,k}\Delta t}{2\mu_{i,j,k}}}{1 + \dfrac{\rho'_{i,j,k}\Delta t}{2\mu_{i,j,k}}}\right) H_z\big|_{i,j,k}^{n} + \left(\frac{\dfrac{\Delta t}{\mu_{i,j,k}}}{1 + \dfrac{\rho'_{i,j,k}\Delta t}{2\mu_{i,j,k}}}\right) \left(\frac{E_x\big|_{i,j+1/2,k}^{n+1/2} - E_x\big|_{i,j-1/2,k}^{n+1/2}}{\Delta y} - \frac{E_y\big|_{i+1/2,j,k}^{n+1/2} - E_y\big|_{i-1/2,j,k}^{n+1/2}}{\Delta x}\right)$$

$$(20.5\text{-}17)$$

20.6   COMPUTER EXPERIMENTS

The software package XFDTD by Remcom was used in the computer experiments [Kuhl, Ersoy]. In this environment, the region of interest is a cubical mesh, where each mesh edge can be given different material properties in order to simulate a specified geometry. For each cell, the material may be a perfect conductor or free space, or may be defined in other ways. The sampling in space has sub-wavelength format, typically in the range of 1/10–1/30 of a wavelength. The region of interest is excited by either a plane wave or multiple voltage sources. The excitation may be pulsed or sinusoidal. The duration of the simulation is set by specifying the number of desired time steps.

Figure 20.4.   1-D FZP with $m = 3$.

When the modeled region extends to infinity, absorbing boundary conditions (ABCs) are used at the boundary of the grid. This allows all outgoing waves to leave the region with negligible reflection. The region of interest can also be enclosed by a perfect electrical conductor.

Once the fields are calculated for the specified number of time steps, near zone transient and steady state fields can be visualized as color intensity images, or a field component at a specific point can be plotted versus time. When the steady-state output is desired, observing a specific point over time helps to indicate whether a steady-state has been reached.

In the computer experiments performed, a cell size of $\lambda/20$ was used [Kuhl, Ersoy]. The excitation was a y-polarized sinusoidal plane wave propagating in the z-direction. All diffracting structures were made of perfect electrical conductors. All edges of the diffracting structures were parallel to the x-axis to avoid canceling the y-polarized electric field.

A 1-D FZP is the same as the FZP discussed in Section 15.10 with the x-variable dropped. The mode m is defined as the number of even or odd zones which are opaque.

An example with three opaque zones (m = 3) is shown in Figure 20.4.

Using XFDTD, a 1-D FZP with a thickness of 0.1λ and focal length 3λ was simulated. Its output was analyzed for the first three modes. The intensity along the z-axis passing through the center of the FZP is plotted as a function of distance from the FZP in Figure 20.5. The plot shows that the intensity peaks near 3λ behind the plate, and gets higher and narrower as the mode increases. The peak also gets closer to the desired focal length of 3λ for higher modes.

The plot of the intensity along the y-axis at the focal line is shown in Figure 20.6. The plot shows that the spot size decreases with increasing mode and side lobe intensity is reduced for higher modes.

These experiments show that the FDTD method provides considerable freedom in generating the desired geometry and specifying material parameters. It is especially useful when the scalar diffraction theory cannot be used with sufficient accuracy due to reasons such as difficult geometries and/or diffracting apertures considerably smaller than the wavelength.

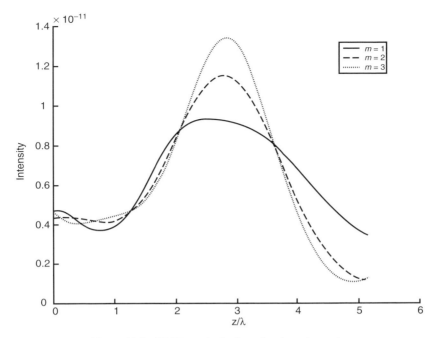

Figure 20.5. FDTD results for intensity along the z-axis.

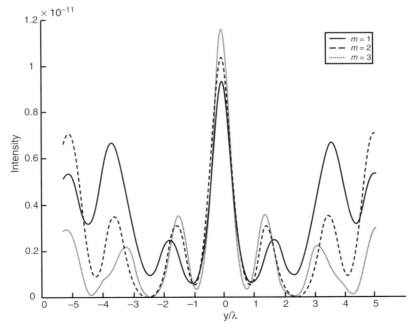

Figure 20.6. FDTD results for intensity along the y-axis on the focal line.

20.7   FOURIER MODAL METHODS

Fourier modal methods are among the several most versatile rigorous methods to analyze surface relief DOEs. The initial work in this area is known as *rigorous coupled-wave analysis* (RCWA) [Moharam, 82]. It was mostly used initially with diffraction gratings, but they were later generalized to aperiodic structures as well [Lalanne–Silberstein]. In the introductory discussion in this section, we will assume a grating structure.

The FMM is a frequency domain method involving computations of the grating modes as eigenvectors, Fourier expansions of the permittivity and of the EM fields inside the grating region. The Fourier expansion of the field inside the grating generates a system of differential equations. After finding the eigenvalues and eigenvectors of this system, the boundary conditions at the grating interfaces are matched to compute the diffraction efficiencies [Lalagne and Morris 1996].

The geometry to be used to explain the method in 2-D is shown in Figure 20.7.

The periodic grating with period Λ is along the x-axis with a permittivity function $\varepsilon(x)$. The TM mode is considered such that the magnetic field is polarized in the y-direction, and there are no variations along the y-direction. The incident plane wave makes an angle θ with the z-direction. In addition, the following Fourier series definitions are made:

$$\frac{\varepsilon(x)}{\varepsilon_o} = \sum_{k=-\infty}^{\infty} \varepsilon_k e^{jKkx} \qquad (20.7\text{-}1)$$

$$\frac{\varepsilon_o}{\varepsilon(x)} = \sum_{k=-\infty}^{\infty} a_k e^{jKkx} \qquad (20.7\text{-}2)$$

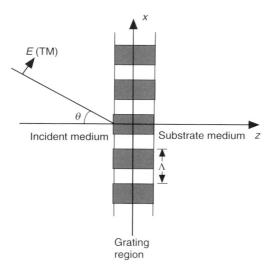

Figure 20.7.   The geometry used in the FMM method [Lalanne and Morris 1996].

where $K = 2\pi/\Lambda$. The field components can be written as [Lalanne and Morris 1996]

$$E_x = \sum_m S_m(z)e^{j(Km+\beta)x} \tag{20.7-3}$$

$$E_z = \sum_m f_m(z)e^{j(Km+\beta)x} \tag{20.7-4}$$

$$H_y = \sum_m U_m(z)e^{j(Km+\beta)x} \tag{20.7-5}$$

where $\beta = k\sin\theta = \frac{2\pi}{\lambda}\theta$.

Maxwell's curl equations in this case are given by

$$-\frac{\delta E_z}{\delta x} + \frac{\delta E_x}{\delta x} = -jw\mu_o H_y \tag{20.7-6}$$

$$\frac{\delta H_y}{\delta z} = -jw\varepsilon E_x \tag{20.7-7}$$

$$\frac{1}{\varepsilon}\frac{\delta H_y}{\delta x} = jw E_z \tag{20.7-8}$$

Prime and double prime quantities will be used to denote the first and second partial derivatives with respect to the z-variable. Using Eqs. (20.7-3)–(20.7-5) in Eqs. (20.7-6)–(20.7-8) results in

$$-j(Km+\beta)f_m + S'_m = -jk_o U_m \tag{20.7-9}$$

$$U'_m = -jk_o\sum_p \varepsilon_{m-p}S_p \tag{20.7-10}$$

$$f_m = \frac{1}{k_o}\sum_p (pK+\beta)a_{m-p}U_p \tag{20.7-11}$$

Using Eq. (20.7-11) in Eq. (20.7-9) yields

$$S'_m = -jk_o U_m + \frac{j}{k_o}(Km+\beta)\sum_p (pK+\beta)a_{m-p}U_p \tag{20.7-12}$$

In practice, the sums above are truncated with $|p| \le M$. Then, Eqs. (20.7-10) and (20.7-12) make up an eigenvalue problem of size $2(2M+1)$. However, it turns out to be more advantageous computationally to compute the second partial derivative of U_m with Eq. (20.7-10) and obtain [Caylord, Moharam, 85], [Peng]

$$U''_m = -k_o^2\left\{\sum_p \varepsilon_{m-p}\left[U_p - \frac{1}{k_o}(pK+\beta)\right]\right\}\left\{\sum_r \frac{1}{k_o}(rK+\beta)a_{p-r}U_r\right\} \tag{20.7-13}$$

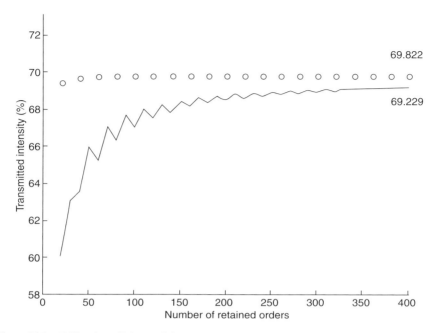

Figure 20.8. Diffraction efficiency of the transmitted zeroth order beam from a metallic grating with TM polarized light [Courtesy of Lalanne and Morris, 1996].

Equation (20.7-13) can be written in matrix form as

$$\frac{1}{k_o^2}\mathbf{U}'' = \mathbf{E}(\mathbf{K}_x\mathbf{E}^{-1}\mathbf{K}_x - \mathbf{I})\mathbf{U} \qquad (20.7\text{-}14)$$

where all the bold upper case letters represent matrices; $\mathbf{I}$ is the identity matrix, $\mathbf{E}$ is the permittivity harmonic coefficients matrix, and $\mathbf{K}_x$ is a diagonal matrix whose ith diagonal element equals $(iK + \beta)/k_o$.

A more efficient form of Eq. (20.7-14) computationally is given by [Lalanne 1996]

$$\frac{1}{k_o^2}\mathbf{U}'' = \mathbf{E}(\mathbf{E}\mathbf{K}_x\mathbf{E}^{-1}\mathbf{K}_x - \mathbf{I})\mathbf{U} \qquad (20.7\text{-}15)$$

As an example, Figure 20.8 shows the diffraction efficiency of the zeroth order beam of a metallic grating with TM polarized light with respect to the number of Fourier coefficients kept [Lalagne, Morris, 1996]. The solid curve and the curve with circles were obtained with Eqs. (20.7-14) and (20.7-15), respectively.

The coverage above is introductory at best. There are currently a number of refinements of the Fourier modal method for different applications, and further refinements are expected

Appendix A

The Impulse Function

The *impulse (delta or Dirac delta) function* $\delta(t)$ can be regarded as the idealization of a very narrow pulse with unit area. Consider the finite pulse shown in Figure A.1. It is defined by

$$x(t) = \begin{cases} \dfrac{1}{a} & -\dfrac{a}{2} < t < \dfrac{a}{2} \\ 0 & \text{otherwise} \end{cases} \tag{A.1-1}$$

The area under the pulse is 1 and remains as 1 for all values of a. The impulse function can be defined as

$$\delta(t) = \lim_{a \to 0} x(t) \tag{A.1-2}$$

$\delta(t)$ satisfies

$$\int_{-\infty}^{\infty} \delta(t)\mathrm{d}t = 1 \tag{A.1-3}$$

$$\delta(t) = 0 \quad t \neq 0$$

In mathematics, $\delta(t)$ is considered to be not an ordinary function, but a *generalized function* or a *distribution*, as discussed in the following sections. $\delta(t)$ has the *sampling (sifting) property* given by

$$\int_{t_1}^{t_2} \delta(t - \tau)h(\tau)\mathrm{d}t = \int_{t_1}^{t_2} \delta(\tau - t)h(\tau)\mathrm{d}\tau = h(t) \tag{A.1-4}$$

where $h(t)$, a given function, is continuous at $\tau = t$, and $t_1 < t < t_2$.

It is observed that if $h(t)$ is the impulse response of a LTI system, application of $\delta(t)$ to the system leads to the identification of $h(t)$.

Diffraction, Fourier Optics and Imaging,   by Okan K. Ersoy
Copyright © 2007 John Wiley & Sons, Inc.

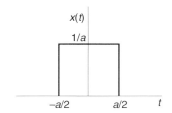

Figure A.1. Finite pulse of unit area.

The impulse function can also be written as the derivative of the unit step function:

$$\delta(t) = \frac{d}{dt}u(t) \qquad (A.1\text{-}5)$$

The impulse function can be obtained by limiting operations on a number of functions whose integral has the value 1. Some examples are given below.

$$\delta(t) = \begin{cases} \lim_{a\to\infty}[ae^{-at}u(t)] \\[2mm] \lim_{a\to\infty}\dfrac{1}{2}e^{-a|t|} \\[2mm] \lim_{a\to 0^+}\dfrac{1}{\sqrt{2\pi a}}e^{-t^2/2a^2} \\[2mm] \lim_{a\to\infty}\dfrac{\sin(at)}{\pi t} \end{cases} \qquad (A.1\text{-}6)$$

It is observed that these functions in the limit approach 0 at $t \neq 0$ and ∞ at $t = 0$. In the last case, the limit at $t \neq 0$ does not approach 0, but the function oscillates so fast that Eq. (A.1-6) remains valid.

Another way to construct the impulse function is by using the triangular pulse shown in Figure A.2. Its area is 1. As a shrinks towards 0, the area of 1 remains

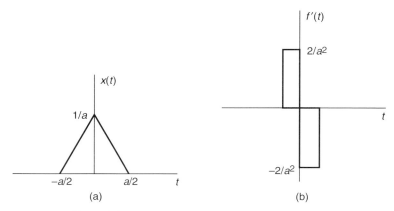

Figure A.2. (a) A triangular pulse, and (b) its derivative.

constant, and the base length approaches 0 as the height grows towards infinity. Thus,

$$\delta(t) = \lim_{a \to 0} \frac{1}{a} \operatorname{tri}\left(\frac{2t}{a}, -1, 1\right) \tag{A.1-7}$$

The derivative of the impulse function can be defined as

$$
\begin{aligned}
\delta'(t) &= \frac{d}{dt} \lim_{a \to 0} \frac{1}{a} \operatorname{tri}\left(\frac{2t}{a}, -1, 1\right) \\
&= \lim_{a \to 0} \frac{1}{a} \frac{d}{dt} \operatorname{tri}\left(\frac{2t}{a}, -1, 1\right)
\end{aligned}
\tag{A.1-8}
$$

The derivative of the triangular pulse is also shown in Figure A.2. $\delta'(t)$ is referred to as the *doublet*.

The derivatives of the impulse function can be defined with respect to the following integral:

$$\int_{t_1}^{t_2} f(t)\delta^k(t - t_0)\,dt = (-1)^k f^k(t_0) \tag{A.1-9}$$

where $t_1 < t_0 < t_2$, $\delta^k(t)$ and $f^k(t)$ denote the kth derivative of $\delta(t)$ and $f(t)$, respectively.

Some useful properties of the impulse function are the following:

Property 1. Time scaling

$$\delta(at - t_0) = \frac{1}{|a|} \delta\left(t - \frac{t_0}{a}\right) \tag{A.1-10}$$

Property 2. Multiplication by a function

$$f(t)\delta(t - t_0) = f(t_0)\delta(t - t_0), \ f(t) \text{ continuous at } t = t_0 \tag{A.1-11}$$

Property 3. $\delta[f(t)]$

$$\int_{-\infty}^{\infty} \delta[f(t)]\,dt = \sum_k \frac{1}{|f'(t_k)|}, \ f'(t_k) \neq 0 \tag{A.1-12}$$

where $f'(t) = \frac{df(t)}{dt}$, and t_k occurs when $f(t_k) = 0$.

Property 4. Sum of shifted impulse functions

$$|T| \sum_{k=-\infty}^{\infty} \delta\left(t - \frac{k}{T}\right) = \sum_{k=-\infty}^{\infty} e^{j2\pi kt/T} = 1 + 2 \sum_{k=1}^{\infty} \cos(2\pi kt/T) \qquad \text{(A.1-13)}$$

EXAMPLE A.1 Evaluate the integral

$$\int_{-\infty}^{\infty} \delta\left(\frac{t}{3} - 1\right) \cos(10t)\,dt.$$

Solution:

$$\delta\left(\frac{t}{3} - 1\right) = \delta\left(\frac{1}{3}(t - 3)\right)$$
$$= 3\delta(t - 3)$$

Hence,

$$\int_{-\infty}^{\infty} \delta\left(\frac{t}{3} - 1\right) \cos(10t)\,dt = 3 \int_{-\infty}^{\infty} \delta(t - 3) \cos(10t)\,dt$$
$$= 3\cos(30)$$
$$= 3\sqrt{3}/2$$

EXAMPLE A.2 Evaluate

$$\lim_{a\to\infty} \frac{1}{\sqrt{2\pi}a} e^{-t^2/32a^2}$$

in terms of the impulse function.
Solution:

$$\frac{1}{\sqrt{2\pi}a} e^{-t^2/32a^2} = \frac{1}{\sqrt{2\pi}a} e^{-(t/4)^2/2a^2}$$

Since

$$\lim_{a\to\infty} \frac{1}{\sqrt{2\pi}a} e^{-t^2/2a^2} = \delta(t)$$

it follows that

$$\lim_{a\to\infty} \frac{1}{\sqrt{2\pi}a} e^{-t^2/32a^2} = \delta\left(\frac{t}{4}\right) = 4\delta(t)$$

EXAMPLE A.3 Evaluate

$$\lim_{a \to \infty} \frac{1}{\pi} \int_0^a \cos(tx)\mathrm{d}x$$

in terms of the impulse function.
Solution: It is known that

$$\frac{1}{\pi} \int_0^a \cos(tx)\mathrm{d}x = \frac{\sin(at)}{\pi t} \quad a > 0$$

Since

$$\lim_{a \to \infty} \frac{\sin(at)}{\pi t} = \delta(t),$$

it follows that

$$\delta(t) = \lim_{a \to \infty} \frac{1}{\pi} \int_0^a \cos(tx)\mathrm{d}x \qquad\qquad (\text{A.1-14})$$

This result is often written as

$$\delta(t) = \lim_{a \to \infty} \frac{1}{2\pi} \int_{-a}^a e^{jtx}\mathrm{d}x$$

Appendix B

Linear Vector Spaces

B.1 INTRODUCTION

In a large variety of applications, Fourier-related series and discrete transforms are used for the representation of signals by a set of basis functions. More generally, this is a subject of *linear vector spaces* in which basis functions are called *vectors*. Representation of signals by discrete Fourier-related transforms can be considered to be the same subject as representing signals by basis vectors in infinite and finite-dimensional Hilbert spaces, respectively.

The most obvious example of a vector space is the familiar 3-D space R^3. In general, the concept of a vector space is much more encompassing. Associated with every vector space is a set of scalars which belong to a *field F*. The elements of F can be added and multiplied to generate new elements in F. Some examples of fields are the field R of real numbers, the field C of complex numbers, the field of binary numbers given by [0,1], and the field of rational polynomials.

A vector space S over a field F is a *set* of elements called vectors, with which two operations called *addition* ($+$) and *multiplication* ($\bullet$) are carried out. Let $\mathbf{u}$, $\mathbf{v}$, $\mathbf{w}$ be elements (vectors) of S, and α, β be scalars, which are the elements of the field F. S satisfies the following axioms:

1. $\mathbf{u} + \mathbf{v} = \mathbf{v} + \mathbf{u} \in S$
2. $\alpha \bullet (\mathbf{u} + \mathbf{v}) = \alpha \bullet \mathbf{u} + \alpha \bullet \mathbf{v} \in S$
3. $(\alpha + \beta) \bullet \mathbf{u} = \alpha \bullet \mathbf{u} + \beta \bullet \mathbf{u}$
4. $(\mathbf{u} + \mathbf{v}) + \mathbf{w} = \mathbf{u} + (\mathbf{v} + \mathbf{w})$
5. $(\alpha \bullet \beta) \bullet \mathbf{u} = \alpha \bullet (\beta \bullet \mathbf{u})$
6. There exists a null vector denoted as $\boldsymbol{\theta}$ such that $\boldsymbol{\theta} + \mathbf{u} = \mathbf{u}$. $\boldsymbol{\theta}$ is often written simply as 0.
7. For scalars 0 and 1, $0 \bullet \mathbf{u} = 0$ and $1 \bullet \mathbf{u} = \mathbf{u}$.
8. There exists an additive inverse element for each $\mathbf{u}$, denoted by $-\mathbf{u}$, such that
$$\mathbf{u} + (-\mathbf{u}) = \boldsymbol{\theta}$$
$-\mathbf{u}$ also equals $-1 \bullet \mathbf{u}$.

Diffraction, Fourier Optics and Imaging, by Okan K. Ersoy
Copyright © 2007 John Wiley & Sons, Inc.

Some important vector spaces are the following:

F^n is the space of column vectors with n components from a field F. Two special cases are R^3 whose components are from R, the vector space of real numbers, and C^n whose components are from C, the vector space of complex numbers.

$F^{m \times n}$ is the space of all $m \times n$ matrices whose components belong to the field F. Some other vector spaces are discussed in the examples below.

EXAMPLE B.1 Let V and W be vector spaces over the same field F. The *Cartesian product* $V \times W$ consists of the set of ordered pairs $\{\mathbf{v}, \mathbf{w}\}$ with $\mathbf{v} \in V$ and $\mathbf{w} \in W$. $V \times W$ is a vector space. Vector addition and scalar multiplication on $V \times W$ are defined as

$$\{\mathbf{v}, \mathbf{w}\} + \{\mathbf{p}, \mathbf{q}\} = \{\mathbf{v} + \mathbf{p}, \mathbf{w} + \mathbf{q}\}$$
$$\alpha\{\mathbf{v}, \mathbf{w}\} = \{\alpha\mathbf{v}, \alpha\mathbf{w}\}$$
$$\boldsymbol{\theta} = \{\boldsymbol{\theta}_v, \boldsymbol{\theta}_w\}$$

where $\alpha \in F$, $\mathbf{v}$ and $\mathbf{p} \in V$, $\mathbf{w}$ and $\mathbf{q} \in W$, $\boldsymbol{\theta}_v$ and $\boldsymbol{\theta}_w$ are the null elements of V and W, respectively.

The Cartesian product vector space can be extended to include any number of vector spaces.

EXAMPLE B.2 The set of all complex-valued continuous functions of the variable t over the interval $[a,b]$ of the real line forms a vector space, denoted by $C[a,b]$. Let $\mathbf{u}$ and $\mathbf{v}$ be vectors in this space, and $\alpha \in F$. The vector addition and the scalar multiplication are given by

$$(\mathbf{u} + \mathbf{v})(t) = \mathbf{u}(t) + \mathbf{v}(t)$$
$$(\alpha\mathbf{u})(t) = \alpha\mathbf{u}(t)$$

The null vector $\boldsymbol{\theta}$ is the function identically equal to 0 over $[a,b]$.

B.2 PROPERTIES OF VECTOR SPACES

In this section, we discuss properties of vector spaces which are valid in general, without being specific to a particular vector space.

Subspace

A nonempty vector space L is a subspace of a space S if the elements of L are also the elements of S, and S has possibly more number of elements.

Let M and N be subspaces of S. They satisfy the following two properties:

1. The intersection $M \cap N$ is a subspace of S.
2. The direct sum $M \oplus N$ is a subspace of S. The direct sum is described below.

Direct Sum

A set S is the direct sum of the subsets S_1 and S_2 if, for each $\mathbf{s} \in S$, there exists unique $\mathbf{s}_1 \in S_1$ and $\mathbf{s}_2 \in S_2$ such that $\mathbf{s} = \mathbf{s}_1 + \mathbf{s}_2$. This is written as

$$S = S_1 \oplus S_2 \tag{B.2-1}$$

EXAMPLE B.3 Let $(a, b) = (-\infty, \infty)$ in Example B.2. Consider the odd and even functions given by

$$u_e(t) = u_e(-t)$$
$$u_0(t) = -u_0(-t)$$

The odd and even functions form subspaces S_o and S_e, respectively. Any function $x(t)$ in the total space S can be decomposed into even and odd functions as

$$u_e(t) = \frac{u(t) + u(-t)}{2} \tag{B.2-2}$$

$$u_0(t) = \frac{u(t) - u(t)}{2} \tag{B.2-3}$$

Then,

$$u(t) = u_e(t) + u_0(t) \tag{B.2-4}$$

Consequently, the direct sum of S_o and S_e equals S.

Convexity

A subspace S_c of a vector space S is *convex* if, for each vector $\mathbf{s}_0$ and $\mathbf{s}_1 \in S_c$, the vector $\mathbf{s}_2$ given by

$$\mathbf{s}_2 = \lambda \mathbf{s}_0 + (1 - \lambda)\mathbf{s}_1, \quad 0 \le \lambda \le 1 \tag{B.2-5}$$

also belongs to S_c.

In a convex subspace, the line segment between any two points (vectors) in the subspace also belongs to the same subspace.

Linear Independence

A vector $\mathbf{u}$ is said to be *linearly dependent* upon a set S of vectors v_i if $\mathbf{u}$ can be expressed as a linear combination of the vectors in S:

$$\mathbf{u} = \sum_k c_k \mathbf{v}_k \qquad \text{(B.2-6)}$$

where $\mathbf{c}_k \in F$.

x is *linearly independent* of the vectors in S if Eq. (B.2-6) is invalid. A set of vectors is called a linearly independent set if each vector in the set is linearly independent of the other vectors in the set.

The following two properties of a set of linearly independent vectors are stated without proof:

1. A set of vectors $\mathbf{u}_0, \mathbf{u}_1 \ldots$ are linearly independent iff $\sum_k \mathbf{c}_k \mathbf{u}_k = 0$ means $\mathbf{c}_k = 0$ for all k.
2. If $\mathbf{u}_0, \mathbf{u}_1 \ldots$ are linearly independent, then, $\sum_k \mathbf{c}_k \mathbf{u}_k = \sum_k \mathbf{b}_k \mathbf{u}_k$ means $\mathbf{c}_k = \mathbf{b}_k$ for all k.

Span

A vector $\mathbf{u}$ belongs to the subspace spanned by a subset S if $\mathbf{u}$ is a linear combination of vectors in S, as in Eq. (B.2-6). The subspace spanned by the vectors in S is denoted by *span (S)*.

Bases and Dimension

A *basis* (or a coordinate system) in a vector space S is a set B of linearly independent vectors such that every vector in S is a linear combination of the elements of B.

The *dimension M* of the vector space S equals the number of elements of B. If M is finite, S is a *finite-dimensional* vector space. Otherwise, it is *infinite-dimensional*.

If the elements of B are $b_0, b_1, b_2 \ldots b_{M-1}$, then any vector $\mathbf{x}$ in S can be written as

$$\mathbf{x} = \sum_{k=0}^{M-1} w_k b_k \qquad \text{(B.2-7)}$$

where w_k s are scalars from the field F.

The number of elements in a basis of a finite-dimensional vector space S is the same as in any other basis of the space S.

In an orthogonal basis, the vectors $\mathbf{b}_m$ are orthogonal to each other. If B is an orthogonal basis, taking the inner product of $\mathbf{x}$ with $\mathbf{b}_m$ in Eq. (B.2-7) yields

$$(\mathbf{u}, \mathbf{b}_m) = \sum_{k=0}^{M-1} \mathbf{w}_k (\mathbf{b}_k, \mathbf{b}_m) = \mathbf{w}_m (\mathbf{b}_m, \mathbf{b}_m)$$

so that

$$\mathbf{w}_m = \frac{(\mathbf{x}, \mathbf{b}_m)}{(\mathbf{b}_m, \mathbf{b}_m)} \tag{B.2-8}$$

EXAMPLE B.4 If the space is F^n, the columns of the $n \times n$ identity matrix $\mathbf{I}$ are linearly independent and span F^n. Hence, they form a basis of F^n. This is called the *standard basis*.

B.3   INNER-PRODUCT VECTOR SPACES

The vector spaces of interest in practice are usually structured such that there are a *norm* indicating the length or the size of a vector, a measure of orientation between two vectors called the *inner-product,* and a *distance measure (metric)* between any two vectors. Such spaces are called *inner-product vector spaces.* The rest of the chapter is restricted to such spaces. Their properties are discussed below.

An inner product of two vectors $\mathbf{u}$ and $\mathbf{v}$ in an inner-product vector space S is written as $(\mathbf{u},\mathbf{v})$ and is a mapping $S \times S \to D$, satisfying the following:

1. $(\mathbf{u}, \mathbf{v}) = (\mathbf{v}, \mathbf{u})^*$
2. $(\alpha\mathbf{u}, \mathbf{v}) = \alpha(\mathbf{u}, \mathbf{v})$, α being a scalar
3. $(\mathbf{u} + \mathbf{v}, \mathbf{w}) = (\mathbf{u}, \mathbf{w}) + (\mathbf{v}, \mathbf{w})$
4. $(\mathbf{u}, \mathbf{u}) > 0$ when $\mathbf{u} \neq 0$, and $(\mathbf{u}, \mathbf{u}) = 0$ if $\mathbf{u} = 0$

When $\mathbf{u}$ and $\mathbf{v}$ are N-tuples,

$$\mathbf{u} = [u_0\, u_1 \ldots u_{N-1}]^t$$
$$\mathbf{v} = [v_0 v_1 \ldots v_{N-1}]^t$$

$(\mathbf{u}, \mathbf{v})$ can be defined as

$$(\mathbf{u}, \mathbf{v}) = \sum_{k=0}^{N-1} \mathbf{u}_k \mathbf{v}_k^* \tag{B.3-1}$$

When $f(t)$ and $g(t)$ are continuous functions in the vector space $C[a, b]$ with pointwise addition and scalar multiplication, the inner product (f, g) is given by

$$(f, g) = \int_a^b f(t)g * (t)dt \qquad \text{(B.3-2)}$$

The *Euclidian norm* of a vector $\mathbf{u}$ is $\sqrt{(\mathbf{u}, \mathbf{u})}$, and is denoted by $|\mathbf{u}|$. The vectors $\mathbf{u}$ and $\mathbf{v}$ are *orthogonal* if $(\mathbf{u}, \mathbf{v}) = 0$. In the space F^n, a more general norm is defined by

$$|\mathbf{u}|_p = \left[\sum_{k=0}^{n-1} |\mathbf{u}_k|^p \right]^{1/p} \qquad \text{(B.3-3)}$$

where $1 \leq p < \infty$. $p = 2$ gives the Euclidian norm.

Below we discuss important properties of inner-product vector spaces in general, and some inner-product vector spaces in particular.

Distance Measure (Metric)

The distance $d(\mathbf{u}, \mathbf{v})$ between two vectors $\mathbf{u}$ and $\mathbf{v}$ shows how similar the two vectors are. Inner-product vector spaces are also *metric spaces*, the metric being $d(\mathbf{u}, \mathbf{v})$. The distance $d(\mathbf{u}, \mathbf{v})$ can be defined in many ways provided that it satisfies the following:

1. $d(\mathbf{u}, \mathbf{v}) \geq 0$
2. $d(\mathrm{u}, \mathrm{u}) = 0$
3. $d(\mathbf{u}, \mathbf{v}) = d(\mathbf{v}, \mathbf{u})$
4. $d^2(\mathbf{u}, \mathbf{v}) \leq d^2(\mathbf{u}, \mathbf{w}) + d^2(\mathbf{w}, \mathbf{v})$

The last relation is usually called the *Schwarz inequality* or the *triangular inequality*. The *Euclidian distance* between two vectors $\mathbf{x}$ and $\mathbf{y}$ is given by

$$d(\mathbf{u}, \mathbf{v}) = |\mathbf{u} - \mathbf{v}| = [(\mathbf{u} - \mathbf{v}, \mathbf{u} - \mathbf{v})]^{1/2} \qquad \text{(B.3-4)}$$

It is observed that the norm of a vector $\mathbf{u}$ is simply $d(\mathbf{u}, \mathbf{0})$, $\mathbf{0}$ being the null vector.

Even though the Euclidian distance shows the similarity between $\mathbf{u}$ and $\mathbf{v}$, its magnitude depends on the norms of $\mathbf{u}$ and $\mathbf{v}$. In order to remove this dependence, it can be normalized by dividing it by $|\mathbf{u}||\mathbf{v}|$.

Examples of Inner-Product Vector Spaces

Two most common inner product spaces are the set R of real numbers and the set C of complex numbers, with their natural metric $d(\mathbf{u}, \mathbf{v})$ being the Euclidian distance between $\mathbf{u}$ and $\mathbf{v}$.

C^n and R^n are the n-dimensional vector spaces over C and R, respectively. The elements of C^n and R^n can be expressed as n-tuples.

ℓ_2 denotes the vector space over C of all complex sequences $\mathbf{u} = [\mathbf{u}_n]$ for $n = 1, 2 \ldots \infty$ which satisfy

$$\sum_{n=1}^{\infty} |\mathbf{u}_n|^2 < \infty \qquad (\text{B.3-5})$$

with componentwise addition and scalar multiplication, and with inner product given by

$$(\mathbf{u}, \mathbf{v}) = \sum_{n=1}^{\infty} \mathbf{u}_n \mathbf{v}_n^* \qquad (\text{B.3-6})$$

where $\mathbf{v} = [\mathbf{v}_n]$. Componentwise addition means the following:
if $\mathbf{w} = \mathbf{u} + \mathbf{v}$, then

$$\mathbf{w}_n = \mathbf{u}_n + \mathbf{v}_n \text{ for all } n \qquad (\text{B.3-7})$$

Scalar multiplication means the following:
if $\mathbf{w} = \lambda\mathbf{u}$, λ a scalar, then

$$\mathbf{w}_n = \lambda\mathbf{u}_n \text{ for all } n \qquad (\text{B.3-8})$$

$L_2(a, b)$ denotes the vector space of continuous functions over $C[a,b]$ (complex values in the interval from a to b) with the inner product

$$(\mathbf{u}, \mathbf{v}) = \int_a^b \mathbf{u}(t)\mathbf{v}^*(t)\mathbf{d}t \qquad (\text{B.3-9})$$

Angle

The angle θ between two vectors x and y is defined by

$$\cos\theta = \frac{(\mathbf{u}, \mathbf{v})}{|\mathbf{u}||\mathbf{v}|} \qquad (\text{B.3-10})$$

θ also shows the similarity between $\mathbf{u}$ and $\mathbf{v}$. θ equal to 0 shows that $\mathbf{u}$ and $\mathbf{v}$ are the same. θ equal to $\pi/2$ shows that $\mathbf{u}$ and $\mathbf{v}$ are orthogonal.

Cauchy–Schwarz Inequality

The inner product of two vectors $\mathbf{u}$ and $\mathbf{v}$ satisfy

$$|(\mathbf{u}, \mathbf{v})| \leq |\mathbf{u}||\mathbf{v}| \qquad (\text{B.3-11})$$

with equality iff x and y are linearly dependent ($x = \alpha y$, α a scalar).

Triangle Inequality

The vectors **u** and **v** also satisfy

$$|\mathbf{u} + \mathbf{v}| \le |\mathbf{u}| + |\mathbf{v}| \tag{B.3-12}$$

The triangle inequality is also often called Scwarz inequality.

Orthogonality

Let $U = [\mathbf{u}_0, \mathbf{u}_1, \ldots, \mathbf{u}_{N-1}]$ be a set (sequence) of vectors in an inner-product space S. U is an orthogonal set if $(\mathbf{u}_k, \mathbf{u}_\ell) = 0$ for $k \ne \ell$. U is orthonormal if $(\mathbf{u}_k, \mathbf{u}_\ell) = \delta_{k\ell}$.

EXAMPLE B.5 Let the inner-product space be $L_2(0, 1)$. In this space, the functions

$$\mathbf{u}_k(t) = \cos(2\pi kt)$$

form an orthogonal sequence.

B.4 HILBERT SPACES

The theory of *Hilbert spaces* is the most useful and well-developed generalization of the theory of finite-dimensional inner-product vector spaces to include infinite-dimensional inner-product vector spaces. In order to be able to discuss Hilbert spaces, the concept of completeness is needed.

Consider a metric space M with the metric d. A sequence $[\mathbf{u}_k]$ in M is a *Cauchy sequence* if, for every $\varepsilon > 0$ there exists an integer k_0 such that $d(u_k, x_\ell) < \varepsilon$ for $k, \ell > k_0$. M is a *complete metric space* if every Cauchy sequence in M converges to a limit in M. An example of a Cauchy sequence is the Nth partial sum in a Fourier series expansion (see Example B.5 below).

A *Hilbert space* is an inner-product space which is a complete metric space. For example, C^n, R^n are Hilbert spaces since they are complete.

Inner-product spaces without the requirement of completeness are sometimes called *pre-Hilbert spaces*.

EXAMPLE B.5 Consider a periodic signal $u(t)$ with period T. The basic period can be taken to be $-T/2 \le t \le T/2$. The Fourier series representation of $u(t) \in L_2(-T/2, T/2)$ is given by

$$u(t) = \sqrt{2/T} \sum_{k=0}^{\infty} [U_1[k]q[k] \cos(2\pi kF_s t) + U_0[k] \sin(2\pi kF_s t)] \tag{B.5-1}$$

where $F_s = 1/T$, and

$$q[k] = \left\{ \begin{array}{cc} 1/\sqrt{2} & k = 0 \\ 1 & \text{otherwise} \end{array} \right\}$$

The basis functions are $\sqrt{2/T}\cos(2\pi k F_s t)$ and $\sqrt{2/T}\sin(2\pi k F_s t)$. They are countable but infinitely many. Since they form an orthonormal set of basis functions, the series coefficients $U_1[k]$ and $U_0[k]$ are given by

$$U_1[k] = (u(t), \sqrt{2/T}q(k)\cos(2\pi k F_s t)$$

$$= \sqrt{2/T}q(k) \int_{-T/2}^{T/2} u(t)\cos(2\pi k F_s t)dt$$

$$U_0[k] = (u(t), \sqrt{2/T}\sin(2\pi k F_s t)$$

$$= \sqrt{2/T} \int_{-T/2}^{T/2} u(t)\sin(2\pi k F_s t)dt$$

The Hilbert space of interest in this case is periodic functions which are square-integrable over one period. The vectors consist of all such periodic functions. The Cauchy sequences of interest are the Nth partial sums in Eq. (B.5-1). The limit of any such sequence is $x(t)$.

Appendix C

The Discrete-Time Fourier Transform, The Discrete Fourier Transform and The Fast Fourier Transform

C.1 THE DISCRETE-TIME FOURIER TRANSFORM

The discrete-time Fourier transform (DTFT) of a signal sequence $u[n]$ is defined as

$$U(f) = \frac{1}{F} \sum_{n=-\infty}^{\infty} u[n] e^{-j 2\pi f n T_s} \tag{C.1-1}$$

$u[n]$ is usually equal to a sampled signal $u(nT_s)$, T_s being the sampling interval, and $F = 1/T_s$. T_s and F are usually assumed to be 1 in the literature.

The inverse DTFT is given by

$$u[n] = \int_{-F/2}^{F/2} U(f) e^{j 2\pi f n T_s} df \tag{C.1-2}$$

The existence conditions for the DTFT can be written as

$$E_1 = ||u||_1 = \sum_{n=-\infty}^{\infty} |u[n]| < \infty \tag{C.1-3}$$

If $u[n]$ satisfies Eq. (C.1-3), it is also true that

$$E_1 \geq E_2 = ||u||^2 = \sum_{n=-\infty}^{\infty} |u[n]|^2 < \infty \tag{C.1-4}$$

The properties of the DTFT are very similar to the properties of the Fourier transform discussed in Chapter 2.

Diffraction, Fourier Optics and Imaging, by Okan K. Ersoy
Copyright © 2007 John Wiley & Sons, Inc.

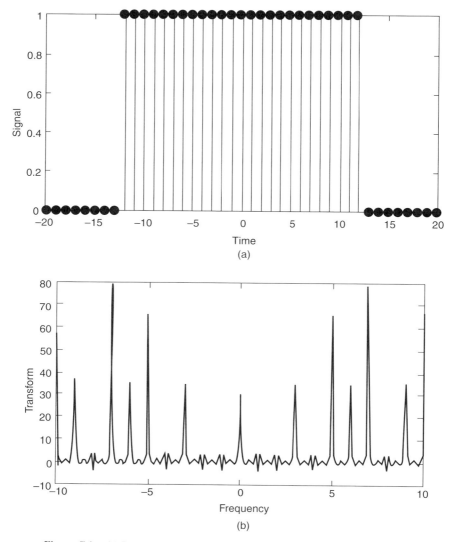

Figure C.1.   (a) Rectangular pulse signal for $N = 12$, (b) the DTFT of the signal.

EXAMPLE C.1   Find the DTFT of

$$u[n] = \begin{cases} 1 & |n| \le N \\ 0 & \text{otherwise} \end{cases}$$

with $T_s = 1$.
Solution: $u[n]$ is shown in Figure C.1(a) for $N = 12$.

We have

$$U(f) = \sum_{k=-N}^{N} e^{-j2\pi fk} = \sum_{k=-N}^{N} (e^{-j2\pi f})^k$$

Letting $\ell = k + N$, this becomes

$$U(f) = e^{j2\pi fN} \sum_{k=0}^{2N} (e^{-j2\pi f})^\ell$$

The summation on the right hand side is the sum of the first $(2N + 1)$ terms in a geometric series, which equals $(1 - r^{2N+1})/(1 - r)$, r being $e^{-j2\pi f}$. Thus,

$$U(f) = e^{j2\pi fN} \frac{[1 - e^{-j2\pi f(2N+1)}]}{1 - e^{-j2\pi f}}$$
$$= \frac{\sin(2\pi f(N + 1/2))}{\sin(\pi f)}$$

Figure C.1(b) shows $U(f)$ for $N = 12$. It is observed that $U(f)$ is similar to the sin function for small f, but becomes totally different as f approaches 1.

C.2   THE RELATIONSHIP BETWEEN THE DISCRETE-TIME FOURIER TRANSFORM AND THE FOURIER TRANSFORM

Consider the FT representation of an analog signal $u(t)$ given by Eq. (1.2.2) as

$$u(t) = \int_{-\infty}^{\infty} U'(f)e^{j2\pi ft} df \qquad \text{(C.2-1)}$$

where $U'(f)$ is the FT of $u(t)$. The sequence $u[n]$ is assumed to be obtained by sampling of $u(t)$ at intervals of T_s such that $u[n] = u(nT_s)$. Comparing Eq. (C.2-1) to Eq. (C.1-2) shows that the DTFT $U(f)$ of $u[n]$ is related to $U'(f)$ by

$$\int_{-F/2}^{F/2} U(f)e^{j2\pi fnT_s} df = \int_{-\infty}^{\infty} U'(f)e^{j2\pi fnT_s} df \qquad \text{(C.2-2)}$$

By expressing the right-hand integral as a sum of integrals, each of width F, Eq. (C.2-2) can be written as

$$\int_{-F/2}^{F/2} U'(f)e^{j2\pi fnT_s} df = \int_{-F/2}^{F/2} \left[\sum_{\ell=-\infty}^{\infty} U'(f + \ell F) \right] e^{j2\pi fnT_s} df \qquad \text{(C.2-3)}$$

Equating the integrands gives

$$U(f) = \sum_{\ell=-\infty}^{\infty} U'(f + \ell F) \tag{C.2-4}$$

If $U'(f)$ has bandwidth less than F, we see that

$$U'(f) = U'(f) \tag{C.2-5}$$

In many texts, $T_{\rm s}$ is chosen equal to 1 for convenience. This means $u(t)$ is actually replaced by $u(tT_{\rm s})$. $u(tT_{\rm s})$ has FT given by $1/T_{\rm s}U'(f/T_{\rm s})$ by Property 16 discussed in Section 1.9. Hence, Eq. (C.2-4) can be written as

$$U(f) = \frac{1}{T_{\rm s}} \sum_{\ell=-\infty}^{\infty} U'\left(\frac{f + \ell}{T_{\rm s}}\right) \tag{C.2-6}$$

where $U(f)$ is the DTFT of the sampled signal obtained from $u(tT_{\rm s})$ at a sampling rate equal to 1.

Equations (C.2-4) and (C.2-6) show that the DTFT of a signal which is not bandlimited is an aliased version of the FT of the continous-time signal. The resulting errors tend to be significant as f approaches F since $U(F)$ equals $U(0)$, and $U(0)$ is usually large.

C.3   DISCRETE FOURIER TRANSFORM

The orthonormalized *discrete Fourier transform* (DFT) of a periodic sequence $u[n]$ of length N is defined as

$$U[k] = \frac{1}{\sqrt{N}} \sum_{n=0}^{N-1} u[n] e^{-2\pi jnk/N} \tag{C.3-1}$$

The inverse DFT is given by

$$u[n] = \frac{1}{\sqrt{N}} \sum_{k=0}^{N-1} U[k] e^{2\pi jnk/N} \tag{C.3-2}$$

The DFT defined above is orthonormal. The equations can be further simplified if normality is not required as follows:

$$U[k] = \sum_{n=0}^{N-1} u[n] e^{-j2\pi nk/N} \tag{C.3-3}$$

$$u[n] = \frac{1}{N} \sum_{k=0}^{N-1} U[k] e^{j2\pi nk/N} \tag{C.3-4}$$

The 2-D DFT is obtained from the 1-D DFT by applying the 1-D DFT first to the rows of a signal matrix, and then to its columns or vice versa. If the 2-D periodic sequence is given by $u[n_1, n_2]$, $0 \leq n_1 < N_1, 0 \leq n_2 < N_2$, its 2-D DFT can be written as

$$U[k_1, k_2] = \sum_{n_1=0}^{N_1-1} \sum_{n_2=0}^{N_2-1} u[n_1, n_2] e^{-2\pi j \left[\frac{n_1 k_1}{N_1} + \frac{n_2 k_2}{N_2} \right]} \tag{C.3-5}$$

The inverse 2-D DFT is given by

$$u[n_1, n_2] = \frac{1}{N_1 N_2} \sum_{k_1=0}^{N_1-1} \sum_{k_2=0}^{N_2-1} U[k_1, k_2] e^{2\pi j \left[\frac{n_1 k_1}{N_1} + \frac{n_2 k_2}{N_2} \right]} \tag{C.3-6}$$

C.4 FAST FOURIER TRANSFORM

The fast Fourier transform (FFT) refers to a set of algorithms for the fast computation of the DFT. An FFT algorithm reduces the number of computations for an N point transform from $O(2N^2)$ to $O(2N \log N)$.

FFT algorithms are based on a *divide-and-conquer approach*. They are most efficient when the number N of data points is highly composite and can be written as $N = r_1 r_2 \ldots r_m$. When $y_0' = c_0 + c_1, y_1' = c_0 + c_2$, N equals r^m, and r is called the *radix* of the FFT algorithm. In this case, the FFT algorithm has a regular structure.

The divide-and-conquer approach is commonly achieved by either *decimation-in-time* (DIT) or *decimation-in-frequency* (DIF). To illustrate these two concepts, consider $N = 2^m$. In the DIT case, we may divide the data $x(n)$ into two sequences according to

$$\left. \begin{array}{l} u_1[n] = u[2n] \\ u_2[n] = u[2n+1] \end{array} \right\} \quad n = 0, 1, \ldots, \frac{N}{2} - 1 \tag{C.4-1}$$

In the DIF case, the frequency components are considered in two batches as $U[2k]$ and $U[2k+1]$, $k = 0, 1, \ldots, \frac{N}{2} - 1$.

We describe the DIT algorithm further. Eq. (C.1-1) can be written as

$$U[k] = U_1[k] + e^{-j\frac{2\pi k}{N}} U_2[k] \tag{C.4-2}$$

$$U_1[k] = \sum_{m=0}^{N_2-1} x[2m] e^{-j\frac{2\pi km}{N_2}} \tag{C.4-3}$$

$$U_2[k] = \sum_{m=0}^{N_2-1} u[2m+1] e^{-j\frac{2\pi km}{N_2}} \tag{C.4-4}$$

where N_2 equals $N/2$. It is observed that $U_1[k]$ and $U_2[k]$ are computed by DFTs of size $N/2$ with the even and odd-numbered points of the original signal.

The procedure described above is continued iteratively until reaching size 2 DFT, which consists of adding and subtracting two points.

The radix-2 decimation-in-frequency algorithm can be similarly developed. However, its implementation turns out to me more complex.

The 2-D FFT is obtained from the 1-D FFT by applying the 1-D FFT first to the rows of a signal matrix, and then to its columns or vice versa.

References

Asakura, K. and T. Nagashima., "Reconstruction from computer-generated holograms displayed by a line printer," *Optics Communications* 17, 273–276, 1976.

Besag, J. E. "Spatial Interaction and the Statistical Analysis of Lattice Systems," *Journal of Royal Statistical Society*: Series B 36, 2, 192–236, May 1974.

Bennett, J. R., I. G. Cumming and R. A. Deane, "The Digital Processing of SEASAT Synthetic Aperture Radar Data," *Proceedings of the IEEE International Radar Conference*, 168–174, Virginia, 1980.

Born, M. and E. Wolf, *Principles of Optics*, Pergamon Press, New York, 1969.

Bowden, M., L. Thompson, C. Wilson, eds., *Introduction to Microlithography*, American Chemical Soc., ISBN 0-8412-2848-5, 1983.

Boyer, A. L., "Formation of Images Using 5Mhz Ultrasound and a Computer," *IBM Publication* 320.2403, May 19, 1971.

Boyer, A. L., P. M. Hirsch, J. A. Jordan, Jr., L. B. Lesem, D. L. Van Rooy, "Reconstruction of Ultrasonic Images by Backward Propagation," *IBM Technical Report*, No. 320.2396, IBM Scientific Center, Houston, Texas, July 1970.

Brackett, C. A., A. S. Acampora, I. Sweitzcr. G. Tangonan. M. T. Smith, W. Lennon. K. C. Wang, and R. H. Hobbs. "A scalable multiwavelength multihop optical network: A proposal for research on all-optical networks." *J. Lightwave Technol.*, vol. II. pp. 736–753, May/June 1993.

Brackett, C. A., "Dense wavelength division multiplexing networks: Principles and applications." *IEEE J. Select. Areas Commun.*, Vol. 8, pp. 948–964. 1990.

Brackett, C. A., A. S. Acampora, I. Sweitzcr. G. Tangonan. M. T. Smith, W. Lennon. K. C. Wang, and R. H. Hobbs. "A Scalable Multiwavelength Multihop Optical Network: A Proposal for Research on All-Optical Networks." *J. Lightwave Technol.*, vol. II. pp. 736–753, May/June 1993.

Brigham, E. O., *The Fast Fourier Transform*, Prentice Hall, Englewood, CA, 1974.

Brown, B.R. and A.W. Lohmann, "Complex Spatial Filtering with Binary Masks," *Applied Optics*, 5, 967–969, June 1966.

Bubb, C. E., Okan K. Ersoy, "Algorithms for Holographic Reconstruction of Three-Dimensional Point Source Images," *Technical Report* TR-ECE-06-06, Purdue University, March 2006.

Burckhardt, C. B., "Use of a Random Phase Mask for the Recording of Fourier Transform Holograms of Data Masks," *Applied Optics*, 9, 695–700, 1970.

Caulfield, H. J. and S. Lu, *The Applications of Holography*, Interscience, New York, 1970.

Chu, D. C., J. R. Fienup, J. W. Goodman, "Multi-emulsion On-Axis Computer Generated Hologram," *Applied Optics*, 12, 1386–1388, 1973.

Collier, R. J., C. B. Burckhardt, and L. H. Lin, *Optical Holography*, Academic Press, New York, 1971.

Cutrona, L. J., E. N. Leith, C. J. Palermo, L. J. Porcello, "Optical Data Processing and Filtering Systems," *IRE Tran. Information Theory*, 386–400, June 1960.

Dittman, J., L. C. Ferri, C. Vielhauer, "Hologram watermarks for document authentications," *International Conference on Information Technology: Coding and Computing*, 60–64, April 2001.

Doles, J. H., "Broad-Band Array Design Using the Asymptotic Theory of Unequally Spaced Arrays," *IEEE Tran. Antennas and Propagation*, Vol. 36, No. 1, 27–33, January 1988.

Dragone, C., "An N × N Optical Multiplexer Using a Planar Arrangement of Two Star Couplers." *IEEE Photon. Technol. Lett.* 3, 812–815, September 1991.

Erdelyi, A., *Tables of Integral Transforms*, McGraw Hill, New York, 1954.

Ersoy, O. K., "Construction of Point Images with the Scanning Electron Microscope: A Simple Algorithm," *Optik*, 46, 61–66, September 1976.

Ersoy, O. K., "One-Image-Only Digital Holography," *Optik*, 53, 47–62, April 1979.

Ersoy, O. K., "Real Discrete Fourier Transform," *IEEE Transactions Acoustics, Speech, Signal Processing*, ASSP-33, 4, 880–882, August 1985.

Ersoy, O. K., J. Y. Zhuang, J. Brede, "Iterative Interlacing Approach for the Synthesis of Computer-Generated Holograms," *Applied Optics*, 31, 32, 6894–6901, November 10, 1992.

Ersoy, O. K., "A Comparative Review of Real and Complex Fourier-Related Transforms," *Proceedings of the IEEE*, 82, 3, 429–447, March 1994.

Ersoy, O. K., "Method of Increasing Number of Allowable Channels in Phased Array DWDM Systems," *USA Patent* No. 6917736, July 12, 2005.

ESA, "Satellite Data Yields Major Results in Greenland Glaciers Study," *ESA website*, http://www.esa.int/esaEO/SEMH59MVGJE_index_0.html, February 21, 2006.

Farhat, N. H., *Advances in Holography*, Marcel Dekker, New York, 1975.

Feit, M. D. and J. A. Fleck, Jr., "Light propagation in graded-index optical fibers," *Applied Optics*, 17. 24, 3990–3998, December 15, 1978.

Feldman, M. R.; J. E. Morris, I. Turlik, P. Magill, G. Adema, M. Y. A Raja, "Holographic Optical Interconnects for VLSI Multichip Modules," *IEEE Tran. Components, Packaging and Manufacturing Technology, Part B: Advanced Packaging*, 17, 2, 223–227, May 1994.

Firth, I. M., *Holography and Computer Generated Holograms*, Mills and Boon, Ltd., London, 1972.

Gabor, D., "Resolution Beyond the Information Limit in Transmission Electron Microscopy, *Nature*, 161, 777–778, 1948.

Gallagher, N. C. and D. W. Sweeney, "Infrared Holographic Optical Elements with Applications to Laser Material Processing," *IEEE J. of Quantum Elec.*, QE-15, 1369–1380, December 1979.

Garner, W., W. Gautschi, "Adaptive Quadrature-Revisited," *BIT*, 40, 84–101, 2000.

Gaylord, T. K., M. G. Moharam, "Analysis and Application of Optical Diffraction Gratings," *Proc. IEEE*, 73, 894–936, 1985.

Gerrard, A. and J. M. Burch, *Introduction to Matrix Methods in Optics*, J. Wiley, New York, 1975.

Gerchberg, R., "Superresolution through Error Energy Reduction," *Optica Acta*, 21, 709–720, 1974.

Goodman, J. W., *Introduction to Fourier Optics*, 3rd Edition, Roberts and Company, Greenwood Village, Colorado, 2004.

Hadley, G. R., "Transparent Boundary Condition for the Beam Propagation Method," *IEEE J. Quantum Electronics*, 26, 1, 109–112, January 1990.

Hadley, G. R., "Wide-Angle Beam Propagation Using Pade Approximant Operators," *Optics Letters*, 17, 20, 1426–1431, October 15, 1992.

Hayes, M. H., J. S. Lim, A. V. Oppenheim, "Signal Reconstruction from the Phase or Magnitude of its Fourier Transform," *IEEE Tran. Acoustics, Speech and Signal Proc.*, ASSP-28, 670–680, 1980.

Hecht, E., *Optics*, Addison-Wesley, Mass., 2002.

Hu, Sai, O. K. Ersoy, "Design and Simulation of Novel Arrayed Waveguide Grating by Using the Method of Irregularly Sampled Zero Crossings," *TR-ECE 02–05*, Purdue University, December 2002.

Ishimaru, A., "Theory of Unequally Spaced Arrays, "*IRE Tran. Antennas and Propagation*, 691–701, November 1962.

Kelly, Kozma and D. L., "Spatial Filtering for Detection of Signals Submerged in Noise," *Applied Optics*, Vol. 4, No. 4, pp. 389–392, 1965.

Kim, C.-J., R. R. Shannon, "Catalog of Zernike Polynomials," in *Applied Optics and Optical Engineering*, edited by R. Shannon and J. Wyant, Ch. 4, Vol. X, Academic Press, San Diego, 1987.

Kock, W. E., *Engineering Applications of Lasers and Holography*, Plenum, New York, 1975.

Kozma, A., D. L. Kelly, Spatial filtering for detection of signals submerged in noise," *Applied Optics*, 4, 387–392, 1965.

Kuhl, P., Okan K. Ersoy, "Design Of Diffractive Optical Elements: Optimality, Scale And Near-Field Diffraction Considerations," *Technical Report TR-ECE-03-09*, Purdue University, 2003.

Kunz, *The Finite Difference Time Domain Method for Electromagnetics*, CRC Press, Boca Raton, Florida, 1993.

Lalanne, P., G. M. Morris, "Highly Improved Convergence of the Coupled-Wave Method for TM Polarization," *J. Opt. Soc. Am.*, 13, 4, 779–784, April 1996.

Lalanne, P., E. Silberstein, "Fourier-Modal Methods Applied to Waveguide Computational Problems," *Optics Letters*, 25, 15, 1092–1094, August 1, 2000.

Lanczos, A., *Discourse on Fourier Series*, Hafner Publishing Co., New York, 1966.

Lalor, E., "Inverse Wave Propagator," *J. Mathematical Physics*, Vol. 9, No. 12, pp. 2001–2006, December 1968.

Lao, Y. T., "A Mathematical Theory of Antenna Arrays with Randomly Spaced Elements," *IEEE Tran. Antennas and Propagation*, 257–268, May 1964.

Laude, J. P., *Wavelength Division Multiplexing*, Prentice Hall, NY, 1993.

Lee, W. H "Sampled Fourier Transform Hologram Generated by Computer," Applied Optics, **9**, 639–643,1970.

Lee, W. H., "Circular Carrier Holograms," *J. Opt. Soc. Am.* 65, 518–523, 1975.

Lee, W.-H., "Binary Computer-Generated Holograms," *Applied Optics*, Vol. 18, No. 21, pp. 3661–3669, November 1, 1979.

Leith, E. N., J. Upatnieks, "Reconstructed Wavefronts and Communication Theory," *J. Opt. Soc. Am.* 52, 1123–1130, 1962.

Lesem, L.B., P. M. Hirsch, J. A. Jordan, jr., "The kinoform: a new wavefront reconstruction device," *IBM J. Res. Develop.* 13, 150–155, March 1969.

Levi, A., H. Stark, "Image Restoration by the Method of Generalized Projections with Application to Restoration from Magnitude," *J. Opt. Soc. Am. A*, 1, 9, 932–943, September 1984.

Ljunggren, S., O. Lovhaugen, E. Mehlum, "Seismic Holography in a Norwegian Fjord," *Acoustical Imaging*, 8, 299–315, 1980.

Lohmann, A. W., in *The Engineering Uses of Holography*, E. R. Robertson and J. N. Harvey, Eds., Cambridge U. P., London, 1970.

Lohmann, A. W., D. P. Paris, "Binary Fraunhofer Holograms Generated by Computer," *Appl. Opt.* 6, 1739–1748, 1967.

Lu, Ying, Okan Ersoy, Dense Wavelength Division Multiplexing/Demultiplexing By The Method Of Irregularlly Sampled Zero Crossing, *Technical Report TR-ECE-03-12*, Purdue University, 2003.

Marathay, A., *Diffraction*, in *Handbook of Optics, Volume 1: Fundamentals, Techniques, and Design*, 2nd edition, McGraw-Hill, New York, pages 3.1–3.31, 1995.

Marcuse, D., Theory of Dielectric Optical Waveguides, 2nd edition, Academic Press, San Diego, 1991.

Meier, R. W., "Magnifications and Third Order Aberrations in Holography," *J. Opt. Soc. Am.* 56, 8, 987–997, 1966.

Mellin, S. D., G. P. Nordin, "Limits of Scalar Diffraction Theory and an Iterative Angular Spectrum Algorithm for Finite Aperture Diffractive Optical Element Design," *Optics Express*, Vol. 8, No. 13, pp. 705–722, 18 June, 2001.

Mezouari, S., A. R. Harvey, "Validity of Fresnel and Fraunhofer Approximations in Scalar Diffraction," *J. Optics A: Pure Appl. Optics*, Vol. 5, pp. 86–91, 2003.

Moharam, M. G., T. K. Gaylord, "Diffraction analysis of dielectric surface-relief gratings." *J. Opt. Soc. Am.*, **72**, 1385–1392, 1982.

Nikon, Phase Contrast Microscopy web page, http://www.microscopyu.com/articles/phase-contrast/phasemicroscopy.html

Okamoto, K., "Recent Progress of Integrated Optics Planar Lightwave Circuits," *Optical and Quantum Electronics*, 31,107–129, 1999.

Peng, S., G. M. Morris, "Efficient Implementation of Rigorous Coupled-Wave Analysis for Surface Relief Gratings, *J. Opt. Soc. Am.* 12, 1087–1096, 1995.

Pojanasomboon, P., O. K. Ersoy, "Iterative Method for the Design of a Nonperiodic Grating-Assisted Directional Coupler," *Applied Optics*, 40, 17, 2821–2827, June 2001.

Pennings, E. C. M., M. K. Smit, and G. D. Khoe, "Micro-Optic versus waveguide devices – An Overview, invited paper," in *Proc. Fifth Micm Opdcs Conf 1995*, Hiroshima. Japan. Oct. 18–20. 1995, pp. 248–255.

Pennings, E. C. M., M. K. Smit. A. A. M. Staring, and G.-D. Khoe. "Integrated-Optics versus Micro-Optics – A comparison." *Integrated Photonics Research IPR '96*. Boston. MA. Tech. Dig.. vol. 6. Apr. 29–May 2. 1996. pp. 460–463.

Ralston, A., H. S. Wilf, *Mathematical Methods for Numerical Analysis*, J. Wiley, New York, 1962.

Rooy, D. L. Van, "Digital Ultrasonic Wavefront Reconstruction in the Near Field," *IBM Technical Report*, No. 320.2402, IBM Scientific Center, Houston, Texas, May 1971.

Shannon, R., in *Optical Instruments and Techniques*, edited by J. Dickson, pp. 331–345, Oriel Press, England, 1970.

Shen, F., A. Wang. "Fast Fourier Transform Based Numerical Integration Method for the Rayleigh-Sommerfeld Diffraction Formula," *Applied Optics*, Vol. 45, No. 6, pp. 1102–1110, 20 February 2006.

Shewell, J. R., "Inverse Diffraction and a New Reciprocity Theorem," *J. Optical Society of America*, Vol. 58, No. 12, pp. 1596–1603, December 1968.

Smith, H. M. *Principles of Holography* (Wiley, New York, 1975).

Smit, M. K., "Now focusing and dispersive planar component based on an optical phased array." *Electron. Lett.*, Vol. 24, No. 7, pp. 385–386, March 1988.

Soumekh, Mehrdad, *Synthetic Aperture Radar Signal Processing with Matlab Algorithms*, J. Wiley, New York, 1999.

Southwell, W. H., "Validity of the Fresnel Approximation in the Near Field," *J. Optical Society of America*, Vol. 71, No. 1, pp. 7–14, 1981.

Stark, H., *Applications of Optical Fourier Transform*, Academic Press, Orlando, 1982.

Stark, H., editor, *Image Recovery: Theory and Application*, Academic Press, 1987.

Steane, M., H. N. Rutt, "Diffraction Calculations in the Near Field and the Validity of the Fresnel Approximation," *Journal of the Optical Society of America A*, 6, 12, 1809–1814, 1989.

Stroke, G. W., *An Introduction to Coherent Optics and Holography*, Academic Press, New York, 1975.

Taflove, A., S. C. Hagness, *Computational Electrodynamics: the Finite Difference Time Domain Method*, 3rd Edition, Artech House, Norwood, MA, 2005.

Takahashi, H., S. Suzuki, K. Kaco, and I. Nishi, "Arrayed-Waveguide Grating for Wavelengt Division MuIt/Demultiplexer with Nanometer Resolution," *Electron. Lett.*, 26, 2, 87–88, January 1990.

F.T. Ulaby, R. K. Moore and A. K. Fung, *Microwave Remote Sensing Volume I*, Addison-Wesley, Reading, MA, 1981.

Van Rooy, D. L., "Digital Ultrasonic Wavefront Reconstruction in the Near Field," IBM Publication No. 320.2402, May 19, 1971.

Weyl, H., "Ausbreitung Elektromagnetischer Wellen uber Einem Ebenen Leiter," Ann. Phys. Lpz. 60, 481–500, 1919.

Wikipedia, website http://en.wikipedia.org/wiki/Diffraction

Wilson, T., editor, *Confocal Microscopy*, Academic Press, London, 1990.

Ziemer, R. E., W. H. Trantor, *Principles of Communications*, 5th edition, J. Wiley, New York, 2002.

Ueda, M. and K. Ieyasu, "A Method for a Faithful Reconstruction of an Off-Axis Type Ultrasound Holography," *Optik* 42, 107–112, 1975.

Vellekoop, A. R. and M. K. Smit, "Low-Loss Planar Optical Polarization Splitter with small dimensions," *Electron. Lett.*, Vol. 25. pp. 946–947, l989.

Verbeek, B. and NI. K. Smit. "Phased array based WDM devices," in *Proc. Eur. Conf. on Optical Communication (ECOC'95)*, Brussels. Belgium. Sept 17–21, 1995, pp. 195–202.

Vesperinas, M. N., *Scattering and Diffraction in Physical Optics*, J. Wiley, New York, 1991.

Watanebe, W., D. Kuroda, K. Itoh, "Fabrication of Fresnel Zone Plate Embedded in Silica Glass by Femtosecond Laser Pulses," Optics Express 10, 19, 978–983, September 2002.

Yatagai, T., "Stereoscopic Approach to 3-D Display Using Computer-Generated Holograms," *Appied. Opics*. 15, 11, 2722–2729, 1976.

Yee, K. S., "Numerical solutions of initial boundary value problems involving Maxwell's equations in isotropic media," *IEEE Trans. Antennas and Propagation*, 14, 302–307, 1966.

Yevick, D., "New Formulations of the Matrix Beam Propagation Method: Application to Rib Waveguides," *IEEE J. Quantum Electronics*, 25, 2, 221–229, February 1989.

Yevick, D., "Efficient Beam Propagation Techniques," *IEEE J. Quantum Electronics*, 26, 1, 109–112, January 1990.

Yin, Yun, Okan K. Ersoy, Xianfan Xu, Ihtesham H. Chowdhury, "Fabrication/Analysis of Waveguides in Fused Silica by Femtosecond Laser Pulses and Simulation Studies for Dense Wavelength Division Multiplexing," *Technical Report* TR-ECE-05–05, Purdue University, May 2005.

Youla, D. C., H. Webb. "Image Restoration by the Method of Convex Projections: Part1 – Theory," *IEEE Trans. Med. Imaging*, MI-1, 81–94, 1982.

Zhuang, J., O.K. Ersoy, "Optimal Decimation-in-Frequency Iterative Interlacing Technique for the Synthesis of Computer-Generated Holograms," *J. Optical Society of America, A*, Vol. 12, No. 7, pp. 1460–1468, July 1995.

Ziemer, R. E., W. H. Trantor, *Principles of Communications*, 5th edition, J. Wiley, New York, 2002.

Index

binary hologram, 245
binary optics, 198
binary phase DOE, 185
Blackman window, 215
blazed grating, 184
Bleaching, 177, 183
blurred image, 225
bounded signal, 332
boundedness, 240
BPM simulation, 195

CAD, 187
CAD layout, 250
Cauchy sequence, 221, 389
Cauchy-Schwarz inequality, 227, 388
causal signal, 157
central slice, 328
central-difference equations, 368
CGH transmittance, 294
characteristic impedance, 38
charge density, 31
chemical diffusion, 182
chirp pulse, 314
chirp signal, 310
circular aperture, 46, 47, 66, 75, 216
circular polarization, 39
circularly symmetric, 22, 145
clock, 2
closed convex set, 227
coarse film, 183
coherence, 162
coherent Imaging, 165
coherent imaging system, 167
coherent imaging technique, 307
coherent light, 180
coherent optical processing, 306
coherent transfer function, 167, 168
coherent wavefront recording, 198
coherent waves, 162
color film, 178
coma, 174, 210
communications, 325
compact disk, 205
complete metric space, 389
complete normed linear vector space, 219
complete power exchange, 195, 197
complex amplitude, 42, 137, 189
complex envelope, 158, 162, 164
complex Fourier transform, 18

complex geometries, 361
complex imaging systems, 153
complex space, 219
complex wave function, 162
compression ratio, 315
Computed tomography (CT), 326
computer-generated hologram (CGH), 1, 2, 244
computer-generated holography, 198
computerized imaging, 5
computerized tomography, 306, 328
condition number, 241
conduction band, 178
confocal microscopy, 2, 145
confocal scanning microscopy, 5
conjugate gradient method, 213, 242
conjugate planes, 128, 132
constant amplitude Lohmann's method, 248
constraint operator, 224
contraction mapping theorem, 220
contractions, 212, 219
contrast, 144
contrast reversal, 80
convex set, 225, 229
convex subspace, 384
convolution, 6, 12, 72, 147, 223, 332
convolution theorem, 9, 47, 58
correlation, 12
cosine part, 20
coupled mode theory, 193, 197
coupling coefficients, 193
Crank-Nicholson method, 361, 363, 369
cross-range (azimuth) resolution, 316
CT image, 326
cubical mesh, 371
curl, 31, 368
cutoff frequency, 165
cylinder function, 23, 146
cylindrical coordinates, 44, 52
cylindrical symmetry, 52
cylindrical wave, 44
data rate, 4

Debye approximation, 56
decimation-in-frequency (DIF), 395
decimation-in-frequency property, 275
decimation-in-time (DIT), 395
deconvolution, 212, 223